Lecture Notes in Computer Science 6970

Commenced Publication in 1973
Founding and Former Series Editors:
Gerhard Goos, Juris Hartmanis, and Jan van Leeuwen

Marina L. Gavrilova   C.J. Kenneth Tan
Mir Abolfazl Mostafavi (Eds.)

Transactions on Computational Science XIV

Special Issue on Voronoi Diagrams
and Delaunay Triangulation

 Springer

Editors-in-Chief

Marina L. Gavrilova
University of Calgary, Department of Computer Science
2500 University Drive N.W., Calgary, AB, T2N 1N4, Canada
E-mail: marina@cpsc.ucalgary.ca

C.J. Kenneth Tan
Exascala Ltd.
Unit 9, 97 Rickman Drive, Birmingham B15 2AL, UK
E-mail: cjtan@exascala.com

Guest Editor

Mir Abolfazl Mostafavi
Université Laval
Centre de Recherche en Géomatique
Département des sciences géomatiques
Québec, QC, G1V 0A6, Canada
E-mail: mir-abolfazl.mostafavi@scg.ulaval.ca

ISSN 0302-9743 (LNCS)   e-ISSN 1611-3349 (LNCS)
ISSN 1866-4733 (TCOMPSCIE)   e-ISSN 1866-4741 (TCOMPSCIE)
ISBN 978-3-642-25248-8   ISBN 978-3-642-25249-5 (eBook)
DOI 10.1007/978-3-642-25249-5
Springer Heidelberg Dordrecht London New York

Library of Congress Control Number: 2011940218

CR Subject Classification (1998): I.3.5, F.2, G.2, E.1, I.4, I.5

Typesetting: Camera-ready by author, data conversion by Scientific Publishing Services, Chennai, India

Printed on acid-free paper

Springer is part of Springer Science+Business Media (www.springer.com)

LNCS Transactions on Computational Science

Computational science, an emerging and increasingly vital field, is now widely recognized as an integral part of scientific and technical investigations, affecting researchers and practitioners in areas ranging from aerospace and automotive research to biochemistry, electronics, geosciences, mathematics, and physics. Computer systems research and the exploitation of applied research naturally complement each other. The increased complexity of many challenges in computational science demands the use of supercomputing, parallel processing, sophisticated algorithms, and advanced system software and architecture. It is therefore invaluable to have input by systems research experts in applied computational science research.

Transactions on Computational Science focuses on original high-quality research in the realm of computational science in parallel and distributed environments, also encompassing the underlying theoretical foundations and the applications of large-scale computation. The journal offers practitioners and researchers the opportunity to share computational techniques and solutions in this area, to identify new issues, and to shape future directions for research, and it enables industrial users to apply leading-edge, large-scale, high-performance computational methods.

In addition to addressing various research and application issues, the journal aims to present material that is validated – crucial to the application and advancement of the research conducted in academic and industrial settings. In this spirit, the journal focuses on publications that present results and computational techniques that are verifiable.

Scope

The scope of the journal includes, but is not limited to, the following computational methods and applications:

- Aeronautics and Aerospace
- Astrophysics
- Bioinformatics
- Climate and Weather Modeling
- Communication and Data Networks
- Compilers and Operating Systems
- Computer Graphics
- Computational Biology
- Computational Chemistry
- Computational Finance and Econometrics
- Computational Fluid Dynamics
- Computational Geometry

- Computational Number Theory
- Computational Physics
- Data Storage and Information Retrieval
- Data Mining and Data Warehousing
- Grid Computing
- Hardware/Software Co-design
- High-Energy Physics
- High-Performance Computing
- Numerical and Scientific Computing
- Parallel and Distributed Computing
- Reconfigurable Hardware
- Scientific Visualization
- Supercomputing
- System-on-Chip Design and Engineering

Editorial

The Transactions on Computational Science journal is part of the Springer series *Lecture Notes in Computer Science*, and is devoted to the gamut of computational science issues, from theoretical aspects to application-dependent studies and the validation of emerging technologies.

The journal focuses on original high-quality research in the realm of computational science in parallel and distributed environments, encompassing the facilitating theoretical foundations and the applications of large-scale computations and massive data processing. Practitioners and researchers share computational techniques and solutions in the area, identify new issues, and shape future directions for research, as well as enable industrial users to apply the techniques presented.

The current volume, edited by Mir Abolfazl Mostafavi, is devoted to Voronoi Diagrams and Delaunay Triangulation. It is comprised of extended versions of nine papers, carefully selected from the papers presented at the International Symposium on Voronoi Diagrams, Quebec City, Canada, June 28–30, 2010. It also contains a Guest Editor's Editorial with reflections on the historical background of the Voronoi diagram.

We would like to extend our sincere appreciation to the Special Issue Guest Editor Mir Abolfazl Mostafavi for his dedication to preparing this high-quality special issue, to all authors for submitting their papers, and to all Associate Editors and referees for their valuable work. We would also like to express our gratitude to the LNCS editorial staff of Springer, in particular Alfred Hofmann, Ursula Barth, and Anna Kramer, who supported us at every stage of the project.

It is our hope that the fine collection of papers presented in this issue will be a valuable resource for Transactions on Computational Science readers and will stimulate further research into the vibrant area of computational science applications.

July 2011

Marina L. Gavrilova
C.J. Kenneth Tan

Voronoi Diagrams and Delaunay Triangulation
Guest Editor's Preface

The Voronoi diagram is a concept that has been around for a long time. In "Le monde de M. Descartes et le traité de la lumière" published in 1644, Descartes used Voronoi-like diagrams to show the disposition of matter in the solar system and its surroundings. However, the first presentation of the concept of the Voronoi diagram appeared in the work of Dirichlet (1850) and was named after the Russian mathematician Georgy Fedoseevich Voronoi (or *Voronoy*), who defined and studied the general n-dimensional case in 1908. Since then, several extensions have originated from these publications. The first extension to the Voronoi diagram was in crystallography, where a set of points was placed regularly in space and the Voronoi cells were labeled as Wirkungsberich (area of influence) (Niggli, 1927). Thiessen (1911) used Voronoi regions, which are referred to as Thiessen polygons, to compute accurate estimates of regional rainfall averages. Another extension of the Voronoi diagram was seen in natural and social sciences for studying market areas (Bogue 1949). The Voronoi diagram in geography is used to analyze 2D point patterns (Boots, 1974). It was only in the early 1970s that a number of algorithms for the efficient construction of the Voronoi diagram were developed. This motivated further developments in several areas using computer science and computational geometry.

Voronoi diagrams continue to be developed and their applications are expanding from natural sciences to engineering and from medical sciences to social sciences. This special issue of the journal Transactions on Computational Science is devoted to the latest advances in Voronoi diagrams and their applications in sciences and engineering. It presents a collection of the best papers that were selected from the papers presented during the 7th International Symposium on Voronoi Diagrams and their applications in sciences and engineering. This edition of the symposium, which was chaired by Mir Abolfazl Mostafavi, Professor at the Department of Geomatics and the Director of the Centre for Research in Geomatics at Laval University, was organized in Quebec City, Canada and was sponsored by the Center for Research in Geomatics (CRG), Laval University and the GEOIDE networks of Centres of Excellence.

This special issue presents nine selected papers presented during the symposium. The papers were extended by the authors and passed through two rounds of reviewing before publication. Several interesting topics are discussed in this issue which include:

- Development of new generalized Voronoi diagrams and algorithms including round-trip Voronoi diagrams, maximal zone diagrams, Jensen-Bregman Voronoi diagrams, hyperbolic Voronoi diagrams, and moving network Voronoi Diagrams.
- New algorithms based on Voronoi diagrams for applications in sciences and engineering, including geosensor networks deployment and optimization and homotopic object reconstruction
- Application of Delaunay triangulation for the modeling and representation of Cosmic Web and rainfall distribution.

The editor expects that readers of the Transactions on Computational Science (TCS) will benefit from the papers presented in this special issue on the latest advances in Voronoi diagrams and Delaunay triangulation and their applications in sciences and engineering.

Acknowledgment

The guest editor of this Special Issue on Voronoi Diagrams and Delaunay Triangulation, Transactions on Computational Science (TCS), Vol. XIV, would like to thank all the authors for preparing and submitting original contributions for this special issue. The guest editor is also grateful to the Program Committee members of ISVD 2010, as well as the reviewers, for their great contribution. Special thanks go to the members of the Steering Committee of ISVD 2010 for their valuable help and support during the preparation of this special issue. The Editor-in-Chief of TCS, Dr. Marina L. Gavrilova, also deserves special thanks for her advice, vision, and support. Finally, the professional help and commitment of the Publisher Springer is greatly acknowledged.

July 2011   Mir Abolfazl Mostafavi
                                                                        Guest Editor

LNCS Transactions on Computational Science – Editorial Board

Table of Contents

Revisiting Hyperbolic Voronoi Diagrams in Two and Higher Dimensions from Theoretical, Applied and Generalized Viewpoints

Toshihiro Tanuma[1], Hiroshi Imai[2], and Sonoko Moriyama[3]

[1] Graduate School of Information Science and Technology,
University of Tokyo, Japan
to-tanuma@is.s.u-tokyo.ac.jp
[2] CS Dept./CINQIE, University of Tokyo
imai@is.s.u-tokyo.ac.jp
[3] Graduate School of Information Sciences, Tohoku University, Japan
moriso@dais.is.tohoku.ac.jp

Abstract. This paper revisits hyperbolic Voronoi diagrams, which have been investigated since mid 1990's by Onishi et al., from three standpoints, background theory, new applications, and geometric extensions.

First, we review two ideas to compute hyperbolic Voronoi diagrams of points. One of them is Onishi's method to compute a hyperbolic Voronoi diagram from a Euclidean Voronoi diagram. The other one is a linearization of hyperbolic Voronoi diagrams. We show that a hyperbolic Voronoi diagram of points in the upper half-space model becomes an affine diagram, which is part of a power diagram in the Euclidean space. This gives another proof of a result obtained by Nielsen and Nock on the hyperbolic Klein model. Furthermore, we consider this linearization from the view point of information geometry. In the parametric space of normal distributions, the hyperbolic Voronoi diagram is induced by the Fisher metric while the divergence diagram is given by the Kullback-Leibler divergence on a dually flat structure. We show that the linearization of hyperbolic Voronoi diagrams is similar to one of two flat coordinates in the dually flat space, and our result is interesting in view of the linearization having information-geometric interpretations.

Secondly, from the viewpoint of new applications, we discuss the relation between the hyperbolic Voronoi diagram and the greedy embedding in the hyperbolic plane. Kleinberg proved that in the hyperbolic plane the greedy routing is always possible. We point out that results of previous studies about the greedy embedding use a property that any tree is realized as a hyperbolic Delaunay graph easily.

Finally, we generalize hyperbolic Voronoi diagrams. We consider hyperbolic Voronoi diagrams of spheres by two measures and hyperbolic Voronoi diagrams of geodesic segments, and propose algorithms for them, whose ideas are similar to those of computing hyperbolic Voronoi diagrams of points.

Keywords: Voronoi diagrams, hyperbolic geometry, divergences, greedy embedding, geodesic segments, spheres.

M.L. Gavrilova et al. (Eds.): Trans. on Comput. Sci. XIV, LNCS 6970, pp. 1–30, 2011.

1   Introduction

Hyperbolic geometry is one of models of non-Euclidean geometry. Hyperbolic geometry relates several topics of mathematics such as group theory, topology (of 3-manifolds in particular), complex variables, and conformal mappings (see [12]). It is interesting not only in theory but also in applications for computer science such as computer graphics, information visualization, and network routing.

A Voronoi diagram is a fundamental concept in computational geometry, and has been investigated in various spaces. Among such spaces, hyperbolic geometry provides deep insights into Voronoi diagrams and their applications as one of models of the non-Euclidean geometry.

The hyperbolic space is represented by models such as the upper half-space (or Poincaré space), denoted by $\mathbb{H}^d$, the Poincaré disk and the Klein disk. Voronoi diagrams in the hyperbolic space, which are called hyperbolic Voronoi diagrams, have been studied since the work of Onishi and Takayama [31]. They considered hyperbolic Voronoi diagrams of n points in the upper half-plane $\mathbb{H}^2$ and propose an $O(n \log n)$ time algorithm. Onishi generalized it to the d-dimensional hyperbolic space $\mathbb{H}^d$ in his thesis [30], showing that a hyperbolic Voronoi diagram of n points in the space can be constructed in $O(\mathrm{CH}(d+1, n))$time, where $\mathrm{CH}(d, n)$ is the time complexity to compute the convex hull of n points in the d-dimensional Euclidean space. An incremental algorithm for hyperbolic Voronoi diagrams in $\mathbb{H}^3$ was given in [28]. In [27], hyperbolic Voronoi diagrams in the Poincaré disk were considered and an incremental algorithm for them was proposed. Hyperbolic Voronoi diagrams in the Klein disk were studied in [26], which showed that the hyperbolic Voronoi diagram in the Klein disk is equal to (part of) a power diagram [4,20].

Hyperbolic Voronoi diagrams have several applications. Nielsen and Nock [26] presented an image browsing application and hyperbolic Voronoi diagrams are used for its operations. A hyperbolic centroidal Voronoi Tessellation [36], which is a special case of a hyperbolic Voronoi diagram, has been studied in connection with computing uniform partitions and remeshing for high-genus surfaces.

Hyperbolic Voronoi diagrams in the upper half-plane $\mathbb{H}^2$ have been applied to information geometry in [32,33,30] since the statistical parametric space of normal distributions with the Fisher metric is almost identical with $\mathbb{H}^2$ [31]. The Fisher metric is a measure for probabilistic distributions in statistical estimation. In information geometry, there is another kind of measures for probabilistic distributions which is called divergences, and Voronoi diagrams by general divergences have been studied. Onishi and Imai studied Voronoi diagrams in the statistical parametric space by the Kullback-Leibler divergence [32,33] and presented unified results in the dually flat space by ∇-divergence or ∇^*-divergence in Onishi [30]. Relations between these Voronoi diagrams for normal distributions were also discussed in [33,30]. Voronoi diagrams in terms of the Bregman divergence were studied in [25].

One of interesting studies on hyperbolic geometry is the greedy embedding for routing on ad hoc wireless networks and sensors. A greedy embedding of a graph in a metric space embeds vertices in the space so that, for any pair of distinct vertices s and t, s has an adjacent vertex in the graph which is closer to t than s in the space. Kleinberg proved every graph has a greedy embedding in the hyperbolic plane [22], while in the Euclidean plane there is a graph which does not admit any greedy embedding [35]. Maymounkov [24] proved that, in the hyperbolic space, no-stretch greedy embeddings of graphs of n vertices requires $\log n$ dimensions and gave a way obtaining greedy embeddings of trees in $\mathbb{H}^3$ with $O(\log^2 n)$ bits. Cvetkovski and Crovella proposed an algorithm to embed dynamic graphs in the hyperbolic plane [13]. Eppstein and Goodrich [15] used more freedom in embedding a normalized tree in the 2-dimensional hyperbolic plane to realize succinctness. The hyperbolic plane thus has advantages in greedy routing.

Various generalizations of Voronoi diagrams in the Euclidean space have been studied. One of them is a Voronoi diagram of line and curved segments in the Euclidean plane. Drysdale and Lee studied Voronoi diagrams of line segments [14]. Yap proposed an $O(n \log n)$ time divide-and-conquer algorithm for n simple curved segments which are line segments or circular arcs [39], and Alt, Cheong and Vigneron proposed an $O(n \log n)$ expected time incremental algorithm for curved segments [1]. There are several types of Voronoi diagrams of spheres (or weighted points). One of them is a power diagram (or also called a Laguerre Voronoi diagram) [4,20]. The power diagram on the sphere was studied by Sugihara [37]. Another of them is called an additively weighted Voronoi diagram, which is defined by an additively weighted distance [4]. Boissonnat and Karavelas [8] gave a tight bound on the worst-case combinatorial complexity of a d-dimensional additively weighted Voronoi cell and an optimal algorithm for it. Boissonnat and Delage [9] proposed an incremental algorithm for d-dimensional additively weighted Voronoi diagrams.

We aim at understanding hyperbolic Voronoi diagrams more to reconfirm its necessity through their fertile characteristics. This paper is an extension of our previous work [38]. We generalize almost results about two-dimensional hyperbolic Voronoi diagrams in [38] to the higher dimensional case, and, in addition, consider several new topics.

Our contributions are as follows:

Computing Hyperbolic Voronoi Diagrams in $\mathbb{H}^d$ and D_P^d from Euclidean Voronoi Diagrams (in section 3): We review Onishi's method [30] to compute the hyperbolic Voronoi diagram from the Euclidean Voronoi diagram. Onishi used this method for the hyperbolic Voronoi diagram in $\mathbb{H}^d$, but we observe that it can be applied to that in the d-dimensional Poincaré disk D_P^d similarly.

Another Proof of Hyperbolic Voronoi Diagrams being Power Diagrams (in section 4): We apply a linearization in the study of hyperbolic convex hulls in $\mathbb{H}^d$ [30] to hyperbolic Voronoi diagrams in $\mathbb{H}^d$. This gives a coordinate system different from the Klein disk discussed in [26]. We show that the d-dimensional hyperbolic Voronoi diagram becomes affine in this new linearized coordinate, and then give another proof that this linearized hyperbolic Voronoi diagram is part of a power diagram in the Euclidean space.

Linearization Related to Statistical Estimation (in section 4): Voronoi diagrams of normal distributions in the space determined by the Fisher metric can be made flat by a linearization similar to the above one. It is observed that this linearization for the Fisher metric of normal distributions has similarities with the linearization of a dually flat space for the Kullback-Leibler divergence for normal distributions in information geometry [3]. This reveals another relation of two linearized Voronoi diagrams in addition to those discussed in [30].

Greedy Embedding and Hyperbolic Voronoi Diagrams (in section 5): We indicate that a hyperbolic Delaunay graph, which is defined as a dual of a hyperbolic Voronoi diagram, relates to the greedy embedding in the hyperbolic plane. We observe that results of previous studies about the greedy embedding in the hyperbolic plane use a property that a hyperbolic Delaunay graph which is a tree can be obtained easily.

Generalization of Hyperbolic Voronoi Diagrams (in section 6): To understand hyperbolic Voronoi diagrams more, we consider their generalizations. Firstly, we consider two types of Voronoi diagrams of spheres in the hyperbolic space. One of them is called an additively weighted hyperbolic Voronoi diagram, which is defined by a measure called an additively weighted hyperbolic distance, and the other of them is a hyperbolic power diagram, which is defined by a different measure called a hyperbolic power. Secondly, we discuss hyperbolic Voronoi diagrams of geodesic segments in $\mathbb{H}^2$. We give algorithms for the three types of Voronoi diagrams. For an additively weighted hyperbolic Voronoi diagrams or a hyperbolic Voronoi diagrams of geodesic segments, we use the same idea. Focusing on boundaries of those hyperbolic Voronoi diagrams of sites, which are spheres or geodesic segments, our algorithm computes them from a Euclidean Voronoi diagram of the same sites by eliminating several faces of the Euclidean Voronoi diagram as in [30,31]. For hyperbolic power diagrams, we apply the linearization in [30] to hyperbolic power diagrams as well as hyperbolic Voronoi diagrams of points. We prove that, by using the linearization, a hyperbolic power diagram becomes affine in the linearized space, and, interestingly, it is part of a power diagram in the Euclidean space. These algorithms demonstrate power of main ideas of algorithms for hyperbolic Voronoi diagrams of points proposed by previous studies [26,30,31].

2   Preliminaries

2.1   Hyperbolic Space and Hyperbolic Voronoi Diagram

We describe several definitions and properties of the hyperbolic space and hyperbolic Voronoi diagrams. See [12] for more details of hyperbolic geometry, and [26,27,28,30,31] for those of hyperbolic Voronoi diagrams.

There are several models representing the hyperbolic space, such as the upper half-space model (or Poincaré space model) $\mathbb{H}^d$, the Poincaré disk model D_P^d, and the Klein disk model D_K^d. Their definitions are as follows:

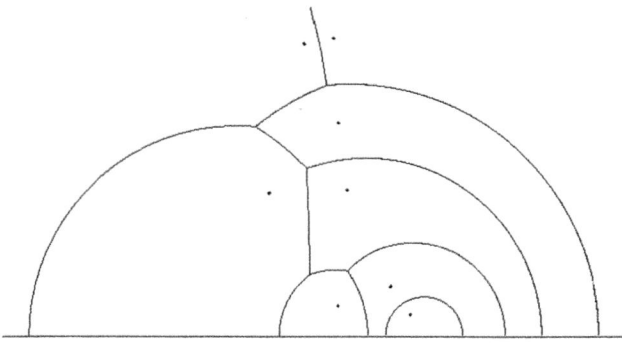

Fig. 1. A hyperbolic Voronoi diagram of 8 points in $\mathbb{H}^2$

1. The upper half-space is the domain $\mathbb{H}^d = \{(x_1, \ldots, x_d) \in \mathbb{R}^d \,|\, x_d > 0\}$ with the Riemannian metric
$$\frac{dx_1^2 + \cdots + dx_d^2}{x_d^2}.$$

2. Poincaré disk is the domain $D_P^d = \{(x_1, \ldots, x_d) \in \mathbb{R}^d \,|\, x_1^2 + \cdots + x_d^2 < 1\}$ with the Riemannian metric
$$4\frac{dx_1^2 + \cdots + dx_d^2}{(1 - x_1^2 - \cdots - x_d^2)^2}.$$

3. Klein disk is the space $D_K^d = \{(x_1, \ldots, x_d) \in \mathbb{R}^d \,|\, x_1^2 + \cdots + x_d^2 < 1\}$ with the Riemannian metric
$$\frac{dx_1^2 + \cdots + dx_d^2}{1 - x_1^2 - \cdots - x_d^2} + \frac{(x_1 dx_1 + \cdots + x_d dx_d)^2}{(1 - x_1^2 - \cdots - x_d^2)^2}.$$

There is an isometric mapping between any pair among these models, and we may mostly focus our discussions on one of them, specifically the upper half-space $\mathbb{H}^d$ in the sequel.

Let $\mathcal{H}$ be a model of the hyperbolic space defined above, and $d_{\mathcal{H}}(\cdot, \cdot)$ be the hyperbolic distance in $\mathcal{H}$. a hyperbolic Voronoi diagram in $\mathcal{H}$ is defined.

Definition 1. *(Hyperbolic Voronoi Diagram) For the distance $d_{\mathcal{H}}(\cdot, \cdot)$, a hyperbolic Voronoi diagram of a finite point set P in $\mathcal{H}$ is a partition of $\mathcal{H}$ by the following hyperbolic Voronoi regions $Vor_{\mathcal{H}}(p)$ $(p \in P)$:*

$$Vor_{\mathcal{H}}(p) = \{x \in \mathcal{H} \,|\, d_{\mathcal{H}}(x, p) \le d_{\mathcal{H}}(x, q) \text{ for } \forall q \in P,\ q \ne p\}.$$

A k-dimensional face of the hyperbolic Voronoi diagram, which is shared by $d - k + 1$ hyperbolic Voronoi regions, is called a Voronoi k-face for k ($0 \le k \le d - 1$). Especially, a Voronoi 0-face is called a Voronoi vertex and a Voronoi $(d - 1)$-face is called a Voronoi facet.

Definition 2. *(Hyperbolic Delaunay Graph) A hyperbolic Delaunay graph of a point set P in $\mathcal{H}$ is a dual graph of its hyperbolic Voronoi diagram such that two vertices, which are points in P, are connected by the geodesic segment if their corresponding Voronoi regions share a Voronoi facet.*

If there is no point which have the same distances from $d+2$ or more points in P, the hyperbolic Delaunay graph is a simiplicial complex.

Now, for each model of the hyperbolic space defined above, we describe several properties of hyperbolic Voronoi diagrams.

$\mathbb{H}^d$ has the following properties:

1. For $p=(p_1,\ldots,p_d)$, $q = (q_1,\ldots,q_d) \in \mathbb{H}^d$, the hyperbolic distance $d_{\mathbb{H}^d}(p_1,p_2)$ is given by

$$d_{\mathbb{H}^d}(p_1,p_2) = \text{arccosh}\left[\frac{\sum_{i=1}^{d-1}(p_i - q_i)^2 + p_d^2 + q_d^2}{2p_d q_d}\right].$$

2. A geodesic in $\mathbb{H}^d$ is expressed as a portion of a circle or a line orthogonal to $x_d = 0$. In $\mathbb{H}^2$, it can be expressed as follows:

$$(x_1 - a)^2 + x_2^2 = r^2\,(a, r \in \mathbb{R},\, r > 0)$$

   or

$$x_1 = c,\quad x_2 > 0\,(c \in \mathbb{R}).$$

3. A hyperbolic hyperplane in $\mathbb{H}^d$ is defined as following surfaces:

$$\sum_{i=1}^{d-1}(x_i - a_i)^2 + x_d^2 = r^2$$

   or

$$\sum_{i=1}^{d-1} a_i x_i + a_d = 0,\, x_d > 0$$

   where $a_i \in \mathbb{R}\,(1 \leq i \leq d)$ and $r > 0$. If $d = 2$, a hyperbolic hyperplane is a geodesic.

4. For a point $a=(a_1,\ldots,a_d) \in \mathbb{H}^d$ and $r > 0$, a set $c = \{p \in \mathbb{H}^d \mid d_{\mathbb{H}^d}(p, a)=r\}$ is expressed as

$$\sum_{i=1}^{d-1}(x_i - a_i)^2 + (x_d - a_d \cosh r)^2 = (a_d \sinh r)^2$$

   and thus it is a Euclidean sphere with center $(a_1,\ldots,a_{d-1}, a_d \cosh r)$ and radius $a_d \sinh r$. The point a is called a hyperbolic center and a point $(a_1,\ldots, a_{d-1}, a_d \cosh r)$ is called a Euclidean center.

Lemma 1. *A bisector of* $p = (p_1, \ldots, p_d)$ *and* $q = (q_1, \ldots, q_d) \in \mathbb{H}^d$ *is a hyperbolic hyperplane and expressed as follows:*

$$\sum_{i=1}^{d-1} \{2(p_i - q_i)x_i - (p_i^2 - q_i^2)\} = 0 \; (\textit{if } p_d = q_d),$$

$$\sum_{i=1}^{d-1} (x_i - \frac{p_i q_d - p_d q_i}{q_d - p_d})^2 + x_d^2 = \sum_{i=1}^{d-1} \frac{(p_i q_d - p_d q_i)^2}{(q_d - p_d)^2} - \sum_{i=1}^{d} \frac{p_i^2 q_d - p_d q_i^2}{q_d - p_d}$$

$$(\textit{otherwise}).$$

Proof. The result can be obtained by
$d_{\mathbb{H}^d}(p, x) = d_{\mathbb{H}^d}(q, x)$, where $x = (x_1, \ldots, x_d)$. □

Onishi and Takayama [31] showed that a hyperbolic Delaunay graph of a finite point set P in $\mathbb{H}^2$ is a subgraph of a Euclidean Delaunay triangulation of P, and a hyperbolic Delaunay graph of P can be computed from a Euclidean Delaunay triangulation of P by eliminating each triangle whose circumcircle is not contained in $\mathbb{H}^2$.

Theorem 1 ([31]). *Let P be a finite point set in $\mathbb{H}^2$. A hyperbolic Delaunay graph on P in $\mathbb{H}^2$ is a subgraph of a Euclidean Delaunay triangulation of P when P is regarded as a point set in the Euclidean plane.*

Onishi [30] generalized the algorithm to the d-dimensional upper half-space, in which some faces of the Euclidean Delaunay triangulation are eliminated. That algorithm is reviewed in section 3.

Theorem 2 ([30]). *A hyperbolic Voronoi diagram of n points in the d-dimensional hyperbolic space can be computed in $O(\mathrm{CH}(d+1, n))$ time, where $\mathrm{CH}(d, n)$ is the time to compute a convex hull of n points in the d-dimensional Euclidean space.*

The hyperbolic Voronoi diagram of points in 2-dimensional Poincaré disk is studied by Nilforoushan and Mohades [27]. It can be generalized to the d-dimensional Poincaré disk, and we give an algorithm computing it in section 3.
D_P^d is more similar to $\mathbb{H}^d$ than the Klein disk model.

1. For $p = (p_1, \ldots, p_d)$, $q = (q_1, \ldots, q_d) \in D_P^d$, the hyperbolic distance $d_{D_P^d}(p_1, p_2)$ is given by

$$d_{D_P^d}(p_1, p_2) = \mathrm{arccosh}[1 + 2\frac{\sum_{i=1}^{d}(p_i - q_i)^2}{(1 - \sum_{i=1}^{d} p_i^2)(1 - \sum_{i=1}^{d} q_i^2)}].$$

2. A geodesic in D_P^d is expressed as a portion of a circle or a line orthogonal to the boundary of D_P^d: $\{(x_1, \ldots, x_d) \mid \sum_{i=1}^{d} x_i^2 = 1\}$. In D_P^2, it can be expressed as follows:

$$x_1^2 + x_2^2 - 2(ax_1 + bx_2) + 1 = 0 \; (a^2 + b^2 > 1)$$

or

$$ax_1 = bx_2.$$

3. A hyperbolic hyperplane in D_P^d is defined as following surfaces:

$$\sum_{i=1}^{d}(x_i^2 - 2a_i x_i) + 1 = 0 \,(\sum_{i=1}^{d} a_i^2 > 1),$$

or

$$\sum_{i=1}^{d} a_i x_i = 0 \,(a_i \in \mathbb{R},\, 1 \le i \le d).$$

If $d = 2$, a hyperbolic hyperplane is a geodesic.

4. For a point $a = (a_1, \ldots, a_d) \in D_P^d$ and $r > 0$, a set $c = \{p \in D_P^d \mid d_{D_P^d}(p, a) = r\}$ is expressed as

$$\sum_{j=1}^{d}\{x_j - \frac{2a_j}{2 + (\cosh r - 1)(1 - \sum_{i=1}^{d} a_i^2)}\}^2 = \frac{(\cosh^2 r - 1)(1 - \sum_{i=1}^{d} a_i^2)}{\{2 + (\cosh r - 1)(1 - \sum_{i=1}^{d} a_i^2)\}^2}$$

and thus it is a Euclidean sphere. The point a is called a hyperbolic center and a center of this sphere as a Euclidean one is called a Euclidean center.

Lemma 2. *A bisector of* $p = (p_1, \ldots, p_d)$ *and* $q = (q_1, \ldots, q_d) \in D_P^d$ *is a hyperbolic hyperplane and expressed as follows:*

$$\sum_{i=1}^{d}(p_i - q_i)x_i = 0 \,\,(\textit{if } \sum_{i=1}^{d} p_i^2 = \sum_{i=1}^{d} q_i^2\,),$$

$$\sum_{j=1}^{d}\{x_j - \frac{p_j(1 - \sum_{i=1}^{d} q_i^2) - q_j(1 - \sum_{i=1}^{d} p_i^2)}{\sum_{i=1}^{d} p_i^2 - \sum_{i=1}^{d} q_i^2}\}^2$$

$$= \sum_{j=1}^{d}\{\frac{p_j(1 - \sum_{i=1}^{d} q_i^2) - q_j(1 - \sum_{i=1}^{d} p_i^2)}{\sum_{i=1}^{d} p_i^2 - \sum_{i=1}^{d} q_i^2}\}^2 - 1$$

(otherwise).

Proof. The result can be obtained by $d_{D_P^d}(p, x) = d_{D_P^d}(q, x)$ where $x = (x_1, \ldots, x_d)$. □

In the Klein disk, a geodesic is an open line segment connecting two points of the disk boundary. This can be extended to higher dimensions. Nielsen and Nock [26] proved that a bisector of two points in the space is a hyperplane, which can be regarded as a power bisector [4,20] of two Euclidean balls, and gave the following theorem:

Theorem 3 ([26]). *A hyperbolic Voronoi diagram of n points in the d-dimensional Klein disk is a power diagram projected on the disk.*

2.2 Voronoi Diagrams of Normal Distributions

We describe Voronoi diagrams in information geometry, especially those of normal distributions. First we describe the Voronoi diagram for the Fisher metric, and then explain that for the divergence in a general form. See [3] for information geometry and [30,32,33] for its Voronoi diagrams for more details.

2.3 Voronoi Diagrams of Normal Distributions

We describe Voronoi diagrams in information geometry, especially those of normal distributions. First we describe the Voronoi diagram for the Fisher metric, and then explain that for the divergence in a general form. See [3] for information geometry and [30,32,33] for its Voronoi diagrams for more details.

Fisher Metric A probability density function with parameters can be identified with a point in the space of those parameters. This space is called a statistical parametric space. For d-dimensional normal distributions, its density function is expressed as

$$p(x;\xi) = 2\pi^{-d/2}|\Sigma|^{-1/2}\exp\{-\frac{1}{2}{}^{t}(x-\mu)\Sigma^{-1}(x-\mu)\}$$

where $\mu = (\mu_1,\ldots,\mu_d) \in \mathbb{R}^d$ and Σ is a variance-covariance matrix. In this paper, we only deal with a special case $\Sigma = \sigma^2 I_d$, where $\sigma > 0$ and I_d is the identity matrix of size d. Then, the statistical parametric space is $\{(\mu_1,\ldots,\mu_d,\sigma) \in \mathbb{R}^{d+1} \mid \sigma > 0\}$.

In statistical estimation theory, the Fisher metric on the space plays an important role, which is determined by the Fisher information matrix:

$$\int \left(\frac{\partial}{\partial \xi_i}\log p(x;\xi)\right)\left(\frac{\partial}{\partial \xi_j}\log p(x;\xi)\right)p(x;\xi)dx.$$

For the space of d-dimensional normal distributions $(d \geq 1)$ with $\Sigma = \sigma^2 I_d$, the Fisher information matrix becomes $(1/\sigma^2)\mathrm{diag}[1,\ldots,1,2d]$, and the metric is given by

$$\frac{d\mu_1^2 + \cdots + d\mu_d^2 + 2d(d\sigma)^2}{\sigma^2}$$

which is almost identical with the metric of $\mathbb{H}^{d+1}$. Therefore, Voronoi diagrams of normal distributions by the Fisher metric can be defined naturally, and are almost identical with hyperbolic Voronoi diagrams in $\mathbb{H}^{d+1}$.

General Divergence. Onishi [30] introduced general divergence Voronoi diagrams based on the framework of information geometry [3] as follows where the tensor notation is used partly. Consider a d-dimensional manifold M with a differentiable strictly convex function $\psi: M \to \mathbb{R}$ over coordinate $\theta = (\theta^i) =$

$(\theta^1, \ldots, \theta^d)$, a dual coordinate $\eta = (\eta_i)$ of M is defined via the Legendre transform as

$$\eta_i = \frac{\partial \psi(\theta)}{\partial \theta^i}$$

with a dual differentiable strictly convex function $\phi \colon M \to \mathbb{R}$ over (η_i) satisfying

$$\psi(\theta(p)) + \varphi(\eta(p)) - \sum_{i=1}^{d} \theta^i(p)\eta_i(p) = 0 \quad (p \in M),$$

where the θ-coordinate and η-coordinate of $p \in P$ are expressed as $\theta(p)$ and $\eta(p)$, respectively. The two coordinate systems (θ^i) and (η_i) are dual, and give flat spaces, thus M is called a dually flat space.

For two points $p_1, p_2 \in M$, the ∇-divergence D and the ∇^*-divergence D^* are defined as follows:

$$D(p_1 \parallel p_2) = \psi(\theta(p_1)) + \varphi(\eta(p_2)) - \sum_{i=1}^{d} \theta^i(p_1)\eta_i(p_2)$$

$$D^*(p_1 \parallel p_2) = \psi(\theta(p_2)) + \varphi(\eta(p_1)) - \sum_{i=1}^{d} \theta^i(p_2)\eta_i(p_1)$$

Let P be a finite point set on M. For a point $p \in P$, its Voronoi region in the ∇-*Voronoi diagram* of P is as

$$Vor(p) = \{x \in M \,|\, D(x \parallel p) \le D(x \parallel q) \text{ for } \forall q \in P \setminus \{p\}\}.$$

Similarly, the Voronoi region of the ∇^*-*Voronoi diagram* of P is defined by replacing in the above definition D by D^*.

Onishi [30] obtained the following, where we state it just for the ∇^*-Voronoi diagram:

Theorem 4 ([30]). *For n points p_k ($k = 1, \ldots, n$) in the d-dimensional dually flat space M with φ over (η_i), add a new axis z and consider the upper envelope of n tangent hyperplanes of $z = \varphi$ at $(\eta_i(p_k), \varphi(p_k))$ with respect to the axis z. Then, the projection of the upper envelope onto $M = (\eta_i)$ is the ∇^*-Voronoi diagram for these n points. This can be computed in $O(\mathrm{CH}(d+1, n))$ time.*

Note that the divergence here is the Bregman divergence [11], whose Voronoi diagram is discussed in [25]. Concerning relations of many divergences, see [2].

Kullback-Leibler Divergence. For probability density functions $p(x; \xi)$, the Kullback-Leibler divergence is

$$D_{\mathrm{KL}}(p(x; \xi), p(x; \xi')) = \int p(x; \xi) \log \frac{p(x; \xi)}{p(x; \xi')} dx.$$

For density function $p(x; \xi) = p(x; \mu_1, \ldots \mu_d, \sigma)$ of d-dimensional normal distributions with $|\Sigma| = \sigma^2 I_d$, defining φ by

$$\varphi(\xi) = \int p(x; \xi) \log p(x; \xi) dx,$$

where $-\varphi$ is the entropy function, and setting η_i by

$$(\eta_1, \ldots, \eta_{d+1}) = (\mu_1, \ldots, \mu_d, \sum_{i=1}^{d} \mu_i^2 + d\sigma^2),$$

the space becomes a dually flat space with its dual coordinate given by

$$(\theta^1, \ldots, \theta^{d+1}) = (\frac{\mu_1}{\sigma^2}, \ldots, \frac{\mu_d}{\sigma^2}, -\frac{d}{2\sigma^2})$$

and the general divergence D coincides with the Kullback-Leibler divergence D_{KL}. The corresponding ∇^*-Voronoi diagram with coordinate (η_i) has many meanings such as clustering and statistical estimation.

3   Computing Hyperbolic Voronoi Diagrams from Euclidean Voronoi Diagrams

In this section, we review Onishi's method [30] to compute a hyperbolic Voronoi diagram of points in $\mathbb{H}^d$ from a Euclidean Voronoi diagram of the same points. We show that this method can be applied to a hyperbolic Voronoi diagram of D_P^d, and thus a hyperbolic Voronoi diagram of n points in D_P^d can be computed in CH$(d+1, n)$ time.

Let $\mathcal{H}^d$ be $\mathbb{H}^d$ or D_P^d, and P be a finite point set in $\mathcal{H}^d$. We assume that there is no (Euclidean) sphere tangent to $d+2$ or more points of P. As we mentioned in section 2, a set of points which have the same distance from a given point in $\mathbb{H}^d$ and D_P^d is expressed as a Euclidean sphere which has two centers. A set of all points on faces of a Euclidean Voronoi diagram of P is a locus of Euclidean centers of spheres tangent to two or more points of P. Similarly, a set of all points on faces of a hyperbolic Voronoi diagram of P is a locus of hyperbolic centers of spheres tangent to two or more points of P. We consider a hyperbolic Voronoi region of a point $p_{(i)} \in P$.

For a point $p \in \mathcal{H}^d \setminus P$, we define $S_P(p)$ as a minimum d-dimensional sphere tangent to one more points of P and whose Euclidean center is p, and g as a mapping which maps a point p such that $S_P(p)$ is contained in $\mathcal{H}^d$ to a hyperbolic center of $S_P(p)$. The following lemma states that, for a point p on a hyperbolic Voronoi region of a point $p_{(i)} \in P$, there is a point on a Euclidean Voronoi region of $p_{(i)}$ corresponding to p and that point is a Euclidean center of $S_P(p)$:

Lemma 3. *Let P be a finite set of points. For each $p_{(i)} \in P$, a region $R(p_{(i)}) = \{p \in Vor_E(p_{(i)}) \mid S_P(p)$ is contained in $\mathcal{H}^d \}$, where $Vor_E(p_{(i)})$ is the Voronoi region of $p_{(i)}$ of a Euclidean Voronoi diagram of P, satisfies that $g(R(p_{(i)}))$ is a hyperbolic Voronoi region of $p_{(i)}$ of a hyperbolic Voronoi diagram of P.*

Proof. Let $p_{(i)} \in P$ be an arbitrary point. For every $p \in R(p_{(i)})$, $g(p)$ is a hyperbolic center of $S_P(p)$, and belongs to $Vor_{\mathcal{H}^d}(p_{(i)})$ since $S_P(p)$ is tangent to $p_{(i)}$ and its inside intersects no points. On the other hand, for every $p \in$

$Vor_{\mathcal{H}^d}(p_{(i)})$, there exists a sphere S such that its hyperbolic center is p, it is tangent to $p_{(i)}$, and no point of P is contained in it. Then, clearly a Euclidean center of S is in $Vor_E(p_{(i)})$.   □

By this lemma, if we eliminate faces of a Euclidean Voronoi diagram of P such that, for any point p on them, $S_P(p)$ is not contained in $\mathcal{H}^d$, we can construct all faces of a hyperbolic Voronoi diagram of P from the retained faces of the Euclidean one.

Now, we discuss what faces of the Euclidean Voronoi diagram of P should be retained or eliminated.

Clearly, any vertex v should not be eliminated if a sphere tangent to $d + 1$ points in P which determines v is contained in $\mathcal{H}^d$. Furthermore, for any k-face f $(1 \leq k \leq d-1)$, if there is a k'-face $(k' < k)$ which is a subface of f and should be retained, f should be also retained, and for any k-face f' $(0 \leq k \leq d - 2)$, there are $d - k$ facets whose intersection is f' and f' should be retained if the $d - k$ facets are retained.

f_{ij} denotes a facet of a Euclidean Voronoi diagram of P determined by two points $p_{(i)}, p_{(j)} \in P$ and $b_H(p_{(i)}, p_{(j)})$ denotes a bisector of $p_{(i)}$ and $p_{(j)}$ by the hyperbolic distance in $\mathcal{H}^d$. Let $p_{(i)} \in P$ and $p_{(i')} \in P$ which is a nearest neighbor of P for $p_{(i)}$ by the hyperbolic distance in $\mathcal{H}^d$. Then, there exists an empty sphere tangent to $p_{(i)}$ and $p_{(i')}$ and contained in $\mathcal{H}^d$ because there exists an empty sphere such that its hyperbolic center is $p_{(i)}$ and it is tangent to $p_{(i')}$. Thus, there is a facet $f_{ii'}$ of a Euclidean Voronoi diagram of P determined by $p_{(i)}$ and $p_{(i')}$, and it should not be eliminated. Let f_{ij} be a facet such that $p_{(j)} \in P \setminus \{p_{(i')}\}$ and all vertices contained by f_{ij} should be eliminated. Then, if f_{ij} is retained, there is a retained facet f_{kl} such that it intersects f_{ij} and $b_H(p_{(i)}, p_{(j)})$ intersects $b_H(p_{(k)}, p_{(l)})$ since portions of $b_H(p_{(i)}, p_{(j)})$ and $b_H(p_{(k)}, p_{(l)})$ become facets in a hyperbolic Voronoi diagram of P.

In the algorithm for hyperbolic diagrams given by Onishi [30], a face lattice of a Euclidean Delaunay triangulation is computed, each face of which corresponds to a face of a Euclidean Voronoi diagram. By eliminating some nodes of it, a face lattice of a hyperbolic Delaunay graph is obtained. Onishi [30] proved that a face lattice of a hyperbolic Delaunay graph of points in $\mathbb{H}^d$ is a subgraph of a face lattice of a Euclidean Delaunay triangulation of the same points, and the all nodes of a face lattice of a Euclidean Delaunay triangulation which does not contained in a face lattice of a hyperbolic Delaunay graph can be eliminated in the linear time of the combinatorial complexity of the Euclidean Delaunay triangulation. We know that these results are also satisfied in D_P^d by the above discussion. Thus, we can obtain the following theorem:

Theorem 5. *A hyperbolic Voronoi diagram of n points in D_P^d can be computed from the Euclidean Voronoi diagram of the same points in $O(\mathrm{CH}(d + 1, n))$ time where $\mathrm{CH}(d, n)$ is the time to compute a convex hull of n points in the d-dimensional Euclidean space.*

We omit to the whole of the algorithm. See [30] for more details.

4 Linearizing Hyperbolic Voronoi Diagrams and Voronoi Diagrams of Normal Distributions

We discuss a linearization of hyperbolic Voronoi diagrams in $\mathbb{H}^d$. We apply a linearization for hyperbolic convex hulls by Onishi [30] to hyperbolic Voronoi diagrams. Furthermore, we observe that this linearization is similar to a linearization for Voronoi diagrams by the Kullback-Leibler divergence for normal distributions.

Before discussing the linearization of hyperbolic Voronoi diagrams, we review several results about hyperbolic convex hulls in $\mathbb{H}^d$ of Onishi's thesis [30].

4.1 Hyperbolic Convex Hulls in $\mathbb{H}^d$ and Its Linearization

A hyperbolic convex hull of a finite point set P is defined as the intersection of all convex sets including all points in P. It is represented as an intersection of half-spaces. Each facet of the hyperbolic convex hull is a hyperbolic hyperplane, since any half-space in $\mathbb{H}^d$, which is bounded by a hyperbolic hyperplane, is convex (Lemma 3.2 of [30]) and an intersection of half-spaces is also convex (Corollary 3.2 of [30]).

A hyperbolic convex hull of P can be computed as a Euclidean convex hull (Theorem 3.2 of [30]) by using a linearization technique. Hyperbolic hyperplanes in $\mathbb{H}^d$ can be linearized by using a transformation l defined by

$$l\colon (x_1, \ldots, x_d) \mapsto (X_1, \ldots, X_d)$$

$$\text{s.t. } X_i = x_i \text{ for } i = 1, 2, \ldots, d-1 \text{ and } X_d = \sum_{i=1}^{d} x_i^2.$$

This transformation l maps each half-space in $\mathbb{H}^d$ to a half-space whose boundary is represented as a Euclidean hyperplane. Hence, the hyperbolic convex hull of P is mapped to the Euclidean convex hull of points mapped from those in P by the transformation l. Stating these results, we obtain the following theorem:

Theorem 6. *The hyperbolic convex hull of n points in the d-dimensional hyperbolic space can be computed in $\mathrm{CH}(d, n)$ time.*

The Klein disk provides an alternative linearization as observed in [19].

4.2 Linearizing Hyperbolic Voronoi Diagrams in $\mathbb{H}^d$

Nielsen and Nock [26] proposed a way to compute a hyperbolic Voronoi diagram of points in D_K^d as a power diagram. A hyperbolic Voronoi diagram in $\mathbb{H}^d$ can be obtained by that in D_K^d by a transformation which preserves the distance. Therefore, a hyperbolic Voronoi diagram in $\mathbb{H}^d$ can be also obtained by computing a power diagram.

We introduce another way to compute a hyperbolic Voronoi diagram of points in $\mathbb{H}^d$ as a power diagram by using the linearization technique induced by the mapping l discussed above. $l(\mathbb{H}^d)$ is a space

$$\{(X_1, \ldots, X_d) \mid X_d > \sum_{i=1}^{d-1} X_i^2\},$$

denoted by L^d, and l is a bijection of $\mathbb{H}^d$ to L^d. We define a distance in L^d as

$$d_{L^d}(P_{(1)}, P_{(2)}) := \operatorname{arccosh}\left[\frac{P_{(1)d} + P_{(2)d} - 2\sum_{i=1}^{d-1} P_{(1)i}P_{(2)i}}{2\sqrt{P_{(1)d} - \sum_{i=1}^{d-1} P_{(1)i}^2}\sqrt{P_{(2)d} - \sum_{i=1}^{d-1} P_{(2)i}^2}}\right].$$

where $P_{(k)} = (P_{(k)1}, \ldots, P_{(k)d})$, $k = 1, 2$. Then, l preserves the distance in $\mathbb{H}^d$ since a property $d_{\mathbb{H}^d}(p_{(1)}, p_{(2)}) = d_L(l(p_{(1)}), l(p_{(2)}))$ is satisfied. Therefore a diagram obtained by mapping a hyperbolic Voronoi diagram in $\mathbb{H}^d$ by l is a Voronoi diagram (which is affine) in L^d, and, contrarily, a diagram obtained by mapping a Voronoi diagram in L^d by

$$l^{-1} \colon (X_1, \ldots, X_d) \mapsto (x_1, \ldots, x_d)$$

$$\text{s.t. } x_i = X_i \text{ for } i = 1, 2, \ldots, d-1 \text{ and } x_d = \sqrt{X_d - \sum_{i=1}^{d-1} X_i^2}$$

is a hyperbolic Voronoi diagram in $\mathbb{H}^d$.

Let P be a finite point set in $\mathbb{H}^d$, and $p_{(1)}$ and $p_{(2)}$ are distinct points in P such that hyperbolic Voronoi regions of $p_{(1)}$ and $p_{(2)}$ are adjacent. By l, a bisector of $p_{(1)}$ and $p_{(2)}$ is mapped to the following hyperplane in L^d:

$$\left(\frac{1}{p_{(2)d}} - \frac{1}{p_{(1)d}}\right)X_d - 2\sum_{i=1}^{d-1}\left(\frac{p_{(2)i}}{p_{(2)d}} - \frac{p_{(1)i}}{p_{(1)d}}\right)X_i + \left(\frac{\sum_{i=1}^{d} p_{(2)d}^2}{p_{(2)d}} - \frac{\sum_{i=1}^{d} p_{(1)d}^2}{p_{(1)d}}\right) = 0,$$

It is easy to prove that this hyperplane is a power bisector of two Euclidean balls, which have a center

$$q_{(k)} = \left(\frac{-p_{(k)1}}{p_{(k)d}}, \ldots, \frac{-p_{(k)d-1}}{p_{(k)d}}, \frac{1}{2p_{(k)d}}\right)$$

and a radius

$$r_k = \sqrt{\frac{4\sum_{i=1}^{d-1} p_{(k)i}^2 + 1}{4p_{(k)d}^2} + \frac{\sum_{i=1}^{d} p_{(k)d}^2}{p_{(k)d}}}$$

for $k = 1, 2$. Therefore, a subgraph which is the intersection of a power diagram of these balls and L^d is a Voronoi diagram in L^d.

Theorem 7. *For a set P of n points in $\mathbb{H}^d$, a Voronoi diagram obtained by mapping a hyperbolic Voronoi diagram of P by the above-defined transformation l is a Euclidean power diagram of balls projected on L^d.*

4.3 Two Measures for Normal Distributions in Information Geometry

In information geometry, the Fisher metric and the Kullback-Leibler divergence are measures for normal distributions. We consider linearization of Voronoi diagrams of normal distributions defined by these two measures. We assume that $\Sigma = \sigma^2 I_d$ for the d-dimensional normal distribution.

As we mentioned in section 2, a Voronoi diagram of normal distributions by the Fisher metric is almost identical with a hyperbolic Voronoi diagram in $\mathbb{H}^{d+1}$. Therefore, it can be also linearized by the transformation l defined before. A space obtained by linearizing the statistical parametric space of normal distributions by l is

$$\{(X_1, \ldots, X_d, Y)|X_i = \mu_i \text{ for } 1 \leq i \leq d, Y = \sum_{i=1}^{d} \mu_i^2 + \frac{\sigma^2}{2d}, \mu_i \in \mathbb{R}, \sigma > 0\},$$

which can be considered as a coordinate system $(\mu_1, \ldots, \mu_d, \sum_{i=1}^{d} \mu_i^2 + \sigma^2/2d)$. On the other hand, a Voronoi diagram of normal distributions by the Kullback-divergence is an affine diagram when it is considered as a ∇^*-Voronoi diagram in the η-coordinate system $(\mu_1, \ldots, \mu_d, \sum_{i=1}^{d} \mu_i^2 + d\sigma^2)$ by the ∇^*-divergence. Therefore, we observe that the Voronoi diagrams by two different measures are affine diagrams in the similar coordinate systems.

It is known that ∇-Voronoi diagrams in the θ-coordinate system by ∇^*-divergence, which is a dual of ∇-divergence, are also affine by results of Onishi and Imai [33] and Onishi's thesis [30]. Recently, it was shown that hyperbolic Voronoi diagrams in the Klein disk are affine by Nielsen and Nock [26]. As we have discussed above, there is a similarity between ∇^*-Voronoi diagrams and Voronoi diagrams obtained by linearizing hyperbolic Voronoi diagrams in $\mathbb{H}^{d+1}$, but there is no similarity between ∇-Voronoi diagrams and hyperbolic Voronoi diagrams in D_K^{d+1}. A significance of a property that those Voronoi diagrams are affine has been not found yet.

5 Hyperbolic Voronoi Diagrams and Greedy Embeddings

Hyperbolic Voronoi diagrams have fertile applications. The previous section sheds a new light on relations between hyperbolic Voronoi diagrams and divergence Voronoi diagrams for statistical estimation, whose investigations were initiated in [31,32] in mid 1990's. There has arose a new application of hyperbolic geometry to greedy routing in the field of sensor network and internet routing since Kleinberg [22] demonstrated that greedy routing is always possible in hyperbolic plane in 2007. From the viewpoint of hyperbolic Voronoi diagrams, Kleinberg's result and subsequent ones use a fact that in the hyperbolic plane it is easy to design a point configuration such that the dual of its hyperbolic Voronoi diagram is a tree. In this section we review recent developments along this line and pose a problem of characterizing duals of hyperbolic Voronoi diagrams.

5.1   Greedy Embeddings and Voronoi Diagrams in a Metric Space

Papadimitriou and Ratajczak [35] investigated fundamental properties and research directions of the greedy embedding, which is defined as follows. Let $\mathcal{S}$ be a metric space (e.g. the Euclidean plane or the hyperbolic plane) and $d_{\mathcal{S}}(\cdot, \cdot)$ is a distance on $\mathcal{S}$. For a graph $G = (V, E)$ with a vertex set V and an edge set E, let $p\colon V \to \mathcal{S}$ be an embedding of G to $\mathcal{S}$. p is a greedy embedding if, for every pair of distinct vertices $s, t \in V$, there exists a vertex u adjacent to s such that $d_{\mathcal{S}}(p(u), p(t)) < d_{\mathcal{S}}(p(s), p(t))$. They show that in the Euclidean plane($= \mathcal{S}$) there is a class of graphs with no greedy embedding.

Greedy embeddings relate to Voronoi diagrams. In the seminal paper by Papadimitriou and Ratajczak [35], they consider a kind of Voronoi region in convex embedding of planar graphs, but more clearly Leong, Liskov and Morris [23] define a Voronoi region of v for its neighbors, and expressed as follows:

$$R_{G,p}(v) = \{x \in \mathcal{S} \mid d_{\mathcal{S}}(x, p(v)) < d_{\mathcal{S}}(x, p(w)) \text{ for } \forall w \in V \text{ s.t. } (v, w) \in E \}.$$

In terms of this Voronoi region, we can restate the greedy embeddability more geometrically.

Fact 1. *Let $G = (V, E)$ be a graph and p is an embedding of G. Then, p is a greedy embedding of G if and only if every vertex v satisfies that $R_{G,p}(v)$ contains no other $p(w)$ for $w \in V \setminus \{v\}$.*

Using geometry further, we obtain the following:

Fact 2. *Let $G = (V, E)$ be an undirected graph and p be an embedding of G to $\mathcal{S}$. If $p(G)$ contains a Delaunay graph of $p(V)$ in $\mathcal{S}$ as its subgraph, there is a greedy embedding of G.*

Bose and Morin [10] showed this in the Euclidean plane (extension to any metric space is trivial), and we may say this is implicit in Papadimitriou and Ratajczak [35] as used and mentioned in Ben, Gotsman and Wormser [7].

5.2   Hyperbolic Voronoi Diagrams and Greedy Embeddings

Kleinberg [22] showed that in the hyperbolic plane there always exists a greedy embedding for any graph. Cvetkovski and Crovella [13] extended it to on-line greedy embedding. It is easy to see that, in each method, the upper half-plane is divided by geodesics nesting in a tree structure, and each cell has a representative point in it and two neighboring cells are divided by a bisector between the corresponding representative points. This is nothing but the hyperbolic Voronoi diagram of those representative points with non-intersecting geodesic bisectors (Voronoi edges). Its Delaunay graph is a tree. In fact, in terms of hyperbolic Voronoi diagrams, their results may be said to use the following fact.

Fact 3. *For any tree with n vertices, there exist infinitely many configurations of n points in the hyperbolic plane whose Delaunay graph is the tree.*

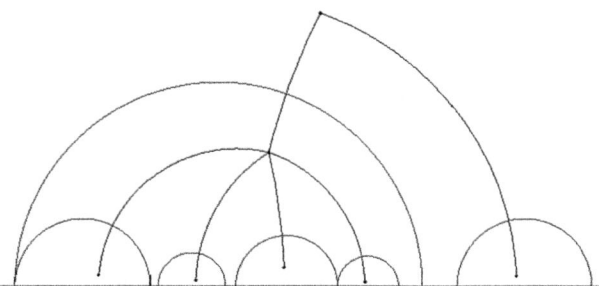

Fig. 2. Realizing a given tree as a Delaunay graph by nesting geodesic half-circles by the tree structure using Lemma 4

This fact is also satisfied in the higher dimensional hyperbolic space. Maymounkov [24] provided greedy embeddings of trees in $\mathbb{H}^3$ with $O(\log^2 n)$ bits.

We show that, actually, a hyperbolic Delaunay graph which is a tree in the d-dimensional hyperbolic space ($d \geq 2$) can be obtained easily. A way we discuss below is similar to an algorithm by Cvetkovski and Crovella, which embeds a spanning tree to D_P^2 so that perpendicular bisectors of edges of the tree do not intersect each other [13].

Lemma 4. *Let p be a point in $\mathbb{H}^d$ and h be a hyperbolic hyperplane in $\mathbb{H}^d$ on which p does not lie. Then, there exists a point $q \in \mathbb{H}^d$ such that q lies in a half-space bounded by h and not containing p and h is a bisector of p and q.*

Proof. Let C_p be a hyperbolic sphere such that its hyperbolic center is p and it is tangent to h at a point p_t. Clearly, there is a sphere C_q with the radius (by the hyperbolic distance) equal to C_p which is tangent to p_t and contained a half-space bounded by h and not containing p. Then, if q is determined as the hyperbolic center of C_q, it satisfies conditions of this lemma. $\square$

Suppose that a point p is below a hyperbolic hyperplane h (i.e., p is contained in the half-sphere determined by h). It is easy to obtain hyperbolic hyperplanes h_i ($i \geq 1$) such that half-spheres determined by them are below h, does not contain p, and does not intersect each other. By Lemma 4, we obtain a point p_i such that h_i is a bisector of p and p_i, and connect all p_i to p as its children. Then, since for all $j \neq k$, h_j and h_k do not intersect each other, an edge connecting p and p_i is an edge of a hyperbolic Delaunay graph. We can repeat this operation for each pair of a geodesic and a point below it, and obtain a hyperbolic Delaunay graph which is a tree. Therefore, we obtain the following theorem.

Theorem 8. *For any $d \geq 2$, any tree can be embedded in $\mathbb{H}^d$ as a hyperbolic Delaunay graph.*

See Figure 2, which is a graph in $\mathbb{H}^2$ obtained by this way.

Combining Fact 2 with Fact 3, we derive a more general understanding of [22,13]. Eppstein and Goodrich [15] pointed out succinct representation of coordinates is necessary in regard of the original *raison d'être* for greedy embeddings, and presented a method satisfying the succinctness in the hyperbolic plane. This method also uses tree-like structures which can be easily handled in the hyperbolic plane.

By the above discussion, the greedy embedding in the hyperbolic plane can be composed by using the property that a hyperbolic Delaunay graph which is a tree can be obtained easily. We expect that there would be more properties of hyperbolic Voronoi diagrams useful for other applications. From this viewpoint, a problem of characterizing hyperbolic Delaunay graphs in practically useful forms would be interesting. A characterization of convex hyperbolic polyhedra and Euclidean Delaunay-related graphs is given in Hodgson, Rivin and Smith [19]. In the Euclidean plane, recognizability of Delaunay graphs was studied algorithmically by Hiroshima, Miyamoto and Sugihara [18].

By Theorem 1 in section 2.1, a hyperbolic Delaunay graph is a subgraph of a Euclidean Delaunay triangulation. Therefore, any graph which cannot be recognized as a hyperbolic Delaunay graph cannot be also recognized as a subgraph of a Euclidean Delaunay triangulation. Furthermore, if a maximal planar graph cannot be recognized as a Euclidean Delaunay triangulation, it cannot be also recognized as a hyperbolic Delaunay graph since the number of its edges is maximal. However, whether other graphs can be recognized as hyperbolic Delaunay graphs is an open problem. In section 2.1, we mention that, by a result of Onishi and Takayama [31], a hyperbolic Delaunay graph of a point set P in $\mathbb{H}^2$ can be obtained from a Euclidean Delaunay graph of P by eliminating each triangle whose circumcircle is not contained in $\mathbb{H}^2$. To understand recognizability of hyperbolic Delaunay graphs, it is necessary to investigate an influence of that elimination.

As we mentioned in section 2.1 and 4, a hyperbolic Voronoi diagram in D_K^2 or L^2 can be considered as a power diagram of Euclidean circles, where L^2 is a space obtained by linearizing $\mathbb{H}^2$. Therefore, the greedy routing in hyperbolic plane is considered as a greedy power routing [7,6] in the Euclidean plane. On the other hand, a greedy power routing in the Euclidean plane cannot be always considered as a greedy routing in the hyperbolic plane since a center and a radius of each Euclidean ball corresponding to a point in D_K^2 or L^2 is determined by the coordinate of the point. It would be interesting to investigate their relation more.

6   Generalization of Hyperbolic Voronoi Diagrams

In this section we give several generalizations of hyperbolic Voronoi diagrams. First, we consider two types of hyperbolic Voronoi diagrams of spheres, one of which is defined by a measure called an additively weighted hyperbolic distance and, the other of which is defined by a different measure called a hyperbolic power. The hyperbolic power can be considered as a generalization of the power in the Euclidean space to in the hyperbolic space.

Secondly, we consider hyperbolic Voronoi diagrams of segments in the 2-dimensional hyperbolic space. Especially, we investigate hyperbolic Voronoi diagrams of geodesic segments.

6.1 Additively Weighted Hyperbolic Voronoi Diagrams of Balls

We define an additively weighted hyperbolic Voronoi diagram and we show that it can be obtained from an additively weighted Euclidean Voronoi diagram of the same balls.

Let $\mathcal{H}^d$ be $\mathbb{H}^d$ or D_P^d, and B be a finite set of balls in $\mathcal{H}^d$. $b = (c, r)$ denotes a ball which has a hyperbolic center $c \in \mathcal{H}^d$ with the radius $r(> 0)$ (which is the hyperbolic distance between c and a point on the boundary of b). We assume that there is no sphere (internal or external) tangent to $d + 2$ or more balls of B in order to simplify the discussion.

For a ball $b = (c, r)$ and a point $p \in \mathcal{H}^d$, an additively weighted hyperbolic distance between p and b, denoted by $d_{\mathcal{H}^d}(p, b)$, is defined as

$$d_{\mathcal{H}^d}(p, b) = d_{\mathcal{H}^d}(p, c) - r.$$

Note that $|d_{\mathcal{H}^d}(p, b)|$ is the minimum hyperbolic distance between p and a point on the boundary of b, because if p does not lie on the boundary of b, there is a sphere internal or external tangent to the boundary of b whose hyperbolic center is p. (If p is not contained in b, such sphere whose hyperbolic center is p is external tangent to b, and if p is contained in b, it is internal tangent to b. Then, $|d_{\mathcal{H}^d}(p, b)|$ is its radius.)

Then an additively weighted hyperbolic Voronoi diagram of B is defined.

Definition 3. *(Additively Weighted Hyperbolic Voronoi Diagram of Balls) Let B be the set of balls defined above. An additively weighted hyperbolic Voronoi diagram of B in $\mathcal{H}^d$ is a partition by the following hyperbolic Voronoi regions of $Vor_{\mathcal{H}^d}(b)$ $(b \in B)$:*

$$Vor_{\mathcal{H}^d}(b) = \{x \in \mathcal{H}^d \mid d_{\mathcal{H}^d}(x, b) \leq d_{\mathcal{H}^d}(x, b') \text{ for } \forall b' \in B, \, b' \neq b\}.$$

Vertices, faces and facets of the additively weighted hyperbolic Voronoi diagram of balls can be defined as well as those of the hyperbolic Voronoi diagram of points.

Any k-face f of the hyperbolic or Euclidean additively weighted Voronoi diagram of B are determined by $d - k + 1$ balls in B, which has the same distance from f, and a point on f is the hyperbolic or Euclidean center of a sphere external or internal tangent to those $d - k + 1$ balls, respectively, or lies on an intersection of the $d - k + 1$ balls. For example, f is a facet of an additively weighted hyperbolic Voronoi diagram and determined by b and b' in B. There are three cases as follows:

1. If a point p on f is not contained in b and b', there is a sphere s external tangent to b and b' whose hyperbolic center is p. Then, the Euclidean center of s is a point lying on a facet of an additively weighted Euclidean Voronoi diagram of B determined by b and b'.
2. If $b \cap b' \neq \emptyset$ and a point p on f lies on boundaries of b and b', p also lies on a facet of an additively weighted Euclidean Voronoi diagram of B determined by b and b'.

3. If $b \cap b' \neq \emptyset$, a point p on f is contained in $b \cap b'$, and p does not lie on boundaries of b and b', there is a sphere s internal tangent to b and b' whose hyperbolic center is p. Then, the Euclidean center of s is a point lying on a facet of an additively weighted Euclidean Voronoi diagram of B determined by b and b'.

Thus, by the same discussion in section 3, we know that the additively weighted hyperbolic Voronoi diagram of B can be computed from the additively weighted Euclidean Voronoi diagram of B. Then, a vertex v of the additively weighted Euclidean Voronoi diagram of B are eliminated, if and only if, v is not an intersection of $d + 1$ balls determining v and the sphere tangent to the $d + 1$ balls is not contained in $\mathcal{H}^d$.

An incremental algorithm computing an additively weighted Euclidean Voronoi diagram of n balls in $O(n^2 \log n + n^{\lceil d/2 \rceil + 1})$ time is given by Boissonnat and Delage [9]. We obtain the following theorem:

Theorem 9. *An additively weighted hyperbolic Voronoi diagram of n balls in $\mathbb{H}^d$ or D_P^d can be computed from an additively weighted Euclidean Voronoi diagram of the same balls in $O(n^2 \log n + n^{\lceil d/2 \rceil + 1})$ time.*

6.2   Hyperbolic Power Diagrams of Balls

We define a hyperbolic Voronoi diagram of balls by a measure different from the hyperbolic distance, which is called a hyperbolic power diagram. We show that a hyperbolic power diagram in $\mathbb{H}^d$ can be linearized by the same way discussed in section 4 and an affine diagram obtained by linearizing a hyperbolic power diagram is part of a power diagram in the Euclidean space.

Let $\mathcal{H}^d$ be $\mathbb{H}^d$ or D_P^d, and B be a finite set of balls (d-dimensional hyperbolic spheres) in $\mathcal{H}^d$. $b = (c, r)$ denotes a ball which has a hyperbolic center $c \in \mathcal{H}^d$ with the radius $r (> 0)$ (which is the hyperbolic distance between c and a point on the boundary of b). For a ball $b = (c, r)$ and a point $p \in \mathcal{H}^d$, a hyperbolic power between p and b, denoted by $pow_{\mathcal{H}^d}(p, b)$, is defined as

$$pow_{\mathcal{H}^d}(p, b) = \frac{\cosh d_{\mathcal{H}^d}(p, c)}{\cosh r}.$$

Note that if $p \notin b$ and q is a tangent point of a tangent geodesic line from p to b, $pow_{\mathcal{H}^d}(p, b) = d_{\mathcal{H}^d}(p, q)$. (Then, a geodesic triangle obtained by connecting p, q and c as its vertices is a right triangle, and, thus, this equation can be obtained by the hyperbolic cosine law.)

By the hyperbolic power, a hyperbolic power diagram of B is defined.

Definition 4. *(Hyperbolic Power Diagram of Balls) Let B be the set of balls defined above. A hyperbolic Power diagram of B in $\mathcal{H}^d$ is a partition by the following hyperbolic power regions of $PR_{\mathcal{H}^d}(b)\,(b \in B)$:*

$$PR_{\mathcal{H}^d}(b) = \{x \in \mathcal{H}^d \mid pow_{\mathcal{H}^d}(x, b) \leq pow_{\mathcal{H}^d}(x, b') \text{ for } \forall b' \in B,\, b' \neq b\}.$$

We discuss a way to compute a hyperbolic power diagram of B and focus our discussions on $\mathbb{H}^d$. In $\mathbb{R}^d$, a power bisector of two balls is a hyperplane. Similarly, in $\mathbb{H}^d$, a hyperbolic power bisector of two balls is a hyperbolic hyperplane.

Lemma 5. *Let $b_1 = (p, r_1)$ and $b_2 = (q, r_2)$ be two balls in $\mathbb{H}^d$, where $p = (p_1, \ldots, p_d)$ and $q = (q_1, \ldots, q_d) \in \mathbb{H}^d$. If the Euclidean centers of b_1 and b_2 are not equal, a hyperbolic power bisector of b_1 and b_2 is a hyperbolic hyperplane, and expressed as follows:*

$$\sum_{i=1}^{d-1} 2(p_i - q_i)x_i - \sum_{i=1}^{d}(p_i^2 - q_i^2) = 0 \; (\text{if } p_d \cosh r_1 = q_d \cosh r_2),$$

$$\sum_{i=1}^{d-1}(x_i - \frac{p_i q_d \cosh r_2 - q_i p_d \cosh r_1}{q_d \cosh r_2 - p_d \cosh r_1})^2 + x_d^2 =$$

$$\sum_{i=1}^{d-1} \frac{(p_i q_d \cosh r_2 - q_i p_d \cosh r_1)^2}{(q_d \cosh r_2 - p_d \cosh r_1)^2} - \sum_{i=1}^{d} \frac{p_i^2 q_d \cosh r_2 - q_i^2 p_d \cosh r_1}{q_d \cosh r_2 - p_d \cosh r_1}$$

(otherwise).

Proof. The result can be obtained by $\mathrm{pow}_{\mathbb{H}^d}(x, b_1) = \mathrm{pow}_{\mathbb{H}^d}(x, b_2)$, where $x = (x_1, \ldots, x_d)$. □

Since hyperbolic power bisectors are expressed as hyperbolic hyperplanes, a hyperbolic power diagram of B in $\mathbb{H}^d$ can be computed from a Euclidean power diagram of balls by the same way discussed in section 4.2.

Let $b_1 = (p_{(1)}, r_1)$ and $b_2 = (p_{(2)}, r_2)$ be two balls of B. By using the linearizing transformation l defined in section 4.1, the hyperbolic power bisector of b_1 and b_2 is mapped to the following hyperplane in $L^d = l(\mathbb{H}^d)$:

$$(\frac{1}{p_{(2)d} \cosh r_2} - \frac{1}{p_{(1)d} \cosh r_1})X_d - 2\sum_{i=1}^{d-1}(\frac{p_{(2)i}}{p_{(2)d} \cosh r_2} - \frac{p_{(1)i}}{p_{(1)d} \cosh r_1})X_i$$

$$+ (\frac{\sum_{i=1}^{d} p_{(2)d}^2}{p_{(2)d} \cosh r_2} - \frac{\sum_{i=1}^{d} p_{(1)d}^2}{p_{(1)d} \cosh r_1}) = 0,$$

It is easy to prove that this hyperplane is a power bisector of two Euclidean balls, which have a center

$$q_{(k)} = (\frac{-p_{(k)1}}{p_{(k)d} \cosh r_k}, \; \ldots \; , \frac{-p_{(k)d-1}}{p_{(k)d} \cosh r_k}, \frac{1}{2p_{(k)d} \cosh r_k})$$

and a radius

$$\tilde{r}_k = \sqrt{\frac{4\sum_{i=1}^{d-1} p_{(k)i}^2 + 1}{4p_{(k)d}^2 \cosh^2 r_k} + \frac{\sum_{i=1}^{d} p_{(k)d}^2}{p_{(k)d} \cosh r_k}}$$

where $k = 1, 2$.

A power diagram of n balls in the Euclidean space has a combinatorial complexity $O(n^{\lceil d/2 \rceil})$, and can be computed in $O(n \log n + n^{\lceil d/2 \rceil})$ time [4]. Therefore, we obtain the following theorem:

Theorem 10. *For a set B of n balls in $\mathbb{H}^d$, an affine diagram obtained by mapping a hyperbolic power diagram of B by the transformation l defined in section 4.1 is a Euclidean power diagram of balls projected on L^d. The hyperbolic power diagram of B has a combinatorial complexity $O(n^{\lceil d/2 \rceil})$, and can be computed in $O(n \log n + n^{\lceil d/2 \rceil})$ time.*

6.3   Hyperbolic Voronoi Diagram of Segments

We discuss hyperbolic Voronoi diagrams of segments in the 2-dimensional upper half-space $\mathbb{H}^2$.

Let S be a finite set of sites in $\mathbb{H}^2$ each of which is a point or an open curved segment such that its endpoints belong to S and it intersects no other sites in S without its endpoints. In order to simplify the discussion, we assume that there is no Euclidean circle tangent to four more sites.

A distance between an open curved segment s and a point $p \in \mathbb{H}^2$ is

$$d_{\mathbb{H}^2}(p, s) = \inf_{q \in s} d_{\mathbb{H}^2}(p, q).$$

Then, we can define hyperbolic Voronoi diagrams of segments as well as Euclidean Voronoi diagrams of segments:

Definition 5. *(Hyperbolic Voronoi Diagram of Segments) Let S be the set of sites defined above. A hyperbolic Voronoi diagram of S in $\mathbb{H}^2$ is a partition by the following hyperbolic Voronoi regions of $Vor_{\mathbb{H}^2}(s)\,(s \in S)$:*

$$Vor_{\mathbb{H}^2}(s) = \{x \in \mathbb{H}^2 \,|\, d_{\mathbb{H}^2}(x, s) \leq d_{\mathbb{H}^2}(x, t) \text{ for } \forall t \in S, t \neq s\}.$$

Any point p on boundaries of hyperbolic Voronoi regions is determined by two or three sites which have the same distance from p. A point which has the same distance from three sites is called a vertex of a hyperbolic Voronoi diagram and a set of points which have the same distance from two sites is called an edge of it. Note that a vertex of a hyperbolic Voronoi diagram is a hyperbolic center of a circle tangent to three sites and a point on an edge of it is a hyperbolic center of a circle tangent to two sites.

We consider what is a hyperbolic center of a Euclidean circle contained in $\mathbb{H}^2$ when its Euclidean center and its radius by the Euclidean distance are given.

Lemma 6. *Let C be a circle such that its Euclidean center is (c_x, c_y), its radius by the Euclidean distance is r and it is contained in $\mathbb{H}^2$. Then, a hyperbolic center of C is $(c_x, \sqrt{c_y^2 - r^2})$.*

Proof. Let (x_h, y_h) be a coordinate of a hyperbolic center of C and r_h be a radius of C by the hyperbolic distance. By the definition of the Euclidean center and the hyperbolic center, the following equations are satisfied:

$$c_x = x_h$$
$$c_y = y_h \cosh r_h$$
$$r = y_h \sinh r_h.$$

By solving the equations about c_x and c_y, we obtain the conclusion.   □

For a point $p \in \mathbb{H}^2 \setminus S$, we define $C_S(p)$ as a minimum Euclidean circle tangent to one more site of S and whose Euclidean center is p, and g as a mapping which maps a point p such that $C_S(p)$ is contained in $\mathbb{H}^2$ to a hyperbolic center of $C_S(p)$.

The following proposition is the extension of the lemma 3 to the set of segments.

Proposition 1. *Let S be the set of sites as defined above. For each $s \in S$, a region $R(s) = \{p \in Vor_E(s) \,|\, C_S(p) \text{ is contained in } \mathbb{H}^2\}$, where $Vor_E(s)$ is the Voronoi region of s of Euclidean Voronoi diagram of S, satisfies that $g(R(s))$ is a hyperbolic Voronoi region of s of a hyperbolic Voronoi diagram of S.*

Proof. Let $s \in S$ be an arbitrary site. For every $p \in R(s)$, $g(p)$ is a hyperbolic center of $C_S(p)$, and belongs to $Vor_{\mathbb{H}^2}(s)$ since $C_S(p)$ is tangent to s and its inside intersects no sites. On the other hand, for every $p \in Vor_{\mathbb{H}^2}(s)$, there exists a circle C such that its hyperbolic center is p, and it is tangent to s and its inside intersects no sites. Then, clearly a Euclidean center of C is in $Vor_E(s)$.   □

6.4   Hyperbolic Voronoi Diagrams of Geodesic Segments

We discuss hyperbolic Voronoi diagrams of S in the case that each site of S is a point in $\mathbb{H}^2$ and a non-intersecting open geodesic segment in $\mathbb{H}^2$ whose endpoints belong to S. First, we consider a hyperbolic distance between a point and a geodesic, and boundaries of a hyperbolic Voronoi region, which consist of bisectors of two sites of S. Then, we show that a hyperbolic Voronoi diagram of S can be computed from the Euclidean Voronoi diagram of S in $O(n \log n)$ time, where $n = |S|$. Furthermore, we consider a hyperbolic Voronoi diagram of geodesic lines, which can be also computed in $O(n \log n)$ time.

We consider a hyperbolic distance between a point and a geodesic segment.

Lemma 7. *Let $p = (p_x, p_y)$ be a point and s be a geodesic in $\mathbb{H}^2$. Then, a hyperbolic distance between p and s is expressed as follows:*

1. *If s is expressed as $(x - a)^2 + y^2 = r^2$ $(a, r \in \mathbb{R}, r > 0)$, then*

$$d_{\mathbb{H}^2}(p, s) = \operatorname{arcsinh}[\frac{1}{2rp_y}|(p_x - a)^2 + p_y^2 - r^2|]$$

2. *If s is expressed as $x = c$ $(c \in \mathbb{R})$, then*

$$d_{\mathbb{H}^2}(p, s) = \operatorname{arcsinh}[\frac{1}{p_y}|p_x - c|]$$

Proof. We only prove the case 1) since a proof of the case 2) is similar to that of the case 1).

Let C be a circle such that its hyperbolic center is p and it tangent to s. Clearly, the hyperbolic distance between p and s, $d = d_{\mathbb{H}^2}(p, s)$ is equal to a radius of C. Let $c = (c_x, c_y)$ and r_C be a Euclidean center and a radius by the

Euclidean distance of C respectively, and $c_s = (a, 0)$ be a Euclidean center of s when s is considered as a Euclidean circle. If $(p_x - a)^2 + p_y^2 > r^2$, the Euclidean distance between c and c_s is equal to $r_c + r$ since c is tangent to s. Thus,

$$\sqrt{(c_x - a)^2 + c_y^2} = r_c + r.$$

By this formula and $c_x = p_x, c_y = p_y \cosh d, r_c = p_y \sinh d$, we can obtain

$$d = \text{arcsinh}[\frac{1}{2rp_y}\{(p_x - a)^2 + p_y^2 - r^2\}].$$

Similarly, if $(p_x - a)^2 + p_y^2 < r^2$, the Euclidean distance between c and c_s is equal to $r - r_c$, and we can obtain

$$d = \text{arcsinh}[\frac{1}{2rp_y}\{r^2 - (p_x - a)^2 - p_y^2\}].$$

Thus,

$$d = \text{arcsinh}[\frac{1}{2rp_y}|(p_x - a)^2 + p_y^2 - r^2|].$$

$\square$

Next, we consider boundaries of a hyperbolic Voronoi region, which consist of bisectors of two geodesic segments and those of a geodesic segment and a point (which can be an endpoint of the geodesic segment). The following lemmas are described about them:

Lemma 8. *Let s_1 and s_2 be geodesics in $\mathbb{H}^2$. Then, bisectors of s_1 and s_2 are geodesics and expressed as follows:*

1. *If s_k $(k = 1, 2)$ is expressed as $(x - a_k)^2 + y^2 = r_k^2$ $(a_k, r_k \in \mathbb{R}, r_k > 0)$, then*

$$\frac{1}{r_1}|(x - a_1)^2 + y^2 - r_1^2| = \frac{1}{r_2}|(x - a_2)^2 + y^2 - r_2^2|.$$

2. *If s_i is expressed as $(x - a_i)^2 + y^2 = r_i^2$ $(a_i, r_i \in \mathbb{R}, r_i > 0)$ and s_j is expressed as $x = c_j$ $(c_j \in \mathbb{R})$, where $i, j = 1, 2, i \neq j$, then*

$$\frac{1}{2r_i}|(x - a_i)^2 + y^2 - r_i^2| = |x - c_j|.$$

3. *If s_k $(k = 1, 2)$ is expressed as $x = c_k$ $(c_k \in \mathbb{R})$, then*

$$x = \frac{c_1 + c_2}{2}.$$

Proof. We can obtain the conclusion by $d_{\mathbb{H}^2}(q, s_1) = d_{\mathbb{H}^2}(q, s_2)$ where $q = (x, y)$.

$\square$

Lemma 9. *Let $p = (p_x, p_y)$ be a point and s be a geodesic in $\mathbb{H}^2$. Then, the bisector of p and s is a solution of the following equations:*

1. If s is expressed as $(x-a)^2 + y^2 = r^2 \, (a, r \in \mathbb{R}, \, r > 0)$, then

$$\frac{1}{r}|(x-a)^2 + y^2 - r^2| = \frac{\sqrt{\{(x-p_x)^2 + (y-p_y)^2\}\{(x-p_x)^2 + (y+p_y)^2\}}}{p_y}.$$

2. If s is expressed as $x = c \, (c \in \mathbb{R})$, then

$$|x - c| = \frac{\sqrt{\{(x-p_x)^2 + (y-p_y)^2\}\{(x-p_x)^2 + (y+p_y)^2\}}}{2p_y}.$$

Proof. We can obtain the conclusion by $d_{\mathbb{H}^2}(q, s) = d_{\mathbb{H}^2}(q, p)$ where $q = (x, y)$, since $d_{\mathbb{H}^2}(q, p)$ can be expressed as

$$d_{\mathbb{H}^2}(q, p) = \operatorname{arcsin}\left[\frac{\sqrt{\{(x-p_x)^2 + (y-p_y)^2\}\{(x-p_x)^2 + (y+p_y)^2\}}}{2yp_y}\right].$$

$\square$

Lemma 10. *Let S be the set of sites defined above, and $s \in S$ be an open geodesic segment whose endpoints are $p_1 = (x_1, y_1), p_2 = (x_2, y_2) \in S$. A boundary of hyperbolic Voronoi regions of s and p_1 of a hyperbolic Voronoi diagram of S is expressed as follows:*

$$\{x - (x_1 + \alpha y_1)\}^2 + y^2 = (1 + \alpha^2)y_1^2$$

or

$$x = x_1$$

where α is a constant depending on x_1, y_1, x_2, y_2.

Proof. Let

$$B = \{p \in \mathbb{H}^2 \mid d_{\mathbb{H}^2}(s, p) = d_{\mathbb{H}^2}(p_1, p)\}$$

and

$$B_E = \{p \in \mathbb{H}^2 \mid C_S(p) \subset \mathbb{H}^2 \text{ and } d_E(s, p) = d_E(p_1, p)\}$$

where $d_E(\cdot, \cdot)$ is the Euclidean distance. By Lemma 1, $B = g(B_E)$ where g is a mapping defined in the previous subsection. B_E is a subset of a line l expressed as $y = \alpha(x - x_1) + y_1$ or $x = x_1$ where α is a constant depending on x_1, y_1, x_2, y_2. If l is expressed as $x = x_1$. B is also expressed as $x = x_1$ clearly.

We assume that l is expressed as $y = \alpha(x - x_1) + y_1$. Since, for each $p \in B_E$, a radius by the Euclidean distance of $C_S(p)$ is equal to the Euclidean distance between p_1 and p, by the definition of g and Lemma 6,

$$g(B_E) = \{(x, \sqrt{y^2 - (x - x_1)^2 - (y - y_1)^2}) \mid y = \alpha(x - x_1) + y_1\}.$$

It is easy to see that $g(B_E)$ is expressed as an equation in this lemma. $\square$

Now, we show that a hyperbolic Voronoi diagram of non-intersected geodesic segments can be computed in $O(n \log n)$ time ($n = |S|$). A hyperbolic Voronoi diagram of a point set can be computed from a Euclidean Delaunay triangulation of it in $O(n \log n)$ time (see section 2.1). We state that a hyperbolic Voronoi diagram of S can be computed from a Euclidean Voronoi diagram of S by the similar way.

A set of all points on vertices and edges of a Euclidean Voronoi diagram of sites is a locus of Euclidean centers of circles tangent to the sites. Similarly, a set of all points on vertices and edges of a hyperbolic Voronoi diagram of sites is a locus of hyperbolic centers of circles tangent to the sites. We consider a hyperbolic Voronoi region of a site $s \in S$. By Lemma 1, every point of $Vor_{\mathbb{H}^2}(s)$ corresponds to a point of $Vor_E(s)$, which is a Voronoi region of s of a Euclidean Voronoi diagram of S. It is clear that for any vertex v of $Vor_{\mathbb{H}^2}(s)$, a Euclidean center of a circle tangent to three sites such that a hyperbolic center of the circle is v is a vertex of $Vor_E(s)$. For any edge e of $Vor_{\mathbb{H}^2}(s)$, a set of all Euclidean centers of circles tangent to two sites, each of which has the same distance from e, is a portion of an edge of $Vor_E(s)$. Thus, we can obtain a hyperbolic Voronoi diagram of S from a Euclidean Voronoi diagram of S by eliminating vertices and portions of edges such that circles determining them are not contained in $\mathbb{H}^2$.

Each vertex v of hyperbolic Voronoi diagrams is determined by three sites s_1, s_2 and s_3 such that $d(s_1, v) = d(s_2, v) = d(s_3, v)$ and each edge e of them is determined by two sites s_1, s_2 such that $d(s_1, p) = d(s_2, p)$ for $\forall p \in e$. $\langle s_1, s_2, s_3 \rangle$ denotes a vertex determined by $s_1, s_2, s_3 \in S$ and $\langle s_1, s_2 \rangle$ denotes an edge determined by $s_1, s_2 \in S$. From now on, we state what vertices and edges should be eliminated from a Euclidean Voronoi diagram of S to obtain a hyperbolic Voronoi diagram of S. For a vertex $v = \langle s_1, s_2, s_3 \rangle$, if a circle tangent to s_1, s_2 and s_3 is not contained in $\mathbb{H}^2$, v should be eliminated. Let $e = \langle s_1, s_2 \rangle$ be an edge. If both or one of endpoints (which are vertices) of e is not an infinite point or should not be eliminated, e is not eliminated. We assume that each endpoint $p_i \in S$ ($i = 1, 2$) of $\langle s_1, s_2 \rangle$ is an infinite point or a point which should be eliminated. Let $C(s_1, s_2, p_i)$ ($i = 1, 2$) be a circle tangent to s_1, s_2 and p_i and $h(s_1, s_2, p_i)$ be a half-plane such that its boundary is a line passing through tangent points $C(s_1, s_2, p_i) \cap s_1$ and $C(s_1, s_2, p_i) \cap s_2$ and it does not contain p_i. Then, if $C(s_1, s_2, p_i) \cap h(s_1, s_2, p_i)$ is not contained in $\mathbb{H}^2$, e should be eliminated, since e is a locus of Euclidean centers of circles tangent to s_1 and s_2, and s_1 and s_2 are in $\mathbb{H}^2$.

Therefore, we obtain the following theorem:

Theorem 11. *Let S be a finite set of sites each of which is a point or a non-intersecting open geodesic segment whose endpoints belong to S, and $n = |S|$. A hyperbolic Voronoi diagram of S has a combinatorial complexity $O(n)$ and can be computed in time $O(n \log n)$.*

Proof. A combinatorial complexity of a Euclidean Voronoi diagram of S is $O(n)$, and each vertex, edge or Voronoi region of a hyperbolic Voronoi diagram of S corresponds to that of a Euclidean Voronoi diagram of S. Thus, a combinatorial complexity of a hyperbolic Voronoi diagram is $O(n)$.

A Euclidean Voronoi diagram $VD_E(S)$ of S can be computed by Yap's divide-and-conquer algorithm [39] in $O(n \log n)$ time. For each vertex and edge of $VD_E(S)$, by using the way discussed above, we can test whether it should be eliminated in a constant time. Since the number of vertices and edges is $O(n)$, we can eliminate vertices and edges, such that for each of them there is no vertex or an edge of a hyperbolic Voronoi diagram of S corresponding it, from $VD_E(S)$ in $O(n)$ time. Thus, a hyperbolic Voronoi diagram of S can be computed in $O(n \log n)$ time.   □

In Theorem 11, each site of S is a point or a geodesic segment, but even if S includes geodesic lines which intersect no other sites, a hyperbolic Voronoi diagram of S can be computed by the same way since geodesic lines are circular arcs whose endpoints are points on x-axis.

In section 5, we mention that a Delaunay graph which is a tree can be obtained easily. Its dual, which is a hyperbolic Voronoi diagram, is a set of non-intersecting geodesic lines, and we can compute a hyperbolic Voronoi diagram of such non-intersecting geodesic lines by using the way discussed above. Since a bisector of two geodesic lines is a geodesic by Lemma 8, each edge of a hyperbolic Voronoi diagram of geodesic lines is a geodesic as well as that of a hyperbolic Voronoi diagram of points. Therefore we can obtain the following theorem:

Theorem 12. *A hyperbolic Voronoi diagram of n non-intersecting geodesic lines has a combinatorial complexity $O(n)$ and can be computed in $O(n \log n)$ time where its edges are again geodesics.*

It is interesting that each edge of a hyperbolic Voronoi diagram of points, whose dual is a hyperbolic Delaunay graph which is a tree, is a geodesic, and each edge of a hyperbolic Voronoi diagram of geodesic lines, which are edges of such a hyperbolic Voronoi diagram of points, is also a geodesic. This property would help the investigation of hyperbolic Delaunay graphs which are trees.

7   Concluding Remarks

We have revealed several new aspects of hyperbolic Voronoi diagrams, and demonstrate their deep structures in theory and applications. In this paper, we have generalized our previous work [38] more, by considering the several results in the higher dimensional hyperbolic space and giving new generalizations of hyperbolic Voronoi diagrams, which are additively weighted hyperbolic Voronoi diagrams and hyperbolic power diagrams.

One of open issues left here is to give a nice characterization of the hyperbolic Delaunay graph in connection with the greedy embedding.

Besides, we are interested in more investigation of hyperbolic power diagrams. In section 6.2, we observed that an affine diagram obtained by linearizing a hyperbolic power diagram is equivalent to part of a power diagram in the Euclidean space. Power diagrams in the Euclidean space relate to other geometric structures such as arrangements and polyhedra [4]. Thus, investigating the relation between hyperbolic power diagrams and other hyperbolic geometric structures may lead to more understanding both hyperbolic and Euclidean geometric structures. In addition, a characterization of a dual graph of the hyperbolic power diagram connecting with the greedy power routing is also an interesting open issue as well as that of a hyperbolic Delaunay graph.

As another research direction, Voronoi diagrams with some Fisher metric in the space of quantum states would be interesting. Voronoi diagrams for divergences have been considered since Oto, Imai and Imai [34] investigated 1-qubit state space with respect to the quantum divergence. Kato et al. [21] surveyed results obtained so far with respect to quantum distances. As underlying geometric structure the diagram has been used to find a quantum channel whose Holevo capacity is attained by 4 signals in [17]. In the quantum case, the Fisher metric is not uniquely determined and several kinds of Fisher-type metrics exist, each being useful in different contexts (e.g., see [16]). Investigating diagrams with respect to such diagrams geometrically would be of theoretical interest.

Acknowledgment. The authors would like to thank Keiko Imai for useful comments on hyperbolic geometry. This work was supported by Special Coordination Funds for Promoting Science and Technology.

References

1. Alt, H., Cheong, O., Vigneron, A.: The Voronoi diagram of curved objects. Discrete & Computational Geometry 34, 439–453 (2005)
2. Amari, S.: α-divergence is unique, belonging to both f-divergence and Bregman divergence classes. IEEE Transactions on Information Theory 55(11), 4925–4931 (2009)
3. Amari, S., Nagaoka, H.: Methods of Information Geometry. Oxford University Press (2000)
4. Aurenhammer, F.: Power diagrams: Properties, algorithms and applications. SIAM Journal of Computing 16(1), 78–96 (1987)
5. Aurenhammer, F.: Voronoi diagrams: a survey of a fundamental geometric data structure. ACM Computing Surveys 23, 345–405 (1991)
6. Ben-Chen, M., Gotsman, C., Gortler, S.J.: Routing with guaranteed delivery on virtual coordinates. In: Proceedings of the 18th Canadian Conference on Computational Geometry, pp. 117–120 (2006)
7. Ben-Chen, M., Gotsman, C., Wormser, C.: Distributed computation of virtual coordinates. In: Proceedings of the 23rd Annual Symposium on Computational Geometry (SCG 2007), pp. 210–219 (2007)
8. Boissonnat, J.-D., Karavelas, M.: On the combinatorial complexity of Euclidean Voronoi cells and convex hulls of d-dimensional spheres. In: 14th ACM-SIAM Symposium on Discrete Algorithms, pp. 305–312 (2003)
9. Boissonnat, J.-D., Delage, C.: Convex Hull and Voronoi Diagram of Additively Weighted Points. In: Proceedings of the 13th Annual European Symposium on Algorithms, pp. 367–378 (2005)
10. Bose, P., Morin, P.: Online Routing in Triangulations. In: Aggarwal, A.K., Pandu Rangan, C. (eds.) ISAAC 1999. LNCS, vol. 1741, pp. 113–122. Springer, Heidelberg (1999)

11. Bregman, L.M.: The relaxation method of finding the common point of convex sets and its application to the solution of problems in convex programming. USSR Computational Mathematics and Mathematical Physics 7(3), 200–217 (1967)
12. Cannon, J.W., Floyd, W.J., Kenyon, R., Parry, W.R.: Hyperbolic Geometry. Flavors of Geometry MSRI Publications 31, 59–115 (1997)
13. Cvetkovski, A., Crovella, M.: Hyperbolic embedding and routing for dynamic graphs. In: Proceedings of the 28th IEEE Conference on Computer Communications (INFOCOM 2009), pp. 1647–1655 (2009)
14. Drysdale, R.L., Lee, D.T.: Generalized Voronoi diagrams in the plane. In: Proc. 16th Allerton Conference on Communication, Control, and Computing, pp. 833–842 (1978)
15. Eppstein, D., Goodrich, M.T.: Succinct Greedy Graph Drawing in the Hyperbolic Plane. In: Tollis, I.G., Patrignani, M. (eds.) GD 2008. LNCS, vol. 5417, pp. 14–25. Springer, Heidelberg (2009)
16. Hayashi, M.: Quantum Information — An Introduction. Springer, Berlin (2006)
17. Hayashi, M., Imai, H., Matsumoto, K., Ruskai, M.B., Shimono, T.: Qubit channels which require four inputs to achieve capacity: implications for additivity conjectures. Quantum Information and Computation 5, 13–31 (2005)
18. Hiroshima, T., Miyamoto, Y., Sugihara, K.: Another proof of polynomial-time recognizability of Delaunay graphs. IEICE Trans. Fundamentals E83-A(4), 627–638 (2000)
19. Hodgson, C.D., Rivin, I., Smith, W.: A characterization of convex hyperbolic polyhedra and of convex polyhedra inscribed in the sphere. Bulletin of the American Mathematical Society 27(2), 246–251 (1992)
20. Imai, H., Iri, M., Murota, K.: Voronoi diagram in the Laguerre geometry and its applications. SIAM Journal on Computing 14, 93–105 (1985)
21. Kato, K., Oto, M., Imai, H., Imai, K.: Computational geometry analysis of quantum state space and its applications. In: Gavrilova, M.L. (ed.) Generalized Voronoi Diagrams, pp. 67–108. Springer, Berlin (2009)
22. Kleinberg, R.: Geographic routing using hyperbolic space. In: Proceedings of the 26th Annual IEEE Conference of the IEEE Computer Communications (INFOCOM 2007), pp. 1902–1909 (2007)
23. Leong, B., Liskov, B., Morris, R.: Greedy virtual coordinates for geographic routing. In: Proceedings of the IEEE International Conference on Network Protocols (ICNP 2007), pp. 71–80 (2007)
24. Maymounkov, P.: Greedy embeddings, trees, and Euclidean vs. Lobachevsky geometry, Online manuscript, M.I.T. (2006),
    http://pdos.csail.mit.edu/petar/papers/maymounkov-greedy-prelim.pdf
25. Nielsen, F., Boissonnat, J.-D., Nock, R.: On Bregman Voronoi diagrams. In: Proceedings of the 18th Annual ACM-SIAM Symposium on Discrete algorithms (SODA 2007), pp. 746–755. SIAM (2007)
26. Nielsen, F., Nock, R.: Hyperbolic Voronoi diagrams made easy. In: Proceedings of the 2010 International Conference on Computational Science and Its Applications (ICCSA 2010), Fukuoka (2010)
27. Nilforoushan, Z., Mohades, A.: Hyperbolic Voronoi Diagram. In: Gavrilova, M.L., Gervasi, O., Kumar, V., Tan, C.J.K., Taniar, D., Laganá, A., Mun, Y., Choo, H. (eds.) ICCSA 2006. LNCS, vol. 3984, pp. 735–742. Springer, Heidelberg (2006)
28. Nilforoushan, Z., Mohades, A., Rezaii, M.M., Laleh, A.: 3D hyperbolic Voronoi diagrams. Computer-Aided Design 42(9), 759–767 (2010)

29. Okabe, A., Boots, B., Sugihara, K., Chiu, S.N.: Spatial Tessellations: Concepts and Applications of Voronoi diagrams, 2nd edn. Wiley Series in Probability and Statistics. Wiley (2000)
30. Onishi, K.: Riemannian Computational Geometry - Convex Hull, Voronoi Diagram and Delaunay-type Triangulation, Doctoral Thesis, Department of Computer Science, University of Tokyo (1998)
31. Onishi, K., Takayama, N.: Construction of Voronoi diagram on the upper half-plane. IEICE Trans. Fundamentals E79-A(4), 533–539 (1996)
32. Onishi, K., Imai, H.: Voronoi diagram in statistical parametric space by Kullback-Leibler divergence. In: Proceedings of the 13th ACM Annual Symposium on Computational Geometry (SCG 1997), pp. 463–465 (1997)
33. Onishi, K., Imai, H.: Voronoi diagrams for an exponential family of probability distributions in information geometry. In: Proceedings of the Japan-Korea Joint Workshop on Algorithms and Computation, Fukuoka, pp. 1–8 (1997)
34. Oto, M., Imai, H., Imai, K.: Computational geometry on 1-qubit quantum states. In: Proceedings of the International Symposium on Voronoi Diagrams in Science and Engineering, Tokyo, pp. 145–151 (2004)
35. Papadimitriou, C., Ratajczak, D.: On a conjecture related to geometric routing. Theoretical Computer Science 344(1), 3–14 (2005)
36. Rong, G., Jin, M., Guo, X.: Hyperbolic centroidal Voronoi tessellation. In: Proceedings of the 14th ACM Symposium on Solid and Physical Modeling, pp. 117–126 (2010)
37. Sugihara, K.: Laguerre Voronoi diagram on the sphere. Journal for Geometry and Graphics 6(1), 69–81 (2002)
38. Tanuma, T., Imai, H., Moriyama, S.: Revisiting Hyperbolic Voronoi Diagrams from Theoretical, Applied and Generalized Viewpoints. In: Proceedings of the International Symposium on Voronoi Diagrams in Science and Engineering 2010, pp. 23–32 (2010)
39. Yap, C.K.: An $O(n \log n)$ algorithm for the Voronoi diagram of a set of simple curve segments. Discrete & Computational Geometry 2, 365–393 (1987)

Mollified Zone Diagrams and Their Computation

Sergio C. de Biasi[1], Bahman Kalantari[1], and Iraj Kalantari[2]

[1] Department of Computer Science,
Rutgers University,
New Brunswick, NJ, USA
{sdebiasi,kalantari}@cs.rutgers.edu
[2] Department of Mathematics,
Western Illinois University,
Macomb, IL, USA
i-kalantari@wiu.edu

Abstract. The notion of the *zone diagram* of a finite set of points in the Euclidean plane is an interesting and rich variation of the classical Voronoi diagram, introduced by Asano, Matoušek, and Tokuyama [1]. In this paper, we define mollified versions of zone diagram named *territory diagram* and *maximal territory diagram*. A zone diagram is a particular maximal territory diagram satisfying a sharp dominance property. The proof of existence of maximal territory diagrams depends on less restrictive initial conditions and is established via Zorn's lemma in contrast to the use of fixed-point theory in proving the existence of the zone diagram. Our proof of existence relies on a characterization which allows embedding any territory diagram into a maximal one. Our analysis of the structure of maximal territory diagrams is based on the introduction of a pair of dual concepts we call *safe zone* and *forbidden zone*. These in turn give rise to computational algorithms for the approximation of maximal territory diagrams. Maximal territory diagrams offer their own interesting theoretical and computational challenges, as well as insights into the structure of zone diagrams. This paper extends and updates previous work presented in [4].

Keywords: computational geometry, Voronoi diagram, zone diagram, mollified zone diagram, territory diagram, maximal territory diagram.

1   Introduction and Motivation

Asano, Matoušek, Tokuyama [1] define a *zone diagram* as a new variation of the classical notion of Voronoi diagram for a given finite set of points in the Euclidean plane. (For a survey of many results on Voronoi diagrams, see [3].) The notion of a zone diagram and its existence is the main motivation behind our defining mollified versions we call *territory diagrams* and *maximal territory diagrams*, as is their study in this article. However, this study is also motivated by an intriguing relationship between approximations to Voronoi diagrams and certain regions of attraction in polynomial root-finding [5,6].

M.L. Gavrilova et al. (Eds.): Trans. on Comput. Sci. XIV, LNCS 6970, pp. 31–59, 2011.
© Springer-Verlag Berlin Heidelberg 2011

Each of these concepts (zone diagram, territory diagram, maximal territory diagram) shall refer to a pair $(P, \mathbf{R})$ where $P = \langle p_1, \ldots, p_n \rangle$ is an n-tuple of points (also called *sites*) in $\mathbb{R}^2$ and $\mathbf{R} = \langle R_1, \ldots, R_n \rangle$ is an n-tuple of subsets of $\mathbb{R}^2$ (also called *regions* or *zones*) whose precise description requires some preliminary definitions.

Definition 1. (*Distance from a point to a region*) *For a set $X \subseteq \mathbb{R}^2$ and a point p in $\mathbb{R}^2$, the distance from p to X is defined as*

$$d(p, X) \equiv \inf \{d(p, z) : z \in X\} , \tag{1}$$

where $d(\cdot, \cdot)$ denotes the Euclidean distance.

Remark 1. It is well-known that if X is convex and closed then $d(p, X)$ is attained at a unique point $z \in X$.

Definition 2. (*Dominance region*) *Given a set $X \subseteq \mathbb{R}^2$ and a point p in $\mathbb{R}^2$, the dominance region of p with respect to X, $\mathrm{dom}(p, X)$, is defined as*

$$\mathrm{dom}(p, X) \equiv \{z \in \mathbb{R}^2 : d(z, p) \leq d(z, X)\} . \tag{2}$$

In other words, $\mathrm{dom}(p, X)$ is the set of all points in the plane closer to p than to X. In particular, $\mathrm{dom}(p, X)$ is closed, convex and necessarily contains p.

In the special case where X is a finite set of points, $\mathrm{dom}(p, X)$ represents the closure of the Voronoi region of the point p with respect to the set $\{p\} \bigcup X$. An elementary geometric proof of the convexity of $\mathrm{dom}(p, X)$ for an arbitrary set X, also generalizing to higher dimensional Euclidean spaces, can be deduced from Fig. 1 which gives a visual proof of the following property.

Proposition 1. *If u, v are two points that are closer to p than to an arbitrary point $z \in X$ (or possibly equidistant to p and z), then so is any convex combination w of u and v.*

Proof. The proof of the proposition follows from the fact that if we draw the perpendicular bisector of the line segment pz, it divides the plane into two half-spaces which are convex, so any two points u and v that are closer to p than to z must lie in the half-space containing p (see Fig. 1). □

Within this framework, zone diagrams are defined in [1] as follows.

Definition 3. (*Zone diagram*) *An n-tuple $\mathbf{R} = \langle R_1, \ldots, R_n \rangle$ of sets in $\mathbb{R}^2$ is a zone diagram for an n-tuple of points $P = \langle p_1, \ldots, p_n \rangle$ in $\mathbb{R}^2$ if for each $i = 1, \ldots, n$ we have*

$$R_i = \mathrm{dom}\left(p_i, \bigcup_{j \neq i} R_j\right) . \tag{3}$$

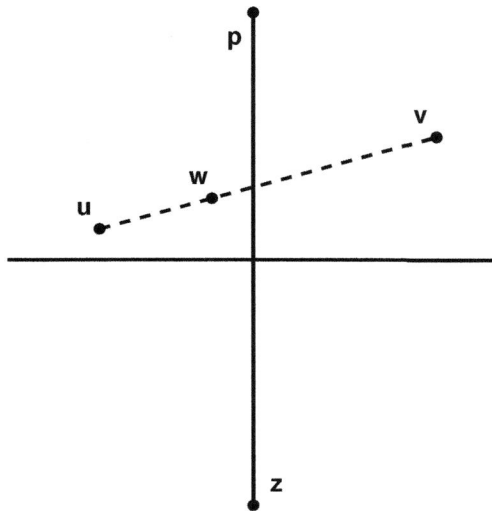

Fig. 1. Verification that if u, v are closer to p than to z so is any of their convex combinations

In other words, R_i is the set of all points in the plane that are closer to p_i than to all $R_j, j \neq i$, that is, $z \in R_i \Leftrightarrow d(z, p_i) \leq d(z, R_j), \ j \neq i$. In particular, each R_i is closed, convex and necessarily contains p. Unlike what happens for Voronoi diagrams, the union of all the dominance regions R_i does not equal the whole plane; a region $N = \mathbb{R}^2 - \bigcup R_i$ (called the *neutral zone*) is left outside all of the R_is such that no point in this region is closer to any p_i than to all $R_j, j \neq i$. It is clear that each R_i must be contained in the Voronoi cell for p_i (see Fig. 2).

Definition 3 turned out to be extremely rich, presenting mathematical challenges (e.g. currently no algebraic description of the border of such regions in known), algorithmic challenges (e.g. currently no tight analysis of the complexity of computing these regions within a given margin of error is known) as well as providing interesting interpretations.

One of the interpretations for zone construction is given in [1] as follows. Consider that n parties are competing for territory, and each party i can incorporate to its region R_i all the points on the plane that it can defend more effectively than any of its enemies can attack it. Each party has a single defense headquarters located at p_i such that it can defend a given point z on the plane with 'reaction time' $d(p_i, z)$. On the other hand, it can launch an attack on any point z on the plane with reaction time of $d(x, z)$ where x is any point in R_i. A zone diagram represents a situation where no party can conquer any additional territory.

In this paper, we consider the following *mollified* variation of that concept. Suppose that in the name of peace and through diplomacy all parties come to an

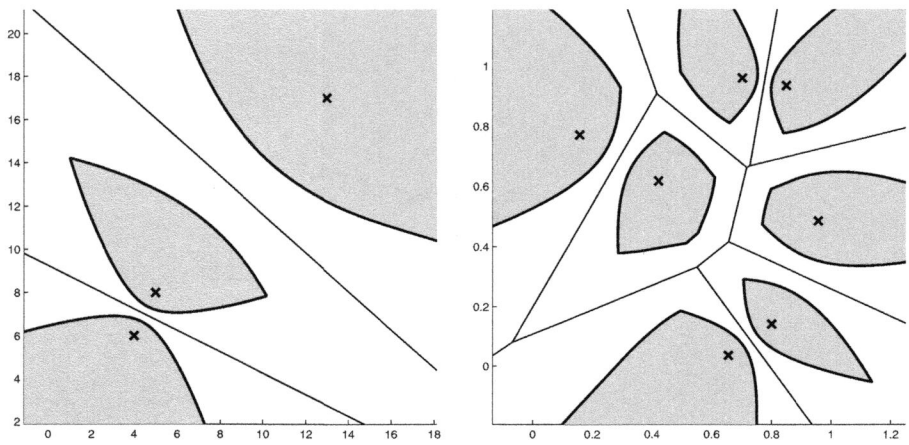

Fig. 2. Example of zone diagrams for 3 sites (*left*) and for 7 sites (*right*). The Voronoi diagrams for the same sites are also shown.

agreement that after a given party claims some territory, it will not be contested by the other parties, and to show good will, each party i agrees not to claim new territory x that is closer to some other party's territory $y \in R_j$ than this territory y is to the party's defense headquarters p_j (since that would allow i to attack and successfully claim the territory y).

In other words, each party i can (as before) only grow its region R_i if this does not produce dominance violations for itself, but now only if this also does not produce dominance violations for any other site. In particular, regions can now only grow and never shrink.

In this situation, a given party will be able to conquer a larger fraction of the originally unclaimed points than in a zone diagram by claiming territory sooner (or more frequently) than other parties.

Additionally, the borders of the resulting regions might contain points that are closer to their home site than to any other region but still unable to expand the region due to the fact that such an expansion would only be possible if accompanied by a corresponding shrinking of some other region(s).

In the following sections, we make these ideas more precise. Territory diagrams as well as maximal territory diagrams are formally defined and then their structure is discussed. Notice that this paper extends and updates the concepts and terminology originally introduced in [4].

2   Mollified Zone Diagrams

2.1   Territory Diagrams

We now define territory diagrams and maximal territory diagrams for a finite n-tuple of points P.

Definition 4. *(Territory diagram)* *Given* $P = \langle p_1, \ldots, p_n \rangle$ *where* $p_i \in \mathbb{R}^2$ *and* $\mathbf{R} = \langle R_1, \ldots, R_n \rangle$ *where* $R_i \subseteq \mathbb{R}^2$ *with* $p_i \in R_i$, *we say* $(P, \mathbf{R})$ *is a territory diagram if for each* $i = 1, \ldots, n$ *we have*

$$R_i \subseteq \mathrm{dom}\left(p_i, \bigcup_{j \neq i} R_j\right). \tag{4}$$

In other words, each R_i must be contained in the set of all points in the plane that are closer to p_i than to all $R_j, j \neq i$, that is, $z \in R_i \Rightarrow d(z, p_i) \leq d(z, R_j), j \neq i$.

Note that this is a mollified version of a zone diagram in which strict equality of each region to the dominance zone of the corresponding site is relaxed to the less restrictive requirement of containment. The borders of each region are in a sense made "softer" than in a zone diagram.

As for zone diagrams, we will call the part of the plane not assigned to any site the *neutral zone* for this particular territory diagram. In the case of a zone diagram, the addition of a point from the neutral zone to any given region will always violate the tight relation of equality for *both* that region as well as for at least one neighboring region; the resulting new association of regions to sites would not be a zone diagram. For a territory diagram, however, issues are subtler, since territory diagrams do not require a perfect match between regions and dominance. Thus if the the addition of a point y from the neutral zone to a region R_i associated to a site p_i does not create any violation of Definition 4, the result is a new territory diagram. But alternatively, the new point y might violate Definition 4 either by being outside the dominance zone for p_i *or* (independently) by forcing a shrinking of some *other* site's dominance zone so that its current region is no longer contained in it. This is a critical point of departure and nuance of main interest in this paper.

Unlike zone diagrams, it does *not* follow from Definition 4 that each R_i must be convex. If every region R_i in $\mathbf{R}$ is convex, then we call $(P, \mathbf{R})$ a *convex territory diagram*.

If every region R_i in $\mathbf{R}$ is radially convex with respect to p_i, then we call $(P, \mathbf{R})$ a *radially convex territory diagram*. (A set X is radially convex with respect to a point $p \in X$ if for any point $a \in X$ the entire line segment between p and a is contained in X.)

Remark 2. It is clear that each R_i in a territory diagram must always be contained in the closure of the Voronoi cell for p_i. It is also clear that the collection of R_is must be pairwise disjoint.

Remark 3. It follows from Definition 4 that all zone diagrams are also territory diagrams (the converse, however, is not true in general).

Definition 5. *(Partial order of territory diagrams)* *Given two territory diagrams* $(P, \mathbf{R})$ *and* $(P, \mathbf{S})$ *for the same tuple of sites* P, *we write* $(P, \mathbf{R}) \preceq (P, \mathbf{S})$ *if* $R_i \subseteq S_i$ *for all* $i = 1, \ldots, n$. *Additionally, we define* $(P, \mathbf{R}) \prec (P, \mathbf{S})$ *if* $(P, \mathbf{R}) \preceq (P, \mathbf{S})$ *but* $\mathbf{R} \neq \mathbf{S}$.

Definition 6. *(Trivial territory diagram) A territory diagram $(P, \mathbf{R})$ is said to be trivial if for the (unique) zone diagram $(P, \mathbf{S})$ we have $(P, \mathbf{R}) \preceq (P, \mathbf{S})$.*

Such a diagram is termed to be trivial because if $(P, \mathbf{S})$ is a zone diagram and for all $i = 1, \ldots, n$ we have $p_i \in R_i$ and $R_i \subseteq S_i$, then it immediately follows that inclusion in the dominance region for each site is preserved and thus $(P, \mathbf{R})$ is always a territory diagram. In other words, if we start from a zone diagram $(P, \mathbf{S})$ and remove arbitrary parts of each S_i, the result is a trivial territory diagram, as long as we do not remove p_i (see Example 1).

Example 1. **(Trivial territory diagram)** The territory diagram

$$
\begin{aligned}
p_1 &= (0, 3) , R_1 = \{(x, y) : x^2 + (y - 3)^2 \leqslant \tfrac{1}{2}\} , \\
p_2 &= (0, -3) , R_2 = \{(x, y) : x^2 + (y + 3)^2 \leqslant \tfrac{1}{2}\} ,
\end{aligned}
\tag{5}
$$

is trivial (see Fig. 3).

Proposition 2. *Nontrivial territory diagrams exist.*

In other words, it is possible to build a territory diagram $(P, \mathbf{R})$ such that one or more R_is are not contained in the corresponding (unique) zone diagram region S_is for the same tuple of sites (see Example 2).

Example 2. **(Nontrivial territory diagram)** The territory diagram

$$
\begin{aligned}
p_1 &= (0, 3) , R_1 = \{(x, y) : x^2 + (y - 3)^2 \leqslant 1\} , \\
p_2 &= (0, -3) , R_2 = \{(x, y) : x^2 + (y + 3)^2 \leqslant \tfrac{5}{2}\} ,
\end{aligned}
\tag{6}
$$

is nontrivial (see Fig. 4).

In this case, R_2 is clearly not contained in the corresponding zone diagram's region for site p_2, since it contains for example the point $(0, -\tfrac{1}{2})$ while the zone diagram region for p_2 must trivially end at the point $(0, -1)$. In spite of that, given the relatively small radius of R_1, all points in R_2 are still closer to p_2 than to p_1, and all the points in R_1 are closer to p_1 than to p_2. Since the region for each site is within the dominance zone induced by the diagram, this is a territory diagram (see Fig. 4).

On the other hand, if region R_1 is expanded to radius $\tfrac{5}{2}$, the result is not a territory diagram anymore, since points on the border of both regions will now be closer to the other region than to their respective sites (see Fig. 5).

2.2   Maximal Territory Diagrams

Definition 7. *(Maximal territory diagram) A territory diagram $(P, \mathbf{R})$ is a maximal territory diagram if it is maximal with respect to the partial order $\prec$, i. e. if there exists no territory diagram $(P, \mathbf{S})$ for the same tuple of sites such that $(P, \mathbf{R}) \prec (P, \mathbf{S})$.*

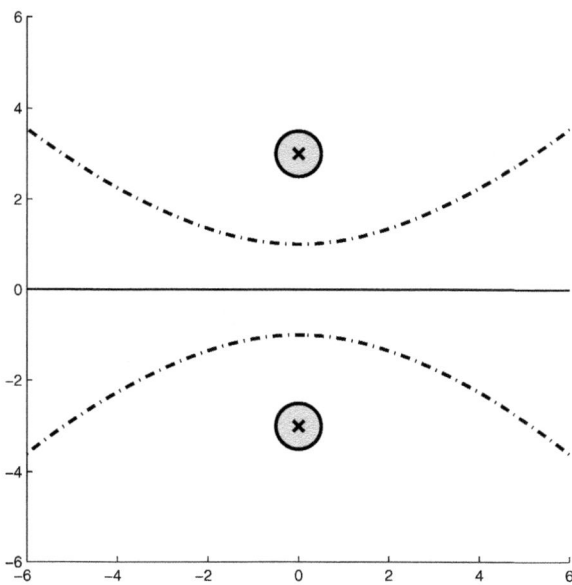

Fig. 3. Example of a trivial territory diagram for 2 sites. The region (*gray*) associated to each site is comfortably within the corresponding region in a zone diagram (*dash-dotted*).

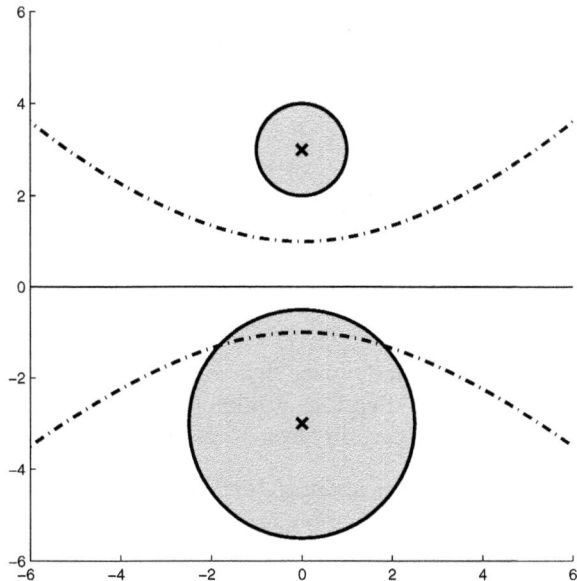

Fig. 4. Example of a nontrivial territory. The regions in this territory diagram (*gray*) are not all contained in the corresponding regions for the zone diagram for the same set of sites (*dash-dotted*).

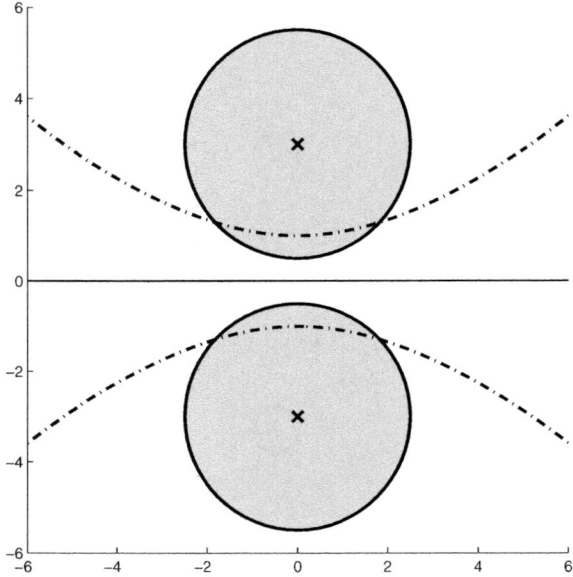

Fig. 5. This is not a territory diagram

If every region R_i in $\mathbf{R}$ is convex, then we call $(P, \mathbf{R})$ a *convex maximal territory diagram*. If every region R_i in $\mathbf{R}$ is radially convex with respect to p_i, then we call $(P, \mathbf{R})$ a *radially convex maximal territory diagram*.

Given two territory diagrams $(P, \mathbf{R})$ and $(P, \mathbf{S})$ such that $(P, \mathbf{R}) \prec (P, \mathbf{S})$ and $(P, \mathbf{S})$ is maximal, we say that $(P, \mathbf{S})$ is a maximal territory diagram for $(P, \mathbf{R})$.

Remark 4. Alternatively, a territory diagram is said to be maximal if each of its regions is maximal, i. e. if each of the regions in the diagram cannot be expanded (replaced by any other region that strictly contains it) without causing a violation of the definition of a territory diagram.

Remark 5. From Definition 7 it follows that each region in a maximal territory diagram is closed.

From Definitions 3 and 5 it follows that every zone diagram is a maximal territory diagram. Since zone diagrams are known to exist [1], this immediately establishes the existence of a certain class of maximal territory diagrams.

Definition 8. *(Elementary maximal territory diagram) Given a tuple of sites P, the maximal territory diagram for which the region associated to each site is exactly equal to its dominance region is called an **elementary maximal territory diagram** and it is equal to the unique zone diagram for P.*

Remark 6. Although every zone diagram is a maximal territory diagram, the converse is not true in general, that is, not all maximal territory diagrams are elementary.

Remark 7. Maximal territory diagrams are not unique. They are not unique for a given set of sites, and in general they are also not unique for a given territory diagram.

Proposition 3. *Non-elementary maximal territory diagrams exist.*

Proof. We prove the proposition by giving an example of a non-elementary maximal territory diagram. □

Example 3. (**Non-elementary maximal territory diagram**) Let $P = \{p_1 = (0, 1), p_2 = (0, -1)\}$. Let R_0 consist of the single point p_1 and R_2 consist of the entire Voronoi cell of p_2 together with its boundary, the x-axis. Then $(P, \mathbf{R})$ where $\mathbf{R} = \langle R_0, R_2 \rangle$ clearly satisfies the definition of a territory diagram for P. However, $\mathbf{R}$ is not maximal since $\mathbf{R}' = \langle R_1, R_2 \rangle$ where R_1 is any line segment with endpoints at $p_1 = (0, 1)$ and $p_y = (0, y), y > 1$ is a territory diagram such that $\mathbf{R} \prec \mathbf{R}'$. We claim that if we pick R_1 to be the union of all such segments (i. e. the half-line $\{(0, y) : y \geqslant 1\}$, see Fig. 6), then with $\mathbf{R} = \langle R_0, R_1 \rangle$ we have that $(P, \mathbf{R})$ is a non-elementary maximal territory diagram.

To verify this claim, we will show that no point u can be added to R_1 to create a larger territory diagram $\mathbf{X}$, that is, one such that $\mathbf{R}' \prec \mathbf{X}$.

Any point $u = (u_x, u_y)$ in the region $F_{2A} = \{(x, y) : 0 < y < 1\}$. Any such point would clearly be closer to the point $v = (u_x, 0) \in R_2$ than p_2 would be to v, thus violating the definition of a territory diagram if u were to be added to R_1.

Next, consider a point $u \neq p_1$ in the region $F_{2B} = \{(x, y), x \neq 0, y \geq 1\}$. Then if we take the line perpendicular to the segment from p_1 to u at point u, it will clearly intersect the x-axis at a point $z = (z_x, 0) \in R_2$ which is closer to u than to p_2, again violating the definition of a territory diagram for R_2 if u were to be added to R_1.

Thus, no additional points in the region $F_2 = F_{2A} \bigcup F_{2B}$ can be added to R_1. Since we already have $R_1 = \mathbb{R}^2 - F_2$, it follows that R_1 is maximal. And since R_2 is already the closure of the Voronoi cell for p_2, it is also maximal.

Let us now consider whether $\mathbf{R}'$ is indeed a territory diagram. The dominance regions for p_1 and p_2 are

$$\begin{aligned} S_1 &= \operatorname{dom}(p_1, R_2) = \{(x, y) : y \geq \tfrac{1}{2}(x^2 + 1)\}, \\ S_2 &= \operatorname{dom}(p_2, R_1) = R_2. \end{aligned} \tag{7}$$

Since it is clearly the case that $R_1 \subseteq S_1$ and $R_2 \subseteq S_2$, and given that $\mathbf{R}'$ is maximal and that $\mathbf{R} \prec \mathbf{R}'$, we conclude that $\mathbf{R}' = \langle R_1, R_2 \rangle$ is a maximal territory diagram for $\mathbf{R} = \langle R_0, R_2 \rangle$ (and in this particular case it is also unique).

Finally, since region R_2 is clearly not within the zone diagram region for p_2, $(P, \mathbf{R})$ is non-elementary.

Both R_1 and R_2 in the previous example are convex, and it is also trivially the case that all zone diagrams contain only convex regions. We might thus speculate that all maximal territory diagrams contain only convex regions. This turns out not to be true.

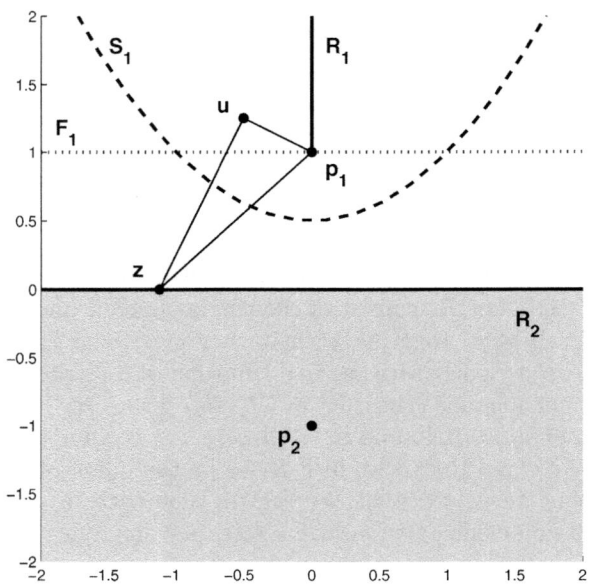

Fig. 6. A non-elementary maximal territory diagram. Notice that for the same pair of sites we could have shown at least one additional maximal territory diagram, namely the elementary maximal territory diagram.

Proposition 4. *Non-convex maximal territory diagrams exist.*

Proof. We prove the proposition by giving an example of a non-convex maximal territory diagram.   □

Example 4. (**Non-convex maximal territory diagram**) Starting from the previous example, keep $P = \{p_1 = (0, 1), p_2 = (0, -1)\}$, but extend R_1 by adding the segment from p_1 to $q = (0, \frac{1}{4})$. This would clearly create dominance violations, namely there would then be points in R_2 closer to R_1 than to p_2; so we drop the original region R_2 and ask what it could be.

Clearly R_2 cannot include any point u such that $d(v, u) < d(v, p_1)$ for some $v \in R_1$. Thus, no points in the region $F_1 = \{u \in \mathbb{R}^2 : \exists v \in R_1, d(v, u) < d(v, p_1)\}$ can be in R_2. We claim that $F_1 = F_{1A} \bigcup F_{1B}$, where (see Fig. 7):

$$\begin{aligned} F_{1A} &= \{(x, y) : y > 1\}, \\ F_{1B} &= \{(x, y) : x^2 + (y - \tfrac{1}{4})^2 < (\tfrac{3}{4})^2\}. \end{aligned} \tag{8}$$

On the other hand, the region S_2 into which R_2 could expand without creating dominance violations for p_2 is simply the closure of the Voronoi cell for p_2 (and q):

$$S_2 = \mathrm{dom}(p_2, R_1) = \mathrm{dom}(p_2, \{q\}) = \{(x, y) : y \leq -\frac{3}{8}\}. \tag{9}$$

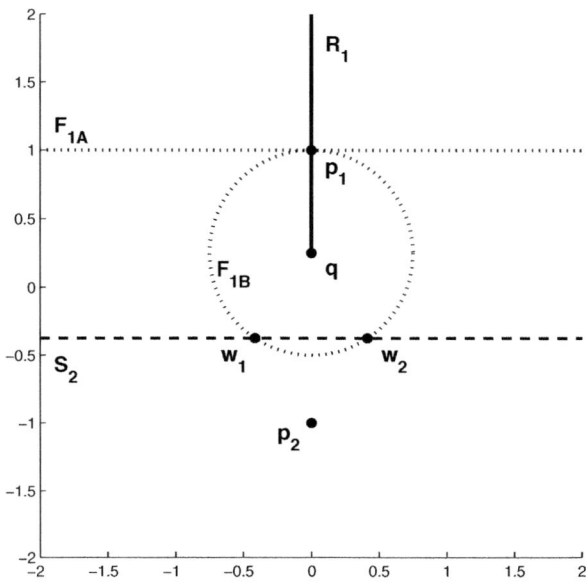

Fig. 7. Building a maximal territory diagram with a non-convex region

There are thus two different restrictions for the growth of R_2: staying inside S_2 and staying outside of F_1. In fact, these are all the restrictions that apply, so we can actually make $R_2 = S_2 - F_1$ without creating any dominance violations.

Would this create a maximal territory diagram? We certainly now have a region R_2 which by construction does not create dominance violations for either p_1 or for p_2 and which is maximal. But is R_1 maximal?

Let us start by noticing that we cannot expand R_1 by adding to it any point $u \in S_2$ since such points would trivially be closer to R_2 than to p_1 and would thus create a dominance violation for p_1.

Next, consider the fact that by construction $d(p_1, q) = d(q, w_1) = d(q, w_2) = d(w_1, p_2) = d(w_2, p_2) = \frac{3}{4}$, and that $d(p_2, q) = \frac{5}{4}$. Thus, we also cannot add any point inside the triangle qw_1w_2 since any such point would be closer to w_1 or w_2 than to p_2, creating a dominance violation for p_2.

Finally, the point u cannot be anywhere else for the same reasons as in the previous example: there would be a point z on the boundary of S_2 either to the left of w_1 or to the right of w_2 such that $d(z, u) < d(z, p_2)$, creating a dominance violation for p_2.

Thus R_1 is indeed maximal, and $\langle R_1, R_2 \rangle$ is a maximal territory diagram (see Fig. 8).

Computing maximal territory diagrams in such an *ad hoc* manner, however, is not very satisfactory, and moreover already we see some patterns emerging. Let us thus build on the insight gained from the previous examples and examine the structure of a maximal territory diagram in greater detail.

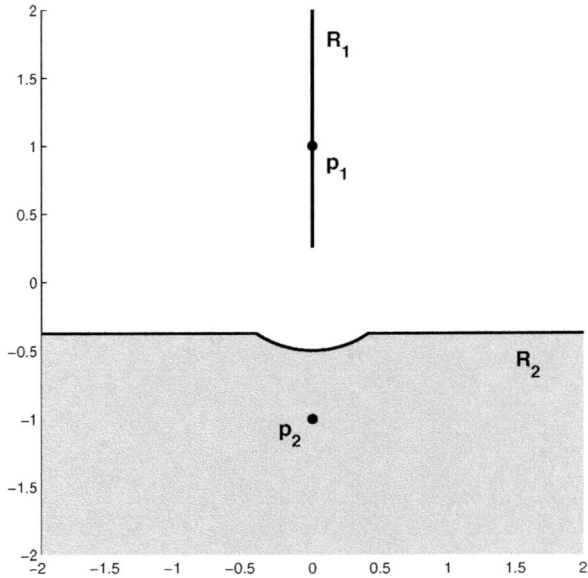

Fig. 8. The resulting non-convex maximal territory diagram

3   The Structure of Territory Diagrams

3.1   Safe Zones

For any territory diagram (and in particular for any maximal territory diagram), it must be the case (see Definition 4) that each region R_i is contained inside the dominance region for its site p_i. For a non-maximal territory diagram, however, this containment is strict, meaning that there are points in the dominance region for at least one site p_i which are not in R_i and which could be safely added to R_i without creating any dominance violations for p_i. This suggests the following definition:

Definition 9. *(Safe zone) The safe zone S_i for a given site p_i in a territory diagram $(P, \mathbf{R})$ is the set of all points that are at least as close to p_i than to any region other than R_i, that is:*

$$S_i = \{z : d(z, p_i) \leq d(z, R_j),\ j \neq i\} . \tag{10}$$

The notion of safe zone is based on the same concept as the **Dom** operator as defined in [1]. In fact, we could have equivalently defined

$$S_i = \mathrm{dom}\left(p_i, \bigcup_{j \neq i} R_j\right) . \tag{11}$$

Remark 8. A territory diagram can thus be equivalently defined as one in which every region R_i is contained in the safe zone S_i for the corresponding site, that is, in which for all sites p_i we have $R_i \subseteq S_i$.

The safe zone S_i for a site p_i does *not* depend in any way on R_i (instead it depends only on the regions associated to other sites) and always contains p_i. In the special case that $R_i = \{p_i\}$ for every i, the safe zone of each site is the closure of its Voronoi cell. Safe zones are always closed and convex but may be unbounded. Finally, the border of each S_i is the bisector curve between p_i and the union of all $R_j, j \neq i$ (see Fig. 9).

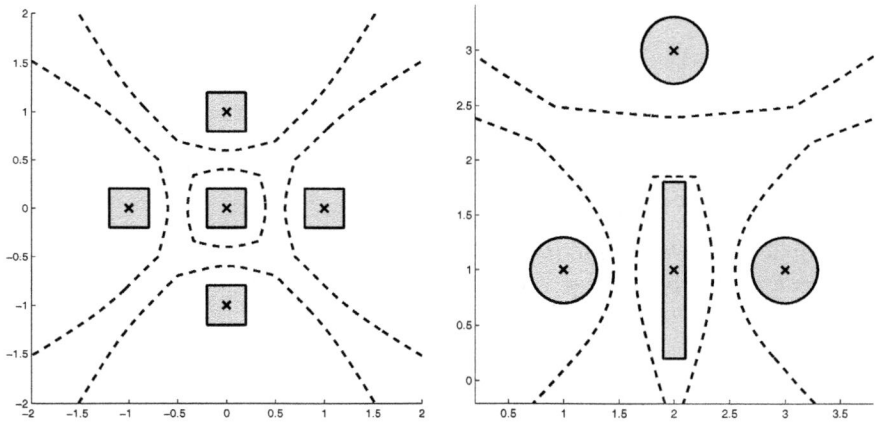

Fig. 9. Safe zone are shown (*dashed*) for two territory diagrams. Notice that any point on the border of a safe zone S_i must be equidistant to p_i and at least one other region $R_{j \neq i}$.

For the maximal territory diagram given in Fig. 6, as we have already discussed, the safe zone for p_1 must have a parabola as its border, since that is the curve that is the bisector for a point and a line, and the safe zone for p_2 is the largest possible: the closure of the Voronoi cell for p_2. For the diagram given in Fig. 8, however, the safe zone for p_1 will have a more complex shape (not shown).

3.2   Forbidden Zones

The notion of safe zone, however, is not enough to describe the parts of the plane into which a given site p_i can expand its region R_i without violating Definition 4. This is because although each site p_i can expand its region R_i inside its safe zone without creating dominance violations for itself, this expansion might create dominance violations for other sites (effectively shrinking their safe zones into their current regions). This is not a problem in the case of iterative computation of zone diagrams, since the other sites can then subsequently shrink their own regions leading to a sequence that converges to a zone diagram. But if we are building a generic maximal territory diagram, regions are not allowed to shrink (as this would violate the partial order of territory diagrams), so we should never in any stage create dominance violations. This suggests an additional concept, complementary to that of safe zones : the part of the plane around a region R_i such that no other region can grow into it without creating a dominance violation for p_i.

Definition 10. (Forbidden zone) *The forbidden zone F_i for a given site p_i with region R_i is the set of all points that are closer to some point $y \in R_i$ than y is to p_i, that is:*

$$F_i = \{z : d(z,y) < d(y, p_i) \text{ for some } y \in R_i\} . \tag{12}$$

The forbidden zone F_i for a site p_i depends *only* on R_i and never contains p_i (see Proposition 7). In the special case that for all sites i we have $R_i = \{p_i\}$, the forbidden zone of each site is the empty set.

Proposition 5. *Definition 10 can be interpreted as including a forbidden open disc of radius $d(y, p_i)$ around each $y \in R_i$, and the forbidden zone F_i for R_i can be interpreted as the union of all such discs, that is :*

$$F_i = \bigcup_{y \in R_i} \{z : d(z,y) < d(y, p_i)\} \tag{13}$$

Proof. Straight from Definition 10, a point z is in F_i if and only if it is inside an open disc with center at some point $y \in R_i$ and radius $d(y, p_i)$. □

For example, in the maximal territory diagram given in Fig. 6, the forbidden zone for p_1 is the (open) region of the plane above the line $y = 1$. It is not difficult to verify that this is the case using Proposition 5. Each open disc with center $y \in R_1$ and radius $d(y, p_1)$ must clearly be entirely within the forbidden zone for p_1 (since any point inside such a disc would be closer to y than to y from p_1). Since in the case of the example in Fig. 6 the boundary of all such discs include p_1 and have an arbitrarily large radius, any (and only) points above the line $y = 1$ will eventually be reached (see Fig. 10, left). On the other hand, the forbidden zone for p_2 contains the whole plane *except* for p_2 and for the points on the half-line $\{(0, y), y > 1\}$ (see Fig. 10, right).

Proposition 6. *The forbidden zone F_i for a site p_i whose region R_i is a line segment ab containing p_i is always equal to two open discs (See Fig. 11, left and middle) with centers at a and b and radii $d(a, p_i)$ and $d(p_i, b)$ respectively. One of the open discs may degenerate into an open disc with radius zero (i. e. the empty set, see Fig. 11, right) if we have either $a = p_i$ or $b = p_i$.*

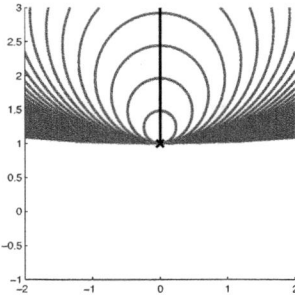 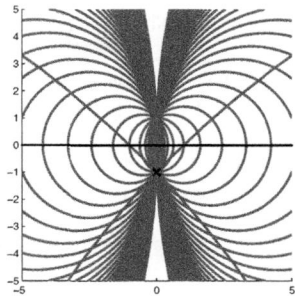

Fig. 10. Sequence of discs approaching the forbidden zones for p_1 (*left*) and for p_2 (*right*). The apparent gap at the bottom right is actually filled; the only region not on the forbidden zone for p_2 is at the top.

Proof. Any point inside either of the two open discs will necessarily be closer to either a or b than to p_i, so F_i must include both discs. On the other hand, any point z in F_i must be closer to some point in the segment ab than to p_i, and thus it must be inside some open disc centered at a point in ab and with radius $d(z, p_i)$. Since all such discs are clearly contained in the ones previously described, F_i cannot contain any point outside of the two discs. Thus we conclude that F_i is equal to the two discs. □

Proposition 7. F_i *never contains* p_i, *that is,* $p_i \notin F_i$.

Proof. Since $d(y, p_i) = 0$ for all $y \in R_i$, it is impossible to find any z such that $d(z, y) < d(y, p_i) = 0$. In particular, if $R_i = \{p_i\}$ then the forbidden zone F_i is empty. □

Proposition 8. F_i *always contains all the points in* R_i, *that is,* $F_i \supseteq R_i - \{p_i\}$. *In other words, the forbidden zone always contains (with the exception of the site itself) the all points in the region associated to the site.*

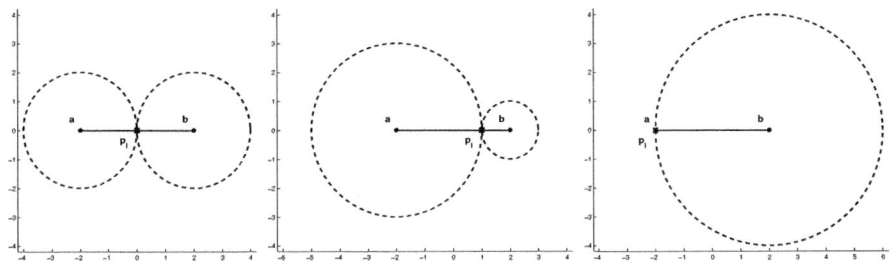

Fig. 11. Forbidden zones when the region is a line segment. (*left,dashed*) The site is at the middle of the segment. (*middle,dashed*) The site is closer to the right end of the segment. (*right,dashed*) The site coincides with the left end of the segment.

Proof. For any point $z \in R_i$ there is a point $y \in R_i$ (that is, z itself) such that $d(z, y) = 0$ which thus satisfies $0 = d(z, y) < d(y, p_i), y \neq p_i$. $\square$

Remark 9. Notice that in general the forbidden zone can be either non-convex (see Fig. 11, left, middle) or convex (see Fig. 11, right).

Theorem 1. *A forbidden zone F_i augmented by its site p_i is always radially convex with respect to p_i.*

Proof. This is a direct consequence of Proposition 5. F_i is a union of discs that are all tangent to p_i, and any such disc is trivially radially convex with respect to p_i. So for any given any point z to be in F_i, it must have come from such a disc, and thus the segment zp_i that connects z to p_i is also now entirely contained in F_i. $\square$

Lemma 1. *If $R_i \subseteq R_i'$ then $F_i \subseteq F_i'$. In other words, if we add more points to a region R_i, the new forbidden zone F_i' for the resulting R_i' will always contain the original forbidden zone F_i.*

Proof. This is a direct consequence of Proposition 5. Since F_i' is trivially equal to the union of all the discs that constituted F_i plus the new discs generated by the points in $R_i' - R_i$, it must contain F_i. $\square$

Theorem 2. *The forbidden zone F_i for any site p_i with region R_i is always the same as the forbidden zone F_i' of p_i with region R_i' where R_i' is the convex hull of R_i (see Fig. 12). In other words,*

$$R_i' = conv(R_i) \Rightarrow F_i' = F_i \tag{14}$$

Proof. Clearly we have $F_i \subseteq F_i'$, since $R_i \subseteq R_i'$ (see Lemma 1). Thus, we just need to show that $F_i' \subseteq F_i$. Consider a point $z \in F_i'$. This means (from Definition 10) that there must be a point $y' \in R_i'$ such that $d(z, y') < d(y', p_i)$. We claim that in this case there is at least one point $y \in R_i$ such that $d(z, y) < d(y, p_i)$, which would mean that $z \in F_i$ whenever $z \in F_i'$ and proving the theorem. We proceed to show that such a y always exists. If $y' \in R_i$ then we can simply pick $y = y'$ and we are done. So assume $y' \notin R_i$. Consider the region of the plane $R_z = \{y : d(y, z) < d(y, p_i)\}$, that is, all points y which if in R_i would put z in F_i. This is exactly the open half-plane of points that are closer to z than they are to p_i. Notice in particular that $y' \in R_z$. But since y' is supposed to be in the convex hull of R_i and since it is not itself in R_i, then clearly there must be at least one other point $y \in R_z$ such that $y \in R_i$. Otherwise, we would have an open half-plane R_z which contains *no* points at all from R_i and still contains a point y' which is supposed to be in the convex hull of R_i, a contradiction (as clearly any open half-plane not containing any points from a set cannot possibly contain any points from the convex hull of that set). $\square$

Corollary 1. *When computing the forbidden zone F_i of a region R_i, it is enough to consider only the border of the convex hull of R_i.*

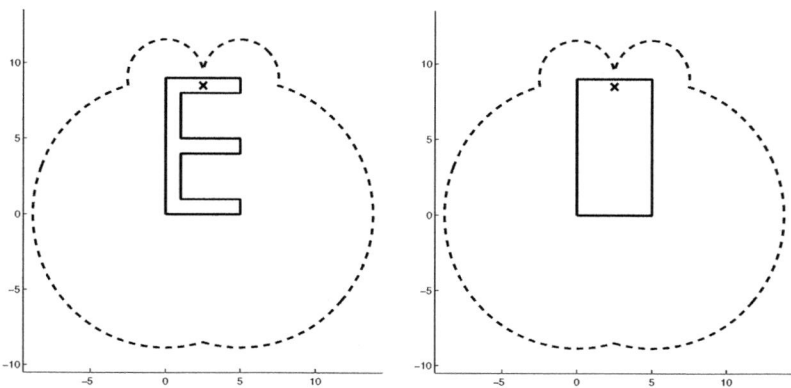

Fig. 12. The forbidden zone for any non-convex region (*left,dashed*) is the same as for the convex hull of the same region (*right,dashed*)

In fact, since for a polygon the edges are all trivially in the convex hull of the vertices, an even stronger result holds in this case :

Corollary 2. *The forbidden zone for any polygonal region depends only on its vertices in convex position, and it is equal to the union of the forbidden zones for these vertices, that is*

$$F_i = \bigcup_{v_j} \{z : d(z, v_j) < d(v_j, p_i)\} \tag{15}$$

where v_j are the vertices of the convex hull of R_i.

Since each vertex has a forbidden zone equal to an open disc, if a site p_i has a region R_i given by a convex polygon with n vertices $v_1, \ldots, v_n$, the forbidden zone for this region is given by the union of n open dics (see Fig. 12, 13). This means that forbidden zones for polygonal regions can be determined not only very efficiently but also exactly : we can easily and quickly compute an algebraic description for their border.

Theorem 3. *The border of the forbidden zone for any polygonal region R with n vertices of which h are in convex position can be computed in time $O(n)$ and it will consist in the piecewise union of h circle segments, each of them centered around one of the vertices in the convex hull of R.*

Proof. Given a simple polygon whose vertices are not all in convex position (but already given in sorted order), we can compute the convex hull in time $O(n)$. Once we have only vertices in convex position (which we know from theorem 2 are the only ones we need to consider), it takes constant time to determine where the circle which describes the border of the disc that corresponds to the forbidden

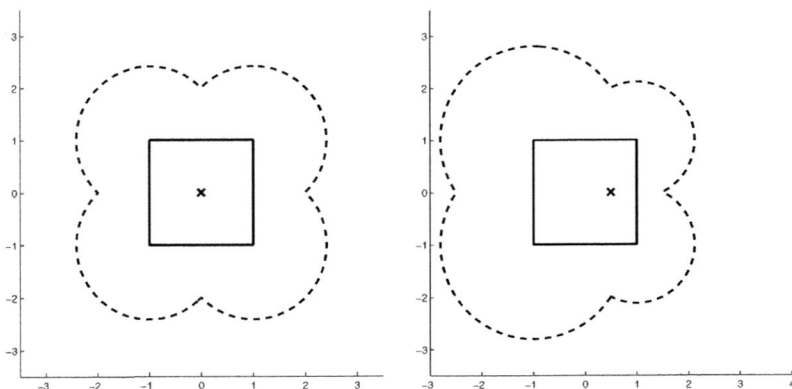

Fig. 13. Forbidden zones for a polygon with 4 vertices (*dashed*), which equal the union of 4 open discs

zone for each vertex intersects the circle around the next vertex, so each vertex can be processed in time $O(1)$. Thus the forbidden zone for a polygonal region can always be computed in time $O(n)$, and the result will be a sequence of h of circle segments around the polygon.   □

Finally, the forbidden zones for shapes that are not polygons can present additional challenges. For example, although it is easy to see that the forbidden zone for a disc whose site is located at its center is also a disc (see Fig. 14, left), if we move the site off-center the resulting shape is not either a single disc or even a finite union of discs (see Fig. 14, right).

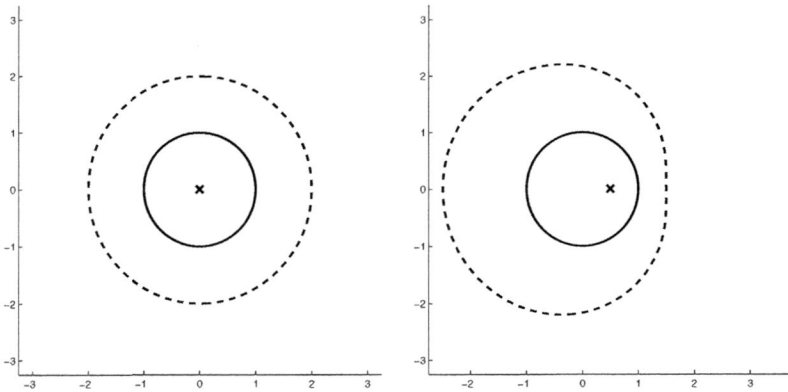

Fig. 14. Forbidden zones for a circle (*dashed*). Notice how location of the site changes the boundary.

3.3   The Structure of Maximal Territory Diagrams

Notice that the concepts of safe zone and forbidden zone are complementary not in terms of the concrete regions that they include but instead in the sense that while the former describes the regions of the plane into which one can expand, the latter describes the regions of the plane into which one cannot.

If we consider the safe zones and forbidden zones for the diagrams given in Figs. 3,4,5, we can immediately determine (See Fig. 15) whether they are territory diagrams and if so, whether they are maximal.

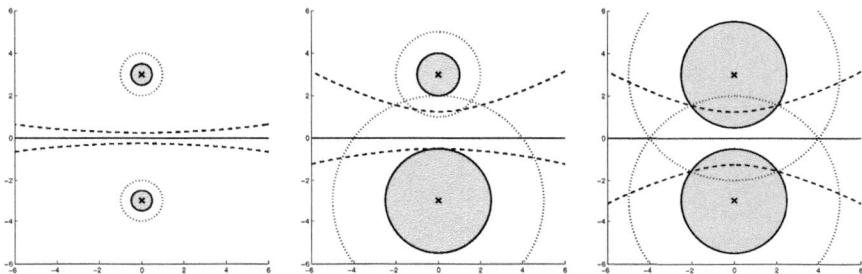

Fig. 15. The safe zones (*dashed*) and forbidden zones (*dotted*) for the diagrams introduced in Figs. 3,4,5 (*left, middle, right*). Notice that in the case of proper territory diagrams (*left, middle*) the regions (*gray*) for each site are completely contained in the site's safe zone, while in those where there are dominance violations (*right*) this does not hold. Notice also that the territory diagrams in (*left, middle*) are clearly not maximal since there are regions inside each site's safe zone which are not inside the other site's forbidden zone and which could thus be added to the site's region without creating dominance violations.

The relevance of safe zones and forbidden zones for the structure of maximal territory diagrams can thus be summarized as follows : in a maximal territory diagram, every region must contain all parts of its site's safe zone that are not in the forbidden zone of any other site.

Theorem 4. *For any maximal territory diagram* $(P, \mathbf{R})$,

$$R_i = S_i - \bigcup_{j \neq i} F_j . \tag{16}$$

Proof. If Equation (16) is satisfied, then clearly $R_i \subseteq S_i$ and from Definition 4 we can conclude that $(P, \mathbf{R})$ is a territory diagram. Moreover, it must be maximal, since if we add any points to any region R_i, they will be either outside S_i, or within $F_i \neq j$, both of which are not permitted. Conversely, if $(P, \mathbf{R})$ is a territory diagram and Equation (16) is not satisfied, then there is at least one point y such that $y \in S_i$ but $y \notin F_i \neq j$, which means that region R_i could be expanded into $R_i \cup \{y\}$, meaning $(P, \mathbf{R})$ is not maximal. □

Notice that in the maximal territory diagrams both in Fig. 6 and Fig. 8 we do have $R_1 = S_1 - F_2$ and $R_2 = S_2 - F_1$.

Since zone diagrams are a special case of maximal territory diagrams, this equality also holds for them. In the case of zone diagrams, however, we have $R_i = S_i$ and thus the safe zones exactly coincide with the diagram regions (see Fig. 16) and are a fixed point for the **Dom** operator (see [1]). In fact, this is precisely the definition of a zone diagram; in all other cases, the regions of a generic maximal territory diagram do not completely fill their corresponding safe zones.

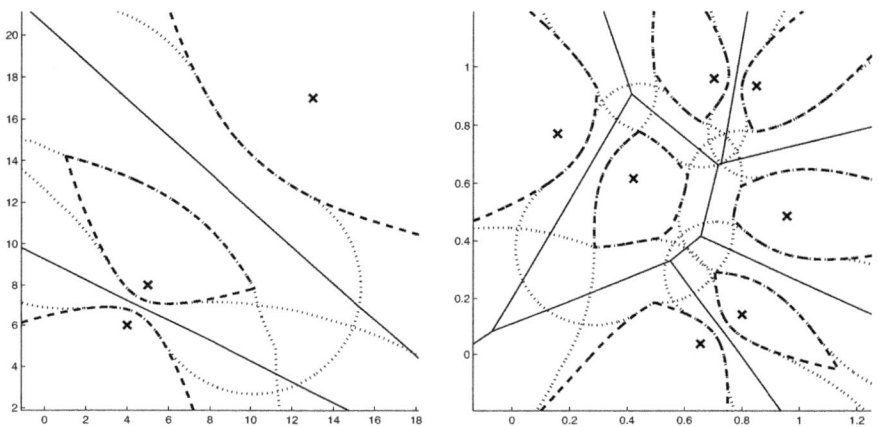

Fig. 16. The safe zones (*dashed*) and forbidden zones (*dotted*) for the zone diagrams introduced in Fig. 2 are shown. The safe zones coincide with the regions in the diagrams. The forbidden zone for each site exactly conforms to the regions (and thus the safe zones) around it without entering them.

3.4   Computation of Maximal Territory Diagrams

No exact algebraic description is currently known for the borders of maximal territory diagrams in general and zone diagrams in particular. Thus, the computation, description and characterization of actual instances of zone diagrams must in general be done using some approximate representation. One that is particularly useful and appropriate is to employ a collection of polygonal discretizations for the border of each region. All but the simplest diagrams shown in this paper were computed using such a representation, and with the exception of the few examples that were designed to be algebraically tractable, they are not exact but approximations within a given margin of error.

An approximation to zone diagrams can be computed using any of the methods described in [1], in particular by iterating the **Dom** operator, and this leads to very robust algorithms with practical applicability.

For maximal territory diagrams, however, we are restricted by the fact that we are not allowed to "overextend" the regions and then later shrink them in search of a fixed point. Thus, the most straightforward and practical way to build non-elementary maximal territory diagrams employs the identity given at the end of the previous section. The basic idea is that given any territory diagram, if we want to find a non-elementary maximal territory diagram for it, we can take each site i in turn and expand its region to be $R_i = S_i - \bigcup_{j \neq i} F_j$. After expanding all sites, the result will by construction be a maximal territory diagram (see Fig. 17). Notice however that the resulting diagram will depend on the order in which the sites are processed (see Fig. 18), as well as on the regions (not just the sites) in the starting territory diagram (see Fig. 19).

This method requires computing safe zones and forbidden zones. Safe zones are essentially determined by bisector curves and thus costly to compute; on the other hand, forbidden zones depend only on one region at a time and are much simpler to determine. Notice that if we consider the computation of safe zones and forbidden zones as primitive operations, this method requires only one iteration for each site.

Another approach to building maximal territory diagrams would be to iteratively expand each region without immediately filling all the available space that it has to grow. The new territory diagram thus obtained could then be again expanded, either by growing the same region or some other region, thereby ascending on the partial order of territory diagrams until eventually reaching a maximal territory diagram.

This variation on the original method allows us to generate and explore a wider range of maximal territory diagrams grown from a given territory diagram. It faces, however, some difficulties; one of them is that computing safe zones is costly, and thus if we perform small incremental changes to each region, the overall cost of recomputing the safe zones at each iteration becomes prohibitive.

In a forthcoming article, we intend to describe and discuss algorithms to compute maximal territory diagrams in more detail.

4   Existence of Maximal Territory Diagrams

In this section, we prove that every territory diagram can be embedded into a maximal territory diagram. We prove this by utilizing Zorn's lemma. In particular, starting with the set of sites as a territory diagram, this proves the existence of a maximal territory diagram. Clearly, in this special case of territory diagram the existence of a maximal territory diagram can be deduced by the existence of the zone diagram. However, Asano, *et al.* make use of the fixed-point theory in [1] to prove the existence of the zone diagram for a given set of sites. They also prove that a zone diagram is unique. The interesting feature of maximal territory diagram is that we may begin with an arbitrary territory diagram.

In order to prove that a territory diagram can be embedded into a maximal territory diagram we give necessary and sufficient conditions for a territory diagram to be a maximal territory diagram. We then use those conditions to give a recipe for extending a given territory diagram that is not maximal into a larger territory diagram. This together with Zorn's lemma are employed to prove to the existence of maximal territory diagram.

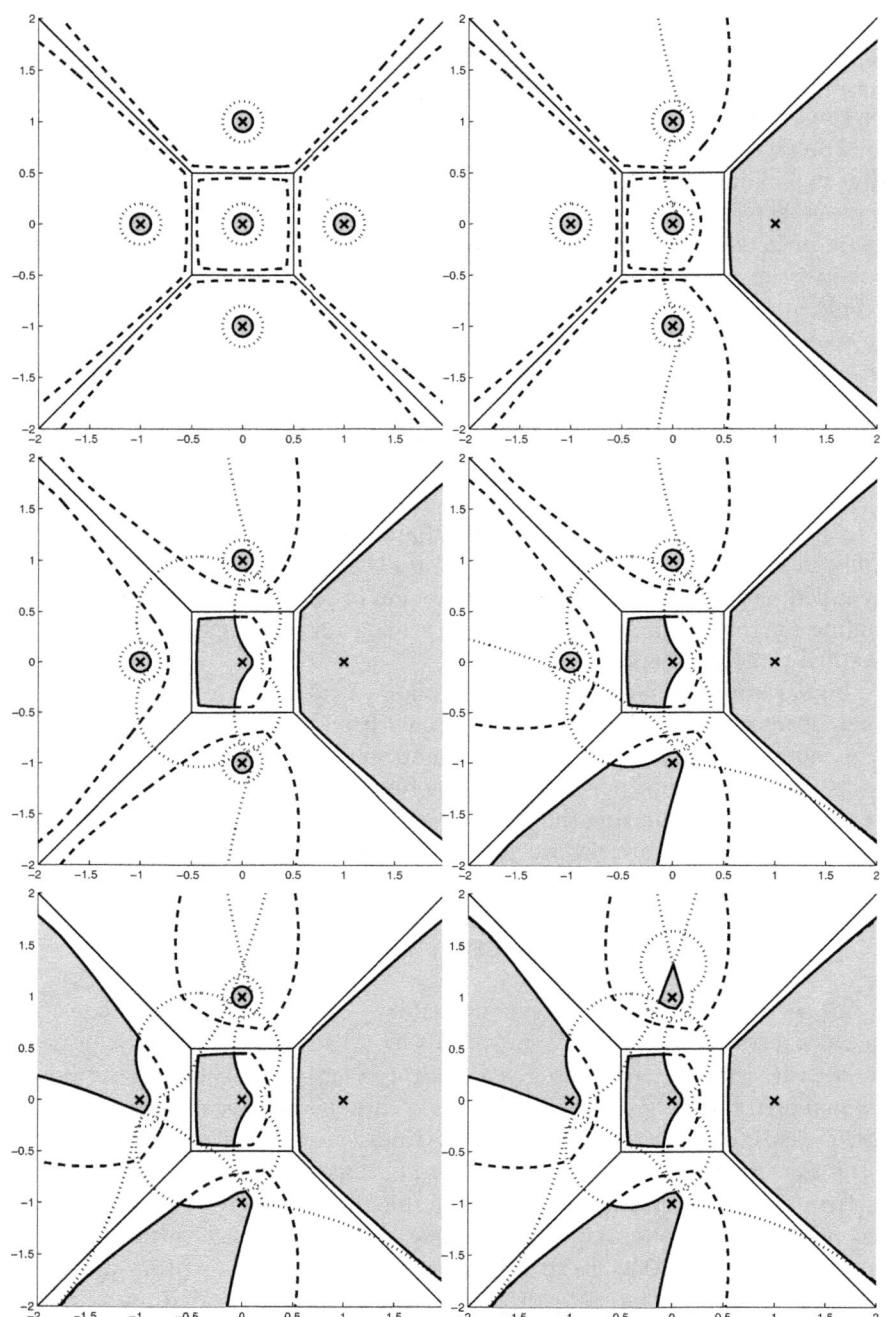

Fig. 17. Building a maximal territory diagram. The original territory diagram is shown at the top, and then one region is expanded in each step to occupy as much of its safe zone (*dashed*) as possible without entering any other region's forbidden zone (*dotted*). Notice that now the safe zones do not coincide with the regions in the diagram.

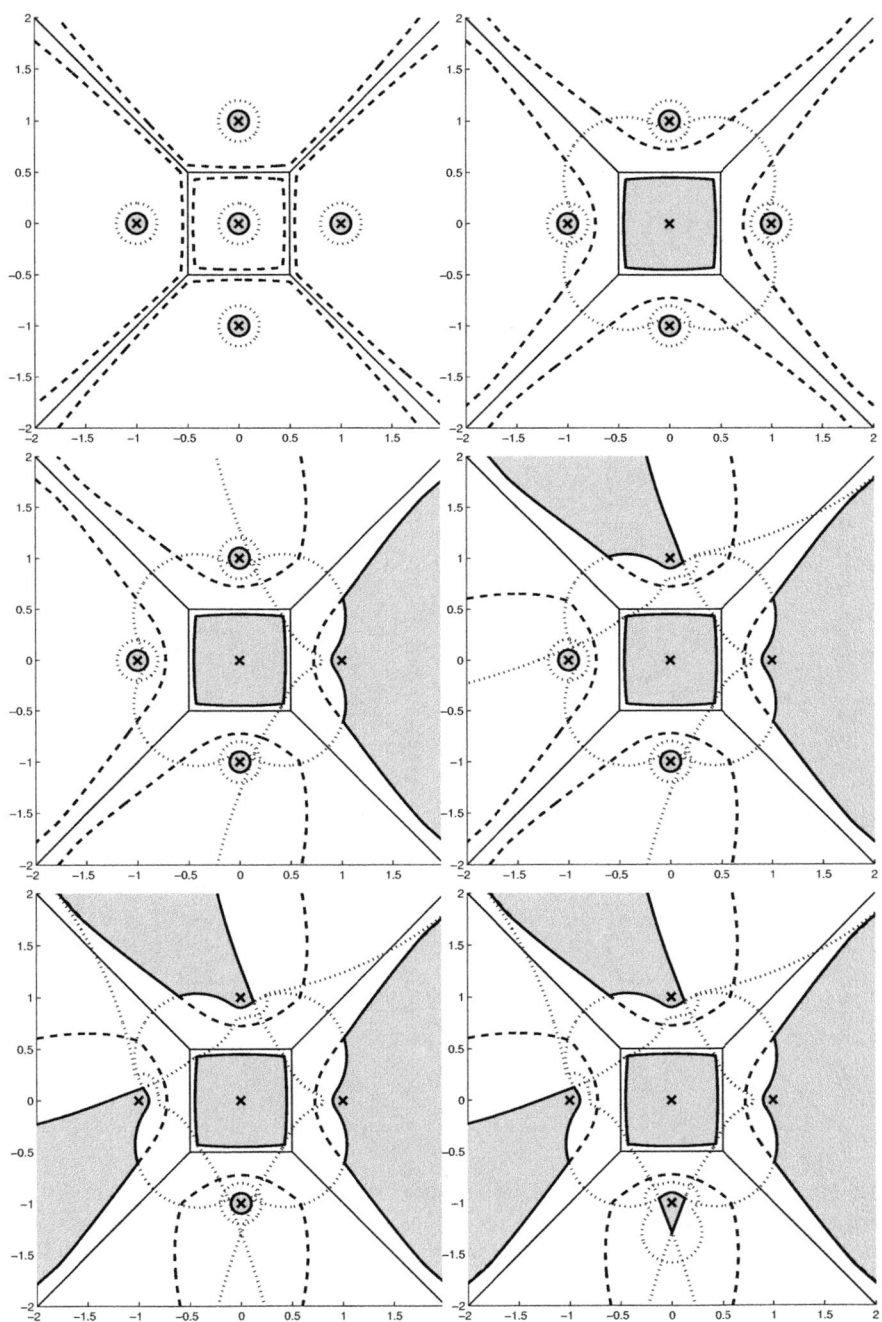

Fig. 18. The regions in the same initial territory diagram as in Fig. 17 are expanded in a different order, resulting in a different maximal territory diagram

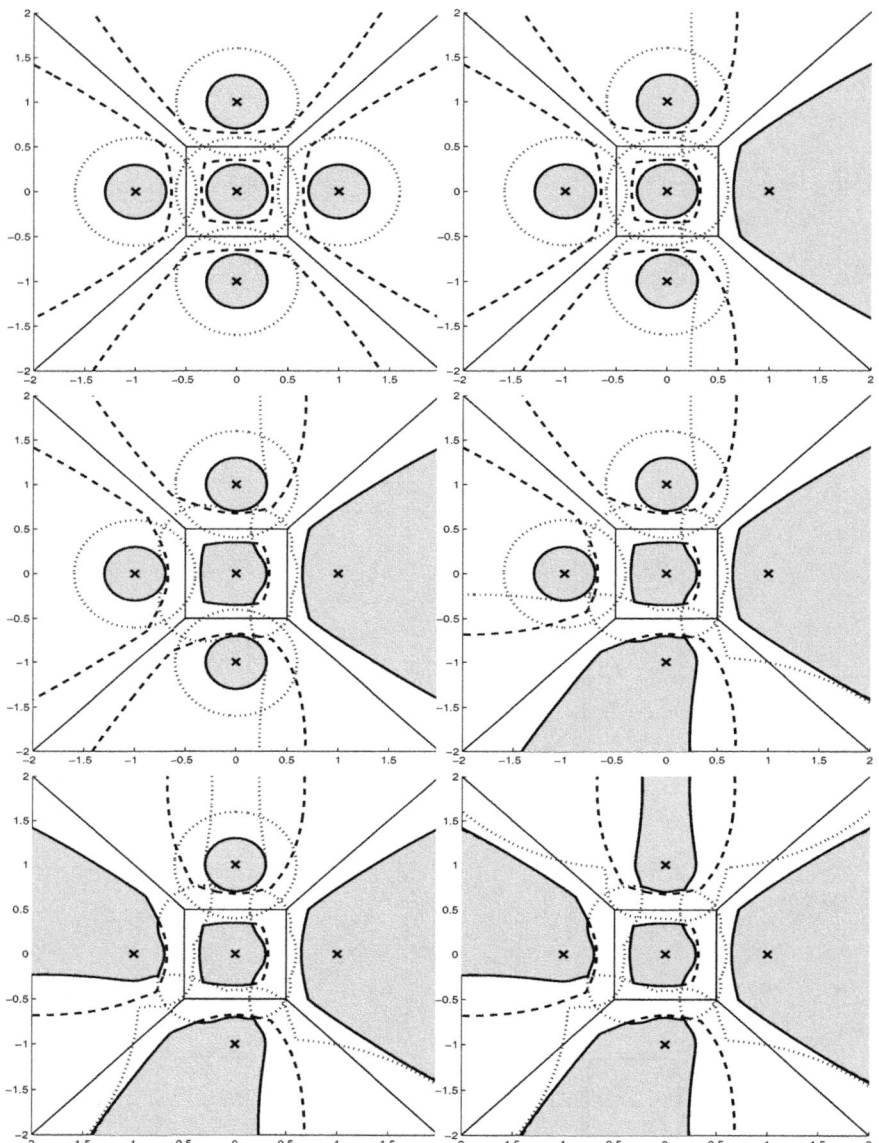

Fig. 19. Starting from different regions results in a different maximal territory diagram even though the sites are the same as in Figs. 17, 18. On the other hand, notice that since all the regions in the initial territory diagram for Figs. 17, 18 are contained in the corresponding regions in the initial territory diagram for this figure, the maximal territory diagram computed here is also a maximal territory diagram for the initial territory diagram for Figs. 17, 18. The converse, however, is not true - the maximal territory diagrams for or Figs. 17, 18 are not a maximal territory diagrams for the initial territory diagram in this figure.

Zorn's lemma states that every partially ordered set in which every chain (i.e. totally ordered subset) has an upper bound contains at least one maximal element. Thus we would immediately conclude existence of a maximal element by establishing the following lemma for chains of territory diagrams.

Lemma 2. *Suppose $(P, \mathbf{R}_\alpha)$ is a chain of territory diagrams where α belongs to some index set I and $\mathbf{R}_\alpha = \langle R_{\alpha,1}, \ldots, R_{\alpha,n} \rangle$. Let $\mathbf{R} = \langle R_1, \ldots, R_n \rangle$, where $R_i = \bigcup_{\alpha \in I} R_{\alpha,i}$, for $i = 1, \ldots, n$. Then $(P, \mathbf{R})$ is a territory diagram. In particular, $\mathbf{R}$ is an upper bound to the chain. If each territory diagram is convex, then $(P, \mathbf{R})$ is convex. If each territory diagram is radially convex, then $(P, \mathbf{R})$ is radially convex.*

Proof. If each territory diagram is convex, then the convexity of each R_i follows since given any two points $u, v \in R_i$ there exists $\alpha, \beta \in I$ such that $u \in R_{\alpha,i}$, $v \in R_{\beta,i}$ and by the chain property either $R_{\alpha,i}$ contains $R_{\beta,i}$ or vice versa. Thus, the segment between u and v is contained, say, in $R_{\alpha,i} \subseteq R_i$.

Similarly, if each territory diagram is radially convex, for any $u \in R_i$, there exists $\alpha \in I$ such that $u \in R_{\alpha,i}$. Thus, the interval from p_i to u is contained in $R_{\alpha,i} \subseteq R_i$; so R_i is radially convex.

Next we prove (essentially similarly to above but indirectly) that $\mathbf{R}$ satisfies Definition 4. Presume otherwise. Then for some i, some $u \in R_i$, some $j \neq i$, and some $z \in R_j$, the following holds:

$$d(u, p_i) > d(u, z) . \tag{17}$$

Hence, there exists $\alpha, \beta \in I$ with $u \in R_{\alpha,i}$ and $z \in R_{\beta,j}$. Since we have a chain of territory diagrams, one of $\mathbf{R}_\alpha$ and $\mathbf{R}_\beta$ contains the other one. Without loss of generality, assume $\mathbf{R}_\beta \preceq \mathbf{R}_\alpha$. But then for the same i and j described above we would have $d(u, p_i) > d(u, z)$ for $u \in R_{\alpha,i}$ and $z \in R_{\alpha,j}$, contradicting the hypothesis that $\mathbf{R}_\alpha$ is a territory diagram. $\square$

4.1   Characterization of Maximal Territory Diagrams

We consider the particular case of convex maximal territory diagrams.

We first need to define some sets.

Let $(P, \mathbf{R})$ be a given convex territory diagram. For each $i = 1, \ldots, n$, set

$$\Delta_i = \mathrm{dom}\left(p_i, \bigcup_{j \neq i} R_j \right) - R_i . \tag{18}$$

The set Δ_i consists of a convex set with some convex part removed (see Fig. 20), which gives a representation of these sets. In a zone diagram each Δ_i is empty. However, in any other maximal territory diagram Δ_i is non-empty for at least one $i \in \{1, \ldots, n\}$.

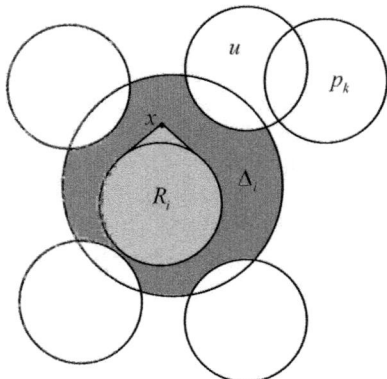

Fig. 20. The region R_i (*light gray*) and its augmentation in the region D_i (*dark gray*)

In Fig. 20, R_i is the light-gray region and Δ_i is the region between R_i and the largest circle. We will explain the dark gray region next.

If R_i is not maximal, then for some index i the region R_i can be extended to R_i', where the difference $R_i' - R_i$ would necessarily be contained in Δ_i. But the points in $R_i' - R_i$ must be outside of the dominance set of other sites. More specifically, if

$$F_i = \bigcup_{k \neq i} \mathrm{dom}\left(p_i, \bigcup_{j \neq i} R_j\right) , \tag{19}$$

then we must have

$$R_i' - R_i \subset \Delta_i - F_i . \tag{20}$$

To visualize the set $\Delta_i - F_i$, we can imagine removing a finite number of convex sets from $\mathrm{dom}(p_i, \bigcup_{j \neq i} R_j)$, where each convex set avoids a certain convex part R_i. But in addition to these finite convex removals, there are many other circular regions that may take away further parts. Specifically, let

$$\begin{aligned} U_i &= \{z \in \Delta_i : \forall k \neq i, \exists u \in F_i \text{ with } d(u,z) \leq d(u,p_k)\} , \\ D_i &= \Delta_i - (F_i \cup U_i) . \end{aligned} \tag{21}$$

In Fig. 20 the set D_i consists of the dark gray region. The white regions removed from Δ_i either belong to the dominance region of other sites, or they belong to a circular neighborhood of points in U_i. The figure shows a point u in the dominance set of p_k which takes a 'bite' from Δ_i, because there are points in Δ_i that are closer to it than to p_k. We now give a characterization of maximality.

Theorem 5. *Let $(P, \mathbf{R})$ be a territory diagram. It is not a maximal territory diagram if and only if there exists a territory diagram $(P, \mathbf{R}')$ satisfying $\mathbf{R} \prec \mathbf{R}'$ where*

$$\mathbf{R}' = \langle R_1, \ldots, R_{i-1}, R_i', R_{i+1}, \ldots R_n \rangle \,, \tag{22}$$

and where R_i', for some i, is the convex hull of R_i and z for some $z \in D_i$.

Proof. Clearly, if there exists a $z \in D_i$, then $R_i' = \text{conv}\{z, R_i\}$ is a strict superset of R_i and $\mathbf{R}'$ is a territory diagram strictly extending $\mathbf{R}$. Thus $\mathbf{R}$ is not maximal.

Suppose $\mathbf{R}$ is not a maximal territory diagram; then for some i, R_i is not a maximal region. Then there exists a territory diagram $(P, \mathbf{S})$ satisfying $\mathbf{R} \prec \mathbf{S}$ with R_i strictly contained in S_i. In particular, there exists $z \in D_i \cap S_i$. But then if we set $R_i' = \text{conv}\{z, R_i\}$, we have that R_i' is a strict superset of R_i so that the territory diagram $(P, \mathbf{R}')$ satisfies $\mathbf{R} \prec \mathbf{R}'$.   □

It may be tempting to expect that a region R_i is maximal if and only if D_i is empty. For instance in the example considered in Fig. 6 such is the case. Let us verify this for both of the regions in the example. As claimed before, it is easy to show that $\text{dom}(p_1, R_2) = \{(x, y) : y \geq \frac{1}{2}(x^2 + 1)\}$, $\text{dom}(p_2, R_1) = R_2$. For this example, $D_2 = \Delta_2 = \emptyset$. Also, $U_1 = \text{dom}(p_1, R_2) - R_1$. Thus, $\Delta_1 = U_1$ and $D_1 = \emptyset$. However, the example in Fig. 21 shows that such is not always the case. In Fig. 21, R_1 is a square centered at p_1. The area between the four circles and R_1 represents D_1. For every point $z \in D_1$ the convex hull of z and R_1 will intersect at least one of the circles in a point different from the four touching points.

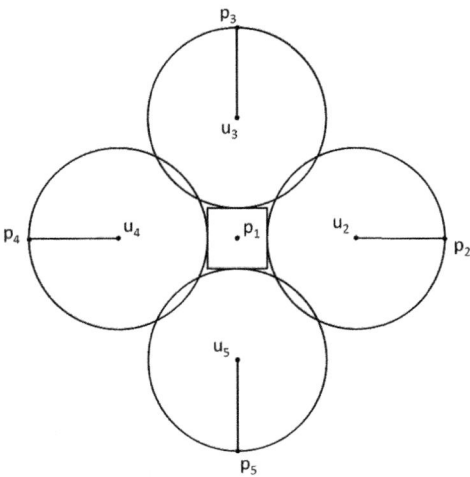

Fig. 21. The region R_1 is maximal while D_1 is nonempty

Definition 11. *Let $(P, \mathbf{R})$ be a territory diagram. With D_i defined as in (21), set*

$$\begin{aligned} Z_i &= \{z : \text{conv}\{z, R_i\} \subset D_i\}, \\ Z(\mathbf{R}) &= \{z \in Z_i : i = 1, \ldots, n\} \,. \end{aligned} \tag{23}$$

Clearly from the definition of a maximal territory diagram and the theorem above we have the next corollary.

Corollary 3. *A territory diagram* $(P, \mathbf{R})$ *is maximal if and only if* $Z(\mathbf{R})$ *is empty.*

4.2   Embedding a Territory Diagram into a Maximal Territory Diagram

We now consider ways in which the characterization theorem for convex maximal territory diagrams would allow embedding a given territory diagram that is not maximal into a maximal territory diagram. From Definition 11 the following easily holds.

Theorem 6. *Let* $(P, \mathbf{R})$ *and* $(P, \mathbf{R}')$ *be territory diagrams satisfying* $\mathbf{R} \prec \mathbf{R}'$. *Then* $Z(\mathbf{R}) \subsetneq Z(\mathbf{R}')$.

Thus if a given territory diagram $(P, \mathbf{R})$ is not maximal, we can compute a point $z \in Z(\mathbf{R})$, say in Z_i, and replace R_i with $\text{conv}\{z, R_i\}$. This gives an incremental improvement. Any maximal convex set that contains z and R_i, and is contained in D_i, results in a maximal region R_i'. (Here, maximality is with respect to inclusion.) Such maximal sets may not be unique. Whether or not such a maximal region can be obtained through countably many operations may well depend on the particular set of points P and the starting territory at hand. In any case, we can now conclude from Lemma 2 and Theorem 6 with our main result of the section :

Theorem 7. *A territory diagram* $(P, \mathbf{R})$, *if not already maximal, can be embedded (i.e.* $\mathbf{R} \prec \mathbf{R}'$*) into a maximal territory diagram* $(P, \mathbf{R}')$.

5   Concluding Remarks and Future Work

Because of their mollified definition in contrast to that of zone diagrams, maximal territory diagrams have their own interesting properties. Their structure and computation is nontrivial and present their own challenging theoretical and computational problems. For instance, one may begin with a particular territory diagram of interest and wish to explore the range of maximal territory diagrams that contain it, possibly subject to achieving some goal or some criterion of optimality (such as maximum area). One might also seek to determine the optimal strategy for determining attributes of a given site in competition with others given limited resources for growth.

Computation of zone diagrams is currently an active area of research that identifies many open problems and challenges (see for example [2]), and finding good approximation algorithms for them as well as determining the complexity, convergence and other properties of such algorithms will probably require new ideas and methods, such as those described in [7]. We believe that the study of maximal territory diagrams both benefits from and contributes to the advancement of such ideas and methods. We also believe that the study of maximal territory diagrams may be a valuable step towards a better understanding of the structure of zone diagrams. Further study of the relation between safe zones and forbidden zones and of the possible forms that they can take is a worthy topic for pursuit both in itself and as a new approach to tackle the difficulties with computing and representing zone diagrams.

Also, analogous to the case of zone diagrams, the results shown here can most likely be extended to higher dimensions, other metrics, and more general spaces. We intend to address these in our future work.

Finally, we also propose the following conjecture that arises in the context of maximal territories and is a generalization of the existence of maximal territories.

Conjecture. Given a set of sites $P = \langle p_1, \ldots, p_n \rangle$ and any $\varepsilon > 0$ there exists a maximal territory diagram $(P, \mathbf{R})$ such that

$$\mu \left(\bigcup_{i=1}^{n} \Delta_i \right) = \varepsilon \,, \tag{24}$$

where μ is the Lebesgue measure. When $\mathbf{R}$ is a zone diagram this measure is zero, but one can imagine that removing a small part of a region would allow adding a small part to another region, hence resulting in a new maximal territory where the corresponding measure of $\bigcup_{i=1}^{n} \Delta_i$ would increase from zero by a small amount.

Acknowlegments. Bahman Kalantari would like to thank Franz Aurenhammer for introducing him to zone diagrams.

References

1. Asano, T., Matoušek, J., Tokuyama, T.: Zone Diagrams: Existence, Uniqueness, and Algorithmic Challenge. SIAM Journal on Computing 37, 1182–1198 (2007)
2. Asano, T., Brimkov, V.E., Barneva, R.P.: Some theoretical challenges in digital geometry, A perspective. Discrete Applied Mathematics 157(16), 3362–3371 (2009)
3. Aurenhammer, F.: Voronoi diagrams - a survey of a fundamental geometric data structure. ACM Computing Surveys 23, 345–405 (1991)
4. de Biasi, S.C., Kalantari, B., Kalantari, I.: Maximal Zone Diagrams and their Computation. In: ISVD 2010, pp. 171–180. IEEE Computer Society (2010)
5. Kalantari, B.: Voronoi Diagrams and Polynomial Root-Finding. In: ISVD 2009, pp. 31–40. IEEE Computer Society (2009)
6. Kalantari, B.: Polynomial Root-Finding Methods Whose Basins of Attraction Approximate Voronoi Diagram. Discrete & Computational Geometry (2011), doi:10.1007/s00454-011-9330-3
7. Reem, D.: An Algorithm for Computing Voronoi Diagrams of General Generators in General Normed Spaces. In: ISVD 2009, pp. 144–152. IEEE Computer Society (2009)

Alpha, Betti and the Megaparsec Universe:
On the Topology of the Cosmic Web

Rien van de Weygaert[1], Gert Vegter[2], Herbert Edelsbrunner[3], Bernard J.T. Jones[1],
Pratyush Pranav[1], Changbom Park[4], Wojciech A. Hellwing[5],
Bob Eldering[2], Nico Kruithof[2], E.G.P. (Patrick) Bos[1], Johan Hidding[1],
Job Feldbrugge[1], Eline ten Have[6], Matti van Engelen[2],
Manuel Caroli[7], and Monique Teillaud[7]

[1] Kapteyn Astronomical Institute, University of Groningen, P.O. Box 800,
9700 AV Groningen, The Netherlands
[2] Johann Bernoulli Institute for Mathematics and Computer Science,
University of Groningen, P.O. Box 407, 9700 AK Groningen, The Netherlands
[3] IST Austria, Am Campus 1, 3400 Klosterneuburg, Austria
[4] School of Physics, Korea Institute for Advanced Study, Seoul 130-722, Korea
[5] Interdisciplinary Centre for Mathematical and Computational Modeling,
University of Warsaw, ul. Pawinskiego 5a, 02-106 Warsaw, Poland
[6] Stratingh Institute for Chemistry, University of Groningen, Nijenborgh 4,
9747 AG Groningen, The Netherlands
[7] Géométrica, INRIA Sophia Antipolis-Méditerranée, route des Lucioles,
BP 93, 06902 Sophia Antipolis Cedex, France

Abstract. We study the topology of the Megaparsec Cosmic Web in terms of
the scale-dependent Betti numbers, which formalize the topological information
content of the cosmic mass distribution. While the Betti numbers do not fully
quantify topology, they extend the information beyond conventional cosmological studies of topology in terms of genus and Euler characteristic. The richer information content of Betti numbers goes along the availability of fast algorithms
to compute them.

For continuous density fields, we determine the scale-dependence of Betti
numbers by invoking the cosmologically familiar filtration of sublevel or superlevel sets defined by density thresholds. For the discrete galaxy distribution, however, the analysis is based on the alpha shapes of the particles. These simplicial
complexes constitute an ordered sequence of nested subsets of the Delaunay tessellation, a filtration defined by the scale parameter, α. As they are homotopy
equivalent to the sublevel sets of the distance field, they are an excellent tool for
assessing the topological structure of a discrete point distribution. In order to develop an intuitive understanding for the behavior of Betti numbers as a function
of α, and their relation to the morphological patterns in the Cosmic Web, we
first study them within the context of simple heuristic Voronoi clustering models.
These can be tuned to consist of specific morphological elements of the Cosmic Web, i.e. clusters, filaments, or sheets. To elucidate the relative prominence
of the various Betti numbers in different stages of morphological evolution, we
introduce the concept of alpha tracks.

M.L. Gavrilova et al. (Eds.): Trans. on Comput. Sci. XIV, LNCS 6970, pp. 60–101, 2011.

Subsequently, we address the topology of structures emerging in the standard LCDM scenario and in cosmological scenarios with alternative dark energy content. The evolution of the Betti numbers is shown to reflect the hierarchical evolution of the Cosmic Web. We also demonstrate that the scale-dependence of the Betti numbers yields a promising measure of cosmological parameters, with a potential to help in determining the nature of dark energy and to probe primordial non-Gaussianities. We also discuss the expected Betti numbers as a function of the density threshold for superlevel sets of a Gaussian random field.

Finally, we introduce the concept of persistent homology. It measures scale levels of the mass distribution and allows us to separate small from large scale features. Within the context of the hierarchical cosmic structure formation, persistence provides a natural formalism for a multiscale topology study of the Cosmic Web.

1   Introduction: The Cosmic Web

The large scale distribution of matter revealed by galaxy surveys features a complex network of interconnected filamentary galaxy associations. This network, which has become known as the *Cosmic Web* [6], contains structures from a few megaparsecs[1] up to tens and even hundreds of megaparsecs of size. Galaxies and mass exist in a wispy web-like spatial arrangement consisting of dense compact clusters, elongated filaments, and sheet-like walls, amidst large near-empty voids, with similar patterns existing at earlier epochs, albeit over smaller scales; see Figure 1 for an illustration of a simulated cosmic mass distribution in the standard LCDM cosmology[2] in a box of size $80\ h^{-1}\mathrm{Mpc}$ [3]. The hierarchical nature of this mass distribution, marked by substructure over a wide range of scales and densities, has been clearly demonstrated [62]. Its appearance has been

---

[1] The main measure of length in astronomy is the parsec. Technically, 1pc is the distance at which we would see the distance Earth-Sun at an angle of 1 arcsec. It is equal to 3.262 lightyears $= 3.086 \times 10^{13}$km. Cosmological distances are substantially larger, so that a megaparsec, with $1\mathrm{Mpc} = 10^6$ pc, is the common unit of distance.

[2] Currently, the LCDM cosmological scenario is the standard - or, "concordance" - cosmological scenario. It seems to be in agreement with a truly impressive amount of observational evidence, although there are also some minor deficiencies. The "L" stands for Λ, the cosmological constant that dominates the dynamics of our Universe, causes its expansion to accelerate and represents in the order of about 73% of its energy content. "CDM" indicates the Cold Dark Matter content of the Universe. Invisible, it appears to form the major fraction of gravitating matter in the Universe, representing some 23% of the cosmic energy density as opposed to the mere 4.4% that we find in the normal baryonic matter we, and the planets and stars, consist of.

[3] The Universe expands according to Hubble's law, stating that the recession velocity of a galaxy is linearly proportional to its distance: $v = \mathrm{H}r$. The Hubble parameter, H, quantifies the expansion rate of the Universe, and is commonly expressed in units of km/s/Mpc. Its present value, the "Hubble constant", is estimated to be $\mathrm{H}_0 \approx 71\ \mathrm{km/s/Mpc}$. Quite often, its value is expressed by a dimensionless number, h, that specifies the Hubble parameter in units of 100 km/s/Mpc.

most dramatically illustrated by the recently produced maps of the nearby cosmos, the 2dFGRS, the SDSS and the 2MASS redshift surveys [12, 28, 58] [4].

The vast Megaparsec Cosmic Web is one of the most striking examples of complex geometric patterns found in nature, and certainly the largest in terms of sheer size. Computer simulations suggest that the observed cellular patterns are a prominent and natural aspect of cosmic structure formation through gravitational instability [42], the standard paradigm for the emergence of structure in our Universe [57, 61]. According to the *gravitational instability scenario*, cosmic structure grows from primordial density and velocity perturbations. These tiny primordial perturbations define a Gaussian density field and are fully characterized by the power spectrum; see Section 8.

1.1 Web Analysis

Over the past decades, we have seen many measures for characterizing different aspects of the large scale cosmic structure: correlation functions (describing the n-point distribution), Minkowski functionals and genus (characterizing the local and global curvature of isodensity surfaces), multi-fractals (summarizing the statistical moments on various scales), and so on.

Despite the multitude of descriptions, it has remained a major challenge to characterize the structure, geometry and topology of the Cosmic Web. Many attempts to describe, let alone identify, the features and components of the Cosmic Web have been of a rather heuristic nature. The overwhelming complexity of both the individual structures as well as their connectivity, the lack of structural symmetries, its intrinsic multiscale nature and the wide range of densities in the cosmic matter distribution has prevented the use of simple and straightforward instruments.

In the observational reality, galaxies are the main tracers of the Cosmic Web, and it is mainly through the measurement of the redshift distribution of galaxies that we have been able to map its structure. Likewise, simulations of the evolving cosmic matter distribution are almost exclusively based upon N-body particle computer calculations, involving a discrete representation of the features we seek to study. Both the galaxy distribution as well as the particles in an N-body simulation are examples of *spatial point processes* in that they are *discretely sampled* and have an *irregular spatial distribution*.

For furthering our understanding of the Cosmic Web, and to investigate its structure and dynamics, it is of prime importance to have access to a set of proper and objective analysis tools. In this contribution, we address the topological and morphological analysis of the large scale galaxy distribution. In particular, we introduce a new measure that is particularly suited to differentiating web-like structures: the homology of the distribution as measured by Betti numbers and their persistence.

---

[4] Because of the expansion of the Universe, any observed cosmic object will have its light shifted redward : its *redshift*, z. According to Hubble's law, the redshift z is directly proportional to the distance r of the object, for $z \ll 1$: $cz = Hr$, in which the constant H is the Hubble parameter. Because it is extremely cumbersome to measure distances r directly, cosmologists resort to the expansion of the Universe and use z as a distance measure. Because of the vast distances in the Universe, and the finite velocity of light, the redshift z of an object may also be seen as a measure of the time at which it emitted the observed radiation.

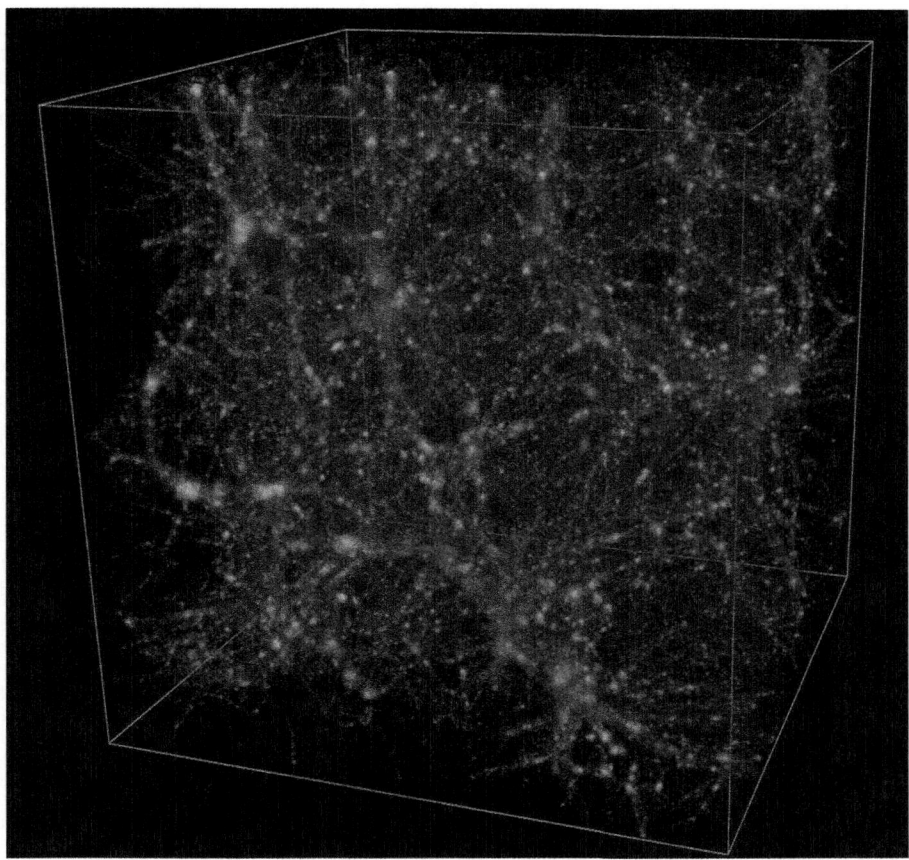

Fig. 1. The Cosmic Web in an LCDM simulation. Shown is the dark matter distribution at the current cosmological epoch in a 384^3 particles N-body simulation of structure formation in a universe with a cosmological constant Λ, accounting for an energy density $\Omega_\Lambda = 0.73$, and cold dark matter, accounting for a total mass density of $\Omega_m = 0.23$. The box has a co-moving size of $80\ h^{-1}$Mpc (i.e. a box co-expanding with the universe and having a current size of $80\ h^{-1}$Mpc). Clearly visible is the intriguing network of filaments and high-density cluster nodes surrounding low-density voids. The mass distribution has a distinct multiscale character, marked by clumps over a wide range of mass scales, reflecting the hierarchical evolution of the cosmic mass distribution. Figure courtesy of Bartosz Borucki.

1.2   Genus and Minkowski Functionals

Gott and collaborators were the first to propose the use of the topology of the megaparsec structure of the Universe [25, 26], to examine whether or not the initial fluctuations had been Gaussian. They introduced the genus of isodensity surfaces of the cosmic mass distribution as a quantitative characterization of cosmic topology in terms of the connectivity of these surfaces. Being an intrinsic topology measure, the genus is relatively insensitive to systematic effects such as non-linear gravitational evolution, galaxy biasing, and redshift-space distortions [41]. This, in fact, is one of the most important advantages of using the genus in studying the patterns in the Megaparsec Universe. This allowed [25], along with a series of subsequent studies, to conclude that the measured genus of the observed large scale structure is consistent with the predictions of cosmo-logical scenarios whose initially Gaussian density and velocity fluctuations correspond to that of the observed power spectrum [23, 24, 36, 39, 40, 67, 71]. Nonetheless, recent work has shown that on small scales, the observed genus is inconsistent with theoreti-cal models [11]. On these scales the - not fully known - details of the galaxy formation process become important.

Additional quantitative information on the morphology of the large scale distribution of galaxies are the Minkowski functionals. Mecke et al. [35] introduced them within a cosmological context as a more extensive characterization of the morphology and ge-ometry of spatial patterns, after which Schmalzing et al. [47, 48] and others developed them into a useful tool for studying cosmological datasets. The Minkowski functionals measure the geometry and shape of manifolds, providing a quantitative characterization of the level sets (the isodensity surfaces) of the matter distribution at a given level of smoothing of the original data. In a d-dimensional space, there are $d + 1$ Minkowski functionals, and the Euler characteristic of isodensity surfaces is one of them. In 3 di-mensions, there are four Minkowski functionals: the volume enclosed by an isodensity surface, the surface area, the integrated mean curvature, and the Euler characteristic (as in (2)). From these, it is possible to generate nonparameteric shape descriptors such as the "typical" thickness, breadth and length of the structures involved at that level of smoothing [45]. There are also algorithms for computing these quantities from discrete point sets [9], and the technique has found wide applications outside of astronomy and physics.

What the Minkowski functionals do not tell about is the topology of the structure as described by the number of blobs, tunnels and voids that are present with changing scale. In that sense, the Betti numbers and their persistence will provide a substantial extension of available topological information. Also, the related use of alpha shapes will provide effective methods of evaluating the Minkowski functionals.

1.3 Homology Groups

Here we advance the topological characterization of the Cosmic Web to a more com-plete description, the homology of the distribution as measured by the scale-dependent Betti numbers of the sample. While a full quantitative characterization of the topology of the cosmic mass distribution is not feasible, its homology is an attractive compromise between detail and speed, providing a useful summary measurement of topology. The ranks of the homology groups, known as the *Betti numbers*, completely characterize the topology of orientable 2-manifolds, such as the isodensity surfaces used in earlier topological analyzes of the Cosmic Web. For a d-dimensional manifold, we have $d + 1$ homology groups, H_p, and correspondingly $d + 1$ Betti numbers, β_p, for $0 \leq p \leq d$. The Betti numbers may be considered as the number of p-dimensional holes. In fact, homology groups can be seen as a useful definition of holes in the cosmic mass distribu-tion. They are fundamental quantities from which the Euler characteristic and genus are derived, including those of the isodensity 2-manifolds. In that sense, they are a powerful generalization of genus analysis.

While the Euler characteristic and the Betti numbers give information about the connectivity of a manifold, two of the other three Minkowski functionals are sensitive to local manifold deformations, and their topological information is limited. The Betti numbers represent a more succinct as well as more informative characterization of the topology.

Following this observation, we will argue and demonstrate that the scale-dependence of Betti numbers makes them particularly suited to analyze and differentiate web-like structures. The Betti numbers measured as a function of scale are a sensitive discriminator of cosmic structure and can effectively reveal differences arising in alternative cosmological models. Potentially, they may be exploited to infer information on important issues such as the nature of dark energy or even possible non-Gaussianities in the primordial density field, as will be corroborated by [68]; see Section 7.

1.4 Alpha Shapes

Most of the topological studies in cosmology depend on some sort of user-controlled smoothing and related threshold to specify surfaces of which the topology may be determined. In cosmological studies, this usually concerns isodensity surfaces and their superlevel and sublevel sets defined on a Gaussian filter scale.

A preferred alternative would be avoid filters and to invoke a density field reconstruction that adapts itself to the galaxy distribution. For our purpose, the optimal technique would be that of the Delaunay Tessellation Field Estimator (DTFE) formalism [10, 46, 64], which uses the local density and shape sensitivity of the Delaunay tessellation to recover a density field that retains the multiscale nature as well as the anisotropic nature of the sampled particle or galaxy distribution. This results in an adaptive and highly flexible representation of the underlying density field, which may then be used to assess its topological and singularity structure; see e.g. [9, 51, 71].

A closely related philosophy is to focus exclusively on the shape of the particle or galaxy distributions itself, and seek to analyze its topology without resorting to the isodensity surfaces and level sets of the corresponding density field. This is precisely where *alpha shapes* enter the stage. They are subsets of a Delaunay triangulation that describe the intuitive notion of the shape of a discrete point set. Introduced by Edelsbrunner and collaborators[18], they are one of the principal concepts from the field of Computational Topology [14, 65, 72].

1.5 Outline

This review contains a report on our project to study the homology and topology of the Cosmic Web. It is a major upgrade of an earlier report [69].

Sections 2 and 3 provide the mathematical background for this review, consisting of a short presentation of homology and Betti numbers, followed by a discussion of alpha shapes and the formalism to infer Betti numbers.

One of the intentions of our study is to understand the information content of scale-dependent Betti numbers with respect to the corresponding geometric patterns observed in the Cosmic Web, i.e. the relative prominence of clusters, filaments and walls. To this end, we first investigate a number of Voronoi clustering models. These models use the geometric structure of Voronoi tessellations as a skeleton of the cosmic mass distribution. In Section 4, we describe these models, and specify the class of Voronoi Element Models and Voronoi Kinematic Models. Section 5 contains an extensive description of the analysis of a set of Voronoi Element and Voronoi Kinematic Models.

Following the analysis of Voronoi clustering models, we turn to the homology analysis in a few cosmological situations of current interest. Section 6 discusses the scale-dependent Betti numbers inferred for the current standard theory of cosmic structure formation, LCDM. The results are obtained from a computer simulation of structure formation in this cosmological scenario. The discussion in Section 6 connects the resulting homological characteristics to the hierarchically evolving mass distribution. In an attempt to investigate and exploit the sensitivity of Betti numbers to key cosmological parameters, we apply our analysis to a set of cosmological scenarios with a different content of dark energy. Section 7 demonstrates that indeed homology may be a promising tool towards inferring information on the dark energy content that dominates the dynamics of our Universe. Equally compelling is the issue of the Betti numbers of superlevel or sublevel sets of Gaussian random fields. To high accuracy, the initial density field out of which all structure in our Universe arose has had a Gaussian character. As reference point for any further assessment of the homology of the evolving cosmic mass distribution, we therefore need to evaluate the homology of Gaussian random fields. This is the subject of Section 8, which focusses on the expected values of the Betti numbers. In addition, it will be a starting point for any related study looking for possible primordial non-Gaussian deviations.

Having shown the potential of homology studies for cosmological purposes, we discuss the prospects of assessing the persistence of cosmological density fields in Section 9. Persistence measures scale levels of the mass distribution and allows us to systematically separate small from large scale features in the mass distribution. It provides us with a rich language to study intrinsically multiscale distributions as those resulting from the hierarchically evolving structure in the Universe. Finally, Section 10 addresses future prospects and relates this review to other work.

2   Homology and Betti Numbers

Homology groups and Betti numbers are concepts from algebraic topology, designed to quantify and compare topological spaces. They characterize the topology of a space in terms of the relationship between the cycles and boundaries we find in the space[5]. For example, if the space is a d-dimensional manifold, $\mathbb{M}$, we have cycles and boundaries of dimension p from 0 to d. Correspondingly, $\mathbb{M}$ has one homology group $H_p(\mathbb{M})$ for each of $d + 1$ dimensions, $0 \leq p \leq d$. By taking into account that two cycles should be considered identical if they differ by a boundary, one ends up with a group $H_p(\mathbb{M})$

---

[5] Assuming the space is given as a simplicial complex, a *p-cycle* is a *p*-chain with empty boundary, where a *p-chain*, γ, is a sum of *p*-simplices. The standard notation is $\gamma = \sum a_i \sigma_i$, where the σ_i are the *p*-simplices and the a_i are the *coefficients*. For example, a 1-cycle is a closed loop of edges, or a finite union of such loops, and a 2-cycle is a closed surface, or a finite union of such surfaces. Adding two *p*-cycles, we get another *p*-cycle, and similar for the *p*-boundaries. Hence, we have a group of *p*-cycles and a group of *p*-boundaries.

whose elements are the equivalence classes of p-cycles[6]. The *rank* of the homology group $H_p(\mathbb{M})$ is the p-th *Betti number*, $\beta_p = \beta_p(\mathbb{M})$.

In heuristic - and practical - terms, the Betti numbers count topological features and can be considered as the number of p-dimensional holes. When talking about a surface in 3-dimensional space, the zeroth Betti number, β_0, counts the components, the first Betti number, β_1, counts the tunnels, and the second Betti number, β_2, counts the enclosed voids. All other Betti numbers are zero. Examples of spaces with Betti numbers which are of interest to us are the sublevel and superlevel density set of the cosmic mass distribution, i.e. the regions whose density is smaller or greater than a specified threshold level.

2.1 Genus and Euler Characteristic

Numerous cosmological studies have considered the *genus* of the isodensity surfaces defined by the megaparsec galaxy distribution [25–27], which specifies the number of handles defining the surface. More formally, it is the maximum number of disjoint closed curves such that cutting along these curves does not increase the number of components. The genus has a simple relation to the Euler characteristic, χ, of the isodensity surface. Consider a 3-manifold subset $\mathbb{M}$ of the Universe and its boundary, $\partial\mathbb{M}$, which is a 2-manifold. With $\partial\mathbb{M}$ consisting of $c = \beta_0(\partial\mathbb{M})$ components, the Gauss-Bonnet Theorem states that the genus of the surface is given by

$$G = c - \frac{1}{2}\chi(\partial\mathbb{M}), \tag{1}$$

where the Euler characteristic $\chi(\partial\mathbb{M})$ is the integrated Gaussian curvature of the surface

$$\chi(\partial\mathbb{M}) = \frac{1}{2\pi} \oint_x \frac{\mathrm{d}x}{R_1(x)R_2(x)}. \tag{2}$$

Here $R_1(x)$ and $R_2(x)$ are the principal radii of curvature at the point x of the surface. The integral of the Gaussian curvature is invariant under continuous deformation of the surface: perhaps one of the most surprising results in differential geometry.

The Euler characteristic of the surface can also be expressed in terms of its Betti numbers, namely, $\chi(\partial\mathbb{M})$ is equal to $\beta_0(\partial\mathbb{M})$ minus $\beta_1(\partial\mathbb{M})$ plus $\beta_2(\partial\mathbb{M})$. This is implied by the Euler-Poincaré Formula, which we will discuss shortly, after introducing triangulation of spaces. Combining these two equations for the Euler characteristic, we get a fundamental relationship between differential geometry and algebraic topology. Returning to the 3-dimensional subset, $\mathbb{M}$, of the Universe, its Euler characteristic is

$$\chi(\mathbb{M}) = \beta_0(\mathbb{M}) - \beta_1(\mathbb{M}) + \beta_2(\mathbb{M}) - \beta_3(\mathbb{M}). \tag{3}$$

---

[6] Correctly defined, the *p-th homology group* is the p-th cycle group modulo the p-th boundary group. In algebraic terms, this is a quotient group. In intuitive terms, this says that two p-cycles are considered the same, or *homologous*, if together they bound a $(p+1)$-chain or, equivalently, if they differ by a p-boundary. Indeed, we do not want to distinguish between two 1-cycles of, say, the torus, if they both go around the hole, differing only in the geometric paths they take to do so.

Whenever $\mathbb{M}$ is a non-exhaustive subset of the connected Universe, its third Betti number vanishes, $\beta_3(\mathbb{M}) = 0$, and its boundary is necessarily non-empty, namely a 2-manifold without boundary. As mentioned above, the Euler characteristic of $\partial\mathbb{M}$ is the alternating sum of Betti numbers, where $\beta_0(\partial\mathbb{M})$ is the number of surface components, $\beta_1(\partial\mathbb{M})$ is twice the genus, and $\beta_2(\partial\mathbb{M}) = \beta_0(\partial\mathbb{M})$. Assuming the Universe is connected like the 3-sphere, we can use Alexander duality and the Mayer-Vietoris sequence to establish a direct relation between the Betti numbers of the 3-manifold with boundary, $\mathbb{M}$, and those of the 2-manifold without boundary, $\partial\mathbb{M}$; see e.g. [15]:

$$\beta_0(\partial\mathbb{M}) = \beta_2(\partial\mathbb{M}) = \beta_0(\mathbb{M}) + \beta_2(\mathbb{M}), \tag{4}$$

$$\beta_1(\partial\mathbb{M}) = 2\beta_1(\mathbb{M}). \tag{5}$$

From this, we infer that the Euler characteristic of the boundary is directly proportional to the Euler characteristic of the 3-manifold:

$$\chi(\partial\mathbb{M}) = \beta_0(\partial\mathbb{M}) - \beta_1(\partial\mathbb{M}) + \beta_2(\partial\mathbb{M}) \tag{6}$$

$$= 2\chi(\mathbb{M}). \tag{7}$$

In the cosmologically interesting situation in which $\partial\mathbb{M}$ is the isodensity surface of either density superlevel or sublevel sets, we find a relation between the genus[7] of the surface and the Betti numbers of the enclosed manifold:

$$G = c - \frac{1}{2}\chi(\partial\mathbb{M}) \tag{8}$$

$$= c - \chi(\mathbb{M}) \tag{9}$$

$$= c - (\beta_0(\mathbb{M}) - \beta_1(\mathbb{M}) + \beta_2(\mathbb{M})). \tag{10}$$

In the analysis described in this paper, we will restrict ourselves to the three Betti numbers, β_0, β_1, and β_2, of the 3-manifolds with boundary defined by the cosmic mass distribution.

2.2   Triangulated Spaces

A practical simplification occurs when we represent a space by a *triangulation*, which is a simplicial complex that retains the topological properties of the space. These are topological spaces assembled from vertices, edges, triangles, tetrahedra, and possibly higher-dimensional simplices[8]. In this situation, homology is defined as described earlier, by comparing chains that are cycles with chains that are boundaries. The availability of simplices has a number of advantages, including the existence of fast algorithms

---

[7] For consistency, it is important to note that the definition in previous topology studies in cosmology [25, 26] slightly differs from this. The genus, g, in these studies has been defined as the number of holes minus the number of connected regions: $g = G - c$. Here, we will refer to g as the *reduced genus*.

[8] Technically, we require that the union of simplices in the simplicial complex is *homeomorphic* to the space it represents, which means that there is a bijective map between the two sets that is continuous and whose inverse is continuous. As an example, consider the boundary of the octahedron, consisting of 6 vertices, 12 edges, and 8 triangles, which forms a triangulation of the 2-dimensional sphere.

to compute homology. Here, we focus on the connection between the number of simplices in the simplicial complex and the Betti numbers of the space.

Suppose $\mathbb{M}$ is a d-dimensional manifold, and K is a triangulation of $\mathbb{M}$. Write n_p for the number of p-dimensional simplices in K. For example, n_0 is the number of vertices, n_1 is the number of edges, and so on. The *Euler characteristic* of K is defined as the alternating sum of these numbers:

$$\chi(K) = \sum_{p=0}^{d} (-1)^p n_p. \tag{11}$$

This is the d-dimensional generalization of the classical Euler characteristic of a polytope:

$$\chi = \#\text{vertices} - \#\text{edges} + \#\text{faces}. \tag{12}$$

For a convex polytope, Euler's Formula states that $\chi = 2$. The *Euler-Poincaré Formula* is a far-reaching generalization of this relation. To state this generalization, we first note that homology is independent of the choice of triangulation, and so are the Betti numbers and the Euler characteristic. The Euler-Poincaré Formula says that the Euler characteristic of a triangulation, which is the alternating sum of simplex numbers, is equal to the alternating sum of Betti numbers of the triangulated space:

$$\chi(K) = \sum_{p=0}^{d} (-1)^p \beta_p(\mathbb{M}). \tag{13}$$

Coming back to the case of a convex polytope in 3-dimensional space, its faces decompose the boundary, which is homeomorphic to the 2-dimensional sphere. We may further decompose the faces into triangles, if necessary, but this makes no difference here. Since this is true for all convex polytopes, their faces are but different triangulations of this same sphere, so the alternating sums must be the same. They are all equal to 2 because this is the Euler characteristic of the 2-dimensional sphere.

2.3 Homology of a Filtration

For the assessment of the topology of a mass or point distribution, a rich source of information is the topological structure of a filtration. Given a space $\mathbb{M}$, a *filtration* is a nested sequence of subspaces:

$$\emptyset = \mathbb{M}_0 \subseteq \mathbb{M}_1 \subseteq \ldots \subseteq \mathbb{M}_m = \mathbb{M}. \tag{14}$$

The nature of the filtrations depends, amongst others, on the representation of the mass distribution. When representing the mass distribution by a continuous density field, $f(\mathbf{x})$, a common practice is to study the sublevel or superlevel sets of the field smoothed on a scale R_s:

$$f_s(\mathbf{x}) = \int f(\mathbf{y}) W_s(\mathbf{y} - \mathbf{x}) \, \mathrm{d}\mathbf{y}, \tag{15}$$

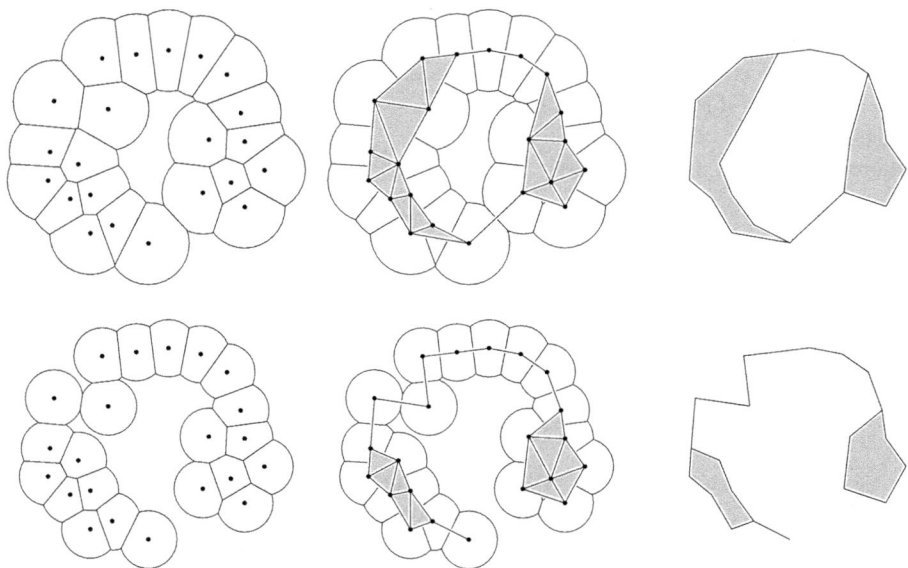

Fig. 2. Illustration of alpha shapes. For two different values of α, we show the relation between the 2-dimensional point distribution, the value of α, and the resulting alpha shape. Around each point in the sample, we draw a circle of radius α. The outline of the corresponding Voronoi tessellation is indicated by the edges (left). All Delaunay simplices *dual* to the decomposition of the union of disks by the Voronoi polygons are shown in black (center). The final resulting alpha shape is shown on the right. Top: large α value. Bottom: small α value.

where $W_s(\mathbf{x} - \mathbf{y})$ is the smoothing kernel. The *sublevel sets* of this field are defined as the regions

$$\mathbb{M}_\nu = \{\mathbf{x} \in \mathbb{M} \mid f_s(\mathbf{x}) \in (-\infty, f_\nu]\} \tag{16}$$

$$= f_s^{-1}(-\infty, f_\nu]. \tag{17}$$

In other words, they are the regions where the smoothed density is less than or equal to the threshold value $f_\nu = \nu\sigma_0$, with σ_0 the dispersion of the density field. When addressing the topology of the primordial Gaussian density field, as in Section 8, the analysis will be based on the filtration consisting of superlevel sets.

The representation of the mass by a discrete particle or galaxy distribution leads to an alternative strategy of filtration, this time in terms of simplicial complexes generated by the particle distribution. In the case of simplicial complexes that are homotopy equivalent to the sublevel sets of either the density or the distance function field, the computation of the homological characteristics of the field is considerably facilitated. Often it is also much easier to visualize the somewhat abstract notion of homology by means of these simplicial complexes.

In our study, we will concentrate on *alpha shapes* of a point set, which are subsets of the corresponding Delaunay triangulation; see Section 3. The alpha shapes are homotopy equivalent to the sublevel sets of the distance field defined by the point distribution, and they constitute a nested sequence of simplicial complexes that forms a topologically useful filtration of the Delaunay triangulation. In the following sections, we will extensively discuss the use of alpha shapes to assess the homology of cosmological particle and galaxy distributions.

3 Alpha Shapes

One of the key concepts in the field of Computational Topology are *alpha shapes*, as introduced by Edelsbrunner and collaborators [18, 20, 37]; see [15] for a recent review. They generalize the convex hull of a point set and are concrete geometric objects that are uniquely defined for a particular point set and a real value α. For their definition, we look at the union of balls of radius α centered on the points in the set, and its decomposition by the corresponding Voronoi tessellation; see the left diagrams in Figure 2. The *alpha complex* consists of all Delaunay simplices that record the subsets of Voronoi cells that have a non-empty common intersection within this union of balls; see the center diagrams of Figure 2. The *alpha shape* is the union of simplices in the alpha complex; see the right diagrams of Figure 2.

Alpha shapes reflect the topological structure of a point distribution on a scale parameterized by the real number α. The ordered set of alpha shapes constitute a filtration of the Delaunay tessellation. The link between alpha shapes and the homology of a point distribution can be appreciated from the fact that tunnels will be formed when, at a certain value of α, an edge is added between two vertices that were already connected, which increases the first Betti number. When new triangles are added, the tunnel may be filled, which decreases the first Betti number. More about this process of growing and shrinking Betti numbers in Section 9, where we discuss the persistence of tunnels and other topological features.

Connections of alpha shapes to diverse areas in the sciences and engineering have developed, including to pattern recognition, digital shape sampling and processing, and structural molecular biology [15]. Applications of alpha shapes have as yet focussed on biological systems, where they have been used in the characterization of the topology and structure of macromolecules. The work by Liang and collaborators [16, 32–34] uses alpha shapes and Betti numbers to assess the voids and pockets in an effort to classify complex protein structures, a highly challenging task given the tens of thousands of protein families involving thousands of different folds. Given the interest in the topology of the cosmic mass distribution [25, 35, 47, 48], it is evident that alpha shapes also provide a highly interesting tool for studying the topology of the galaxy distribution and the particles in N-body simulations of cosmic structure formation. Directly connected to the topology of the point distribution itself, it avoids the need of any user-defined filter kernels.

3.1 Definition

Figure 2 provides an impression of the concept by illustrating the process of defining the alpha shape, for two different values of α. If we have a finite point set, S, in 3-dimensional space and its Delaunay triangulation, we may identify all simplices – vertices, edges, triangles, tetrahedra – in the triangulation. For a given non-negative

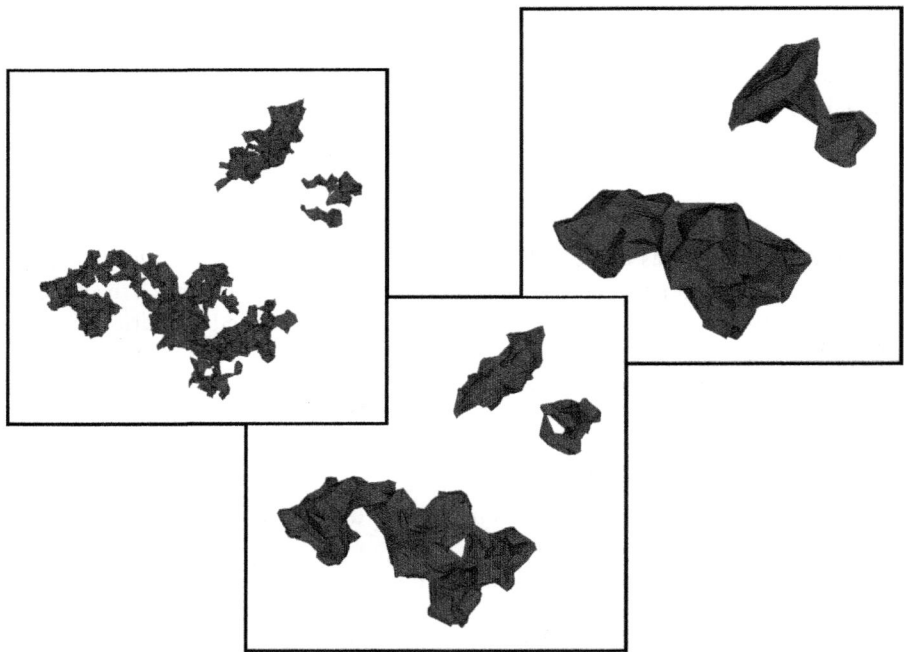

Fig. 3. Illustration of 3-dimensional alpha shapes for three different values of α = 0.001, 0.005, 0.01. Clearly visible are the differences in connectivity and topological structure of the different alpha shapes.

value of α, the *alpha complex* consists of all simplices in the Delaunay triangulation that have an empty circumsphere with radius less than or equal to α. Here "empty" means that the open ball bounded by the sphere does not include any points of S. For an extreme value $\alpha = 0$, the alpha complex merely consists of the vertices of the point set. There is also a smallest value, $\alpha_{\max}$, such that for $\alpha \geq \alpha_{\max}$, the alpha complex is the Delaunay triangulation and the alpha shape is the convex hull of the point set.

As mentioned earlier, the alpha shape is the union of all simplices in the alpha complex. It is a polytope in a fairly general sense: it can be concave and even disconnected. Its components can be three-dimensional clumps of tetrahedra, two-dimensional patches of triangles, one-dimensional strings of edges, and collections of isolated points, as well as combinations of these four types. The set of real numbers leads to a family of shapes capturing the intuitive notion of the overall versus fine shape of a point set. Starting from the convex hull gradually decreasing α, the shape of the point set gradually shrinks and starts to develop enclosed voids. These voids may join to form tunnels and larger voids. For negative α, the alpha shape is empty. An intuitive feel for the evolution of the topological structure may be obtained from the three different alpha shapes of the same point set in 3-dimensional space shown in Figure 3.

It is important to realize that alpha shapes are not triangulations of the union of balls in the technical sense. Instead, they are simplicial complexes that are homotopy equivalent to the corresponding union of balls with radius α. While this is a nontrivial observation that follows from the Nerve Theorem [17], it is weaker than being homeomorphic, which would be necessary for being a triangulation. Nevertheless, the homotopy equivalence implies that the alpha shape and the corresponding union of balls have the same Betti numbers.

Although the alpha shape is defined for all real numbers α, there are only a finite number of different alpha shapes for any finite point set. In other words, the alpha shape process is never continuous: it proceeds discretely with increasing α, marked by the addition of new Delaunay simplices once α exceeds the corresponding threshold.

3.2 Computing Betti Numbers

Following the description above, one may find that alpha shapes are intimately related to the topology of a point set. Indeed, they form a direct way of characterizing the topology of a point distribution. The complete description of its homology in terms of Betti numbers may therefore be inferred from the alpha shapes.

For simplicial complexes, like Delaunay tessellations and alpha complexes, the Betti numbers can be defined on the basis of the p-simplices. We illustrate this for a simplicial complex in 3-dimensional space. Cycling through all its simplices, we base the calculation on the following straightforward considerations. When a vertex is added to the alpha complex, a new component is created and β_0 is increased by 1. Similarly, if an edge is added, it connects two vertices, which either belong to the same or to different components of the current complex. In the former case, the edge creates a new tunnel, so β_1 is increased by 1. In the latter case, two components get connected into one, so β_0 is decreased by 1. If a triangle is added, it either completes a void or it closes a tunnel. In the former case, β_2 is increased by 1, and in the latter case, β_1 is decreased by 1. Finally, when a tetrahedron is added, a void is filled, so β_2 is lowered by 1. Following this procedure, the algorithm has to include a technique for determining whether a p-simplex belongs to a p-cycle. For vertices and tetrahedra, this is rather trivial. On the other hand, for edges and triangles, we use a somewhat more elaborate procedure, involving the classical computer science concept of a union-find data structure [13].

Turning our attention to software that implements these algorithmic ideas, we resort to the Computational Geometry Algorithms Library, CGAL [9]. Within this context, Caroli and Teillaud recently developed an efficient code for the calculation of two-dimensional and three-dimensional alpha shapes in periodic spaces. We use their software for the computation of the alpha shapes of our cosmological models. In the first stage of our project, which concerns the analysis of Voronoi clustering models, we computed the Betti numbers of alpha shapes with a code developed within our own project. Later, for the analysis of the cosmological LCDM models, we were provided with an optimized code written by Manuel Caroli.

---

[9] CGAL is a C++ library of algorithms and data structures for computational geometry, see www.cgal.org.

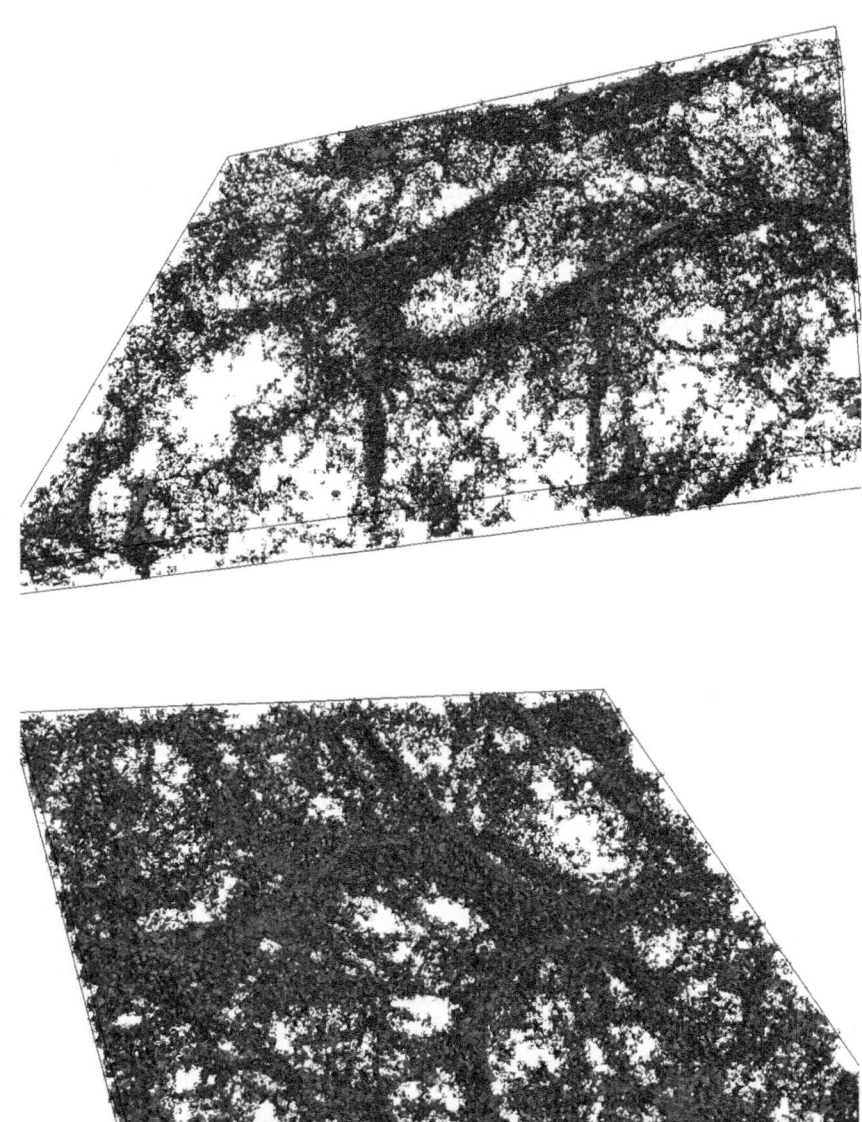

Fig. 4. Examples of alpha shapes of the LCDM GIF simulation. Shown are central slices through the complete alpha shape for two different values of α, viewed from different angles. The sensitivity to the structure and topology of the matter distribution in the Cosmic Web is clearly visible when comparing the lower value of α in the top panel with the higher value of α in the bottom panel.

3.3 Alpha Shapes of the Cosmic Web

In a recent study, Vegter et al. computed the alpha shapes for a set of GIF simulations of cosmic structure formation [22]. It concerns a 256^3 particles GIF N-body simulation, encompassing a LCDM ($\Omega_m = 0.3$, $\Omega_\Lambda = 0.7$, $H_0 = 70$ km/s/Mpc) density field within a (periodic) cubic box with length 141 h^{-1}Mpc and produced by means of an adaptive P^3M N-body code [30].

Figure 4 illustrates the alpha shapes for two different values of α, by showing sections through the GIF simulation. The top panel uses a small value of α, the bottom uses a high value. The intricacy of the web-like patterns is very nicely followed. The top configuration highlights the interior of filamentary and sheet-like features, and reveals the interconnection between these structural elements. The bottom configuration covers an evidently larger volume, which it does by connecting finer features in the Cosmic Web. Noteworthy are the tenuous filamentary and planar extensions into the interior of the voids.

These images testify of the potential power of alpha shapes in analyzing the web-like cosmic matter distribution, in identifying its morphological elements, their connections and, in particular, their hierarchical character. However, to understand and properly interpret the topological information contained in these images, we need first to assess their behavior in simpler yet similar circumstances. To this end, we introduce a set of heuristic spatial matter distributions, Voronoi clustering models.

4   Voronoi Clustering Models

In this section, we introduce a class of heuristic models for cellular distributions of matter using Voronoi tessellations as scaffolds, know as *Voronoi clustering models* [59, 63, 70]. They are well suited to model the large scale clustering of the morphological elements of the Cosmic Web, defined by the stochastic yet non-Poissonian geometrical distribution of the matter and the related galaxy population forming *walls*, *filaments*, and *clusters*.

The small-scale distribution of galaxies within the various components of the cosmic skeleton involves the complicated details of highly nonlinear interactions of the gravitating matter. This aspect may be provided by elaborate physical models and/or N-body computer simulations, but this would distract from the purpose of the model and destroy its conceptual simplicity. In the Voronoi models, we complement a geometrically fixed Voronoi tessellation defined by a small set of nuclei with a heuristic prescription for the location of particles or model galaxies within the tessellation. We distinguish two complementary approaches: the *Voronoi element* and the *Voronoi evolution models*. Both are obtained by moving an initially random distribution of N particles toward the faces, lines, and nodes of the Voronoi tessellation. The Voronoi element models do this by a heuristic and user-specified mixture of projections onto the various geometric elements of the tessellation. The Voronoi evolution models accomplish this via a gradual motion of the galaxies from their initial, random locations towards the boundaries of the cells.

The Voronoi clustering models identify the geometric elements of a 3-dimensional Voronoi tessellations with the morphological component of the Cosmic Web. To

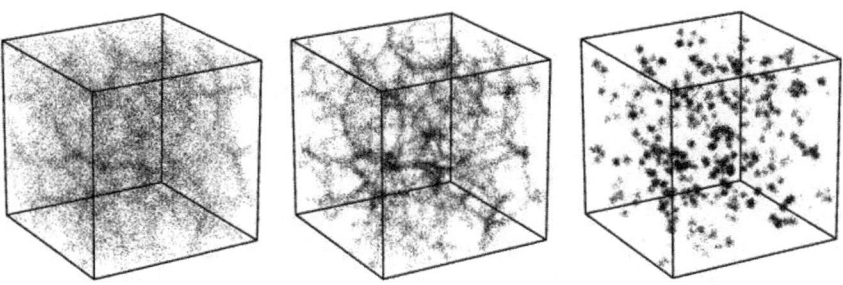

Fig. 5. Three different Voronoi element models, shown in a cubic setting. From left to right: a wall-dominated Voronoi universe, a filamentary Voronoi universe, and a cluster-dominated Voronoi universe.

describe this relationship, we will adhere to the terminology listed in Table 1, which also lists the terminology to the corresponding types of holes in the alpha shape of the particle distribution[10].

4.1   Voronoi Element Models

These are fully heuristic, user-specified spatial galaxy distributions within the cells, faces, lines, and nodes of a Voronoi tessellation. They are obtained by projecting the initially randomly distributed N model galaxies onto the relevant Voronoi face, line, or node. We can also retain a galaxy within the Voronoi cell in which it is located. The Voronoi element models are particularly apt for studying systematic properties of spatial galaxy distributions confined to one or more structural elements of nontrivial geometric spatial patterns.

Pure Voronoi element models place their galaxies exclusively inside or near either faces, lines, or nodes. In contrast, *mixed* models allow combinations in which distributions inside or near cells, faces, lines, and nodes are superimposed. These include particles located in four distinct structural components:

- *field* particles located in the interior of Voronoi cells,
- *wall* particles within and around the Voronoi faces,
- *filament* particles within and around the Voronoi lines,
- *cluster* particles within and around the Voronoi nodes.

The characteristics of the spatial distributions in the mixed models can be varied and tuned according to desired fractions of galaxies of each type. These fractions are free parameters that can be specified by the user; see Figure 5.

---

[10] For consistency and clarity, throughout this paper we adopt the following nomenclature for the spatial components of the Cosmic Web, of Voronoi tessellations and of Delaunay tessellations. The cosmic web consists of *voids, walls, filaments, and clusters*. Voronoi tessellations consist of *cells, faces, lines, and nodes*. Finally, Delaunay tessellations (and alpha shapes) involve *tetrahedra, triangles, edges, and vertices*.

Table 1. Identification of morphological components of the Cosmic Web (lefthand column) with geometric Voronoi tessellation elements (central column). The righthand column list the terminology for the corresponding types of holes in the alpha shape of the particle distribution.

Cosmic Web	Voronoi	Alpha
voids, field	cell	void
walls, sheets, superclusters	face	tunnel
filaments, superclusters	line	gap
clusters	node	

4.2 Voronoi Evolution Models

The second class we consider are the *Voronoi evolution models*. They provide web-like galaxy distributions mimicking the outcome of realistic cosmic structure formation scenarios. They are based upon the notion that voids play a key organizational role in the development of structure, causing the universe to resemble a soapsud of expanding bubbles [29]. While the galaxies move away from the void centers, and stream towards the walls, filaments, and clusters, the fractions of galaxies in or near the cells, faces, lines, and nodes evolve continuously. The details of the model realization depends on the specified time evolution.

Within the class of Voronoi evolution models, the most representative and most frequently used are the *Voronoi kinematic models*. Forming the idealized and asymptotic description of the outcome of a hierarchical gravitational structure formation process, they simulate the asymptotic web-like galaxy distribution implied by the hierarchical void formation process by assuming a single-size dominated void population. Within a void, the mean distance between galaxies increases with time. Before a galaxy enters an adjacent cell, the velocity component perpendicular to the otherwise crossed face disappears. Thereafter, the galaxy continues to move within the face. Before it enters the next cell, the velocity component perpendicular to the otherwise crossed edge disappears. The galaxy continues along a filament and, finally, comes to rest at a node. The resulting evolutionary progression within the Voronoi kinematic model proceeds from an almost featureless random distribution towards a distribution in which matter ultimately aggregates into conspicuous compact cluster-like clumps.

The steadily increasing contrast of the various structural features is accompanied by a gradual shift in the topology of the distribution. The virtually uniform and featureless particle distribution at the beginning ultimately unfolds into a highly clumped distribution of clusters. This evolution involves a gradual progression via a wall-like through a filamentary towards an ultimate cluster-dominated matter distribution.

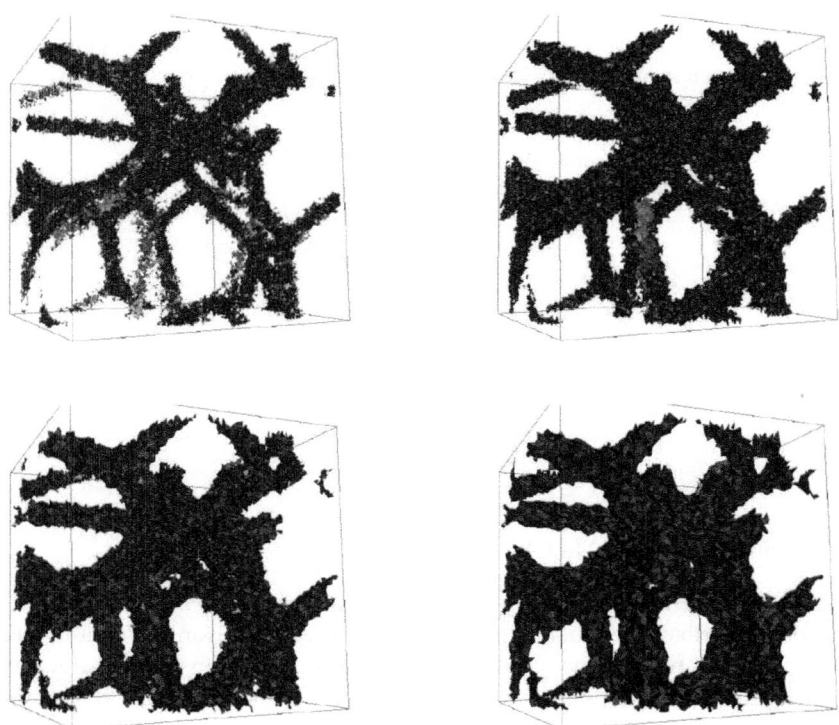

Fig. 6. Four alpha shapes of a Voronoi filament model consisting of $200,000$ particles in a periodic box of size $50\ h^{-1}\mathrm{Mpc}$ with 8 Voronoi cells. We use colors to highlight different components. From top left to bottom right: $10^4 \cdot \alpha = 0.5, 1.0, 2.0, 4.0$.

5 Topological Analysis of Voronoi Universes

In this section, we study the systematic behavior of the Betti numbers of the alpha shapes of Voronoi clustering models; see [22, 66]. For each point sample, we investigate the alpha shape for the full range of the α parameter. We generate six Voronoi clustering models, each consisting of $200,000$ particles within a periodic box of size $50\ h^{-1}\mathrm{Mpc}$. Of each model, we make two realizations, one with 8 and the other with 64 nuclei. We start with two pure Voronoi element models: a wall model, and a filament model. In addition, we study four Voronoi kinematic models, ranging from a mildly evolved to a strongly evolved configuration. In each case, the clusters, filaments, and walls have a finite Gaussian width of $1.0\ h^{-1}\mathrm{Mpc}$.

An impression of the sequence of alpha shapes may be gained from the four panels in Figure 6. For the smallest value of α, we see that the simplices delineate nearly all the filaments in the particle distribution. As α increases, going from the top-left panel down to the bottom right panel, we find that the alpha shape fills in the walls. For even larger values of α, the alpha shape includes the large Delaunay simplices that cover the interior of the Voronoi cells. It is a beautiful illustration of the way in which alpha shapes define, as it were, naturally evolving surfaces that are sensitive to every detail of the morphological structure of the cosmic matter distribution.

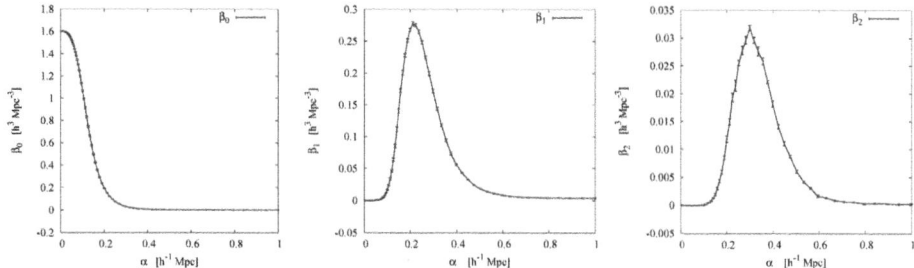

Fig. 7. Topological analysis of the Voronoi filament model illustrated by its alpha shape shown in the upper left panel of Figure 6. From left to right: β_0, β_1, and β_2 counting the components, tunnels, and voids of the alpha shape. The realization contains 64 nuclei or cells.

5.1 Filament Model Topology

We take the Voronoi filament model as a case study, investigating its topology by following the behavior of the three Betti numbers as functions of the parameter α.

Figure 7 shows the relation between the Betti numbers of the alpha shape and the value of α. The zeroth Betti number, β_0, counts the components. Equivalently, $\beta_0 - 1$ counts the gaps between the components. In the current context, the latter interpretation is preferred as we focus on the holes left by the alpha shape. Starting with $\beta_0 = 200,000$ at $\alpha = 0$, the zeroth Betti number gradually decreases to 1 as the components merge into progressively larger entities.

The first Betti number, β_1, counts the tunnels. At first, it increases steeply when edges are added to the alpha complex, some of which bridge gaps while others form tunnels. After β_1 reaches its maximum for α roughly equal to 0.25 times 10^{-4}, the number of tunnels decreases sharply as triangles enter the alpha complex in large numbers.

The second Betti number, β_2, counts the voids in the alpha shape. In the case of the Voronoi filament model, its behavior resembles that of β_1: a peaked distribution around a moderate value of α. It increases when the entering triangles are done closing tunnels and start creating voids. However, when α is large enough to add tetrahedra, these voids start to fill up and β_2 decreases again, reaching zero eventually. Notice that in the given example of the Voronoi filament model, β_2 reaches its maximum for α roughly equal to 0.45 times 10^{-4}, which lies substantially beyond the peak in the β_1 distribution; see also Figure 8.

5.2 Void Evolution

Having assessed one particular Voronoi clustering model in detail, we turn our attention to the differences between the models. While this is still the subject of ongoing research,

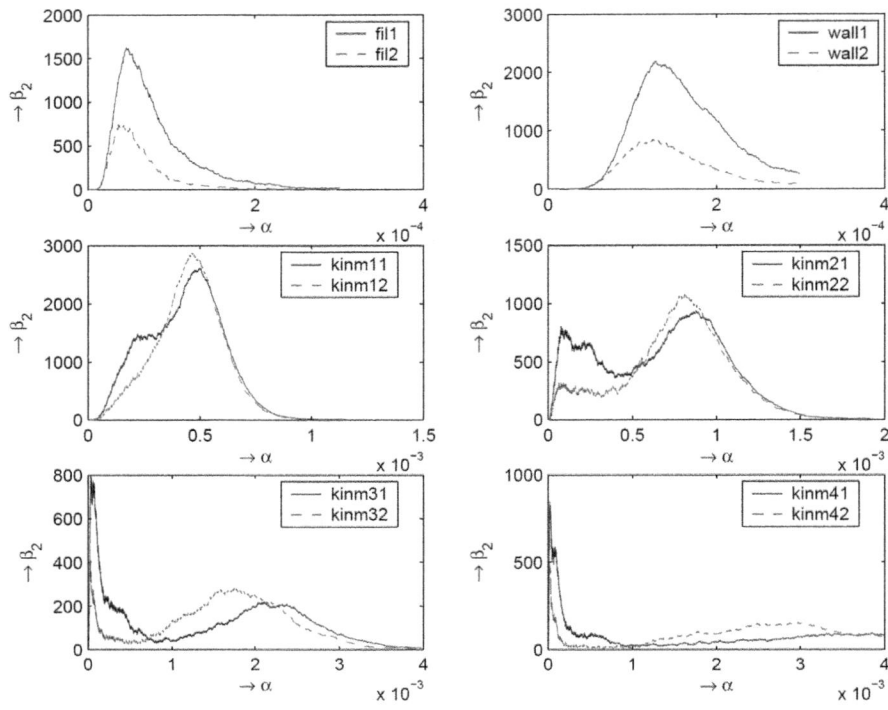

Fig. 8. The dependence of the second Betti number, β_2, on α, for six different Voronoi clustering models. Top left: Voronoi filament model. Top right: Voronoi wall model. Centre left to bottom right: four stages of the Voronoi kinematic model, going from a moderately evolved model dominated by walls (center left) to a highly evolved model dominated by filaments and clusters (bottom right). Blue lines: realizations with 8 nuclei or cells. Red lines: realizations with 64 nuclei or cells.

we find substantial differences on a few particular aspects. For example, β_0 decreases monotonically with α for all models, but the range over which the number of gaps is substantially positive, and the rate with which it decrease are highly sensitive to the underlying distribution. In fact, the approximate derivative, $\partial\beta_0/\partial\alpha$, contains interesting features. Examples are a minimum and a varying width, both potentially interesting for discriminating between the underlying topologies.

Most interesting is the difference in behavior of the second Betti number. As one may infer from Figure 8, substantial differences between models can be observed. This concerns the values and range over which β_2 reaches maxima, as well as new systematic behavior. For kinematic models, we find two or more peaks, each corresponding to different morphological components of the particle distribution. It is revealing to follow the changes in the dependence of β_2 on α, as we look at different evolutionary stages. The panels from center left to bottom right in Figure 8 correspond to four different stages of evolution. The center left panel shows the dependence of β_2 on α for a moderately evolved matter distribution, which is dominated by walls. The center right panel shows the dependence for a stage at which walls and filaments are approximately

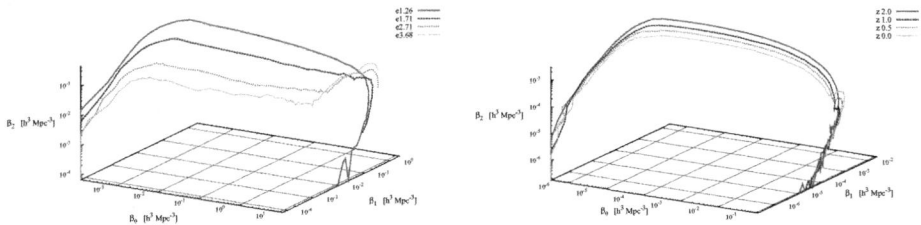

Fig. 9. Left: the alpha tracks for four stages of the kinematic Voronoi clustering model, evolving from almost uniform (red) to a distribution in which most matter resides in cluster nodes (orange). The wall-like (blue) and filamentary (magenta) distributions form intermediate cases. Right: the alpha tracks at different cosmic epochs in an evolving LCDM mass distribution. The redshift, z, is indicated in the top right corner, with the track at an early epoch ($a = 0.17$, $z = 5.0$) bearing resemblance to the early Voronoi kinematic model, and the track at the current epoch ($a = 1.0$, $z = 0.0$) resembling the Voronoi kinematic model at an intermediate distribution.

equally prominent. In the bottom left panel, the filaments represent more than 40% of the mass, while the walls and the gradually more prominent clusters each represent around 25% of the particles. The final, bottom right panel shows the dependence for a highly evolved mass distribution, with clusters and filaments each representing around 40% of the particles.

The different morphological patterns of the Voronoi kinematic models are reflected in the behavior of β_2. In the center left panel, we find a strong peak at α roughly equal to 5 times 10^{-4}, with a shoulder at lower values. The peak reflects the voids inside the walls in the distribution, while the shoulder finds its origin in the somewhat smaller voids inside the filaments. The identity of the peaks becomes more clear when we turn to the two-peak distribution in the center right panel. The strong peak at α roughly equal to 10^{-4} is a manifestation of the strongly emerged filaments in the matter distribution. As the shift to filaments and clusters continues, we see the rise of a third peak at a much smaller value of α; see the bottom two panels. This clearly corresponds to the voids in the high density cluster regions.

5.3   Alpha Tracks

In Figure 9, we synthesize the homology information by tracing curves in the 3-dimensional *Betti space*, whose coordinates are the three Betti numbers. Each curve is parametrized by α, with points $(\beta_0(\alpha), \beta_1(\alpha), \beta_2(\alpha))$ characterizing the homology of the alpha shape at this given scale. We call each curve an *alpha track*, using it to visualize the combined evolution of the Betti numbers for a given Voronoi cluster model. Note that the evolution of the Euler characteristic, $\chi(\alpha)$, is the projection of an alpha track onto the normal line of the plane $\beta_0 - \beta_1 + \beta_2 = 0$. The richer structure of the full alpha track illustrates why patterns with similar genus may still have a substantially different topology.

The left frame of Figure 9 shows the alpha tracks for the four Voronoi kinematic models. We notice that β_0 varies over a substantially larger range of values than β_1 and

β_2. The tracks reveal that the few components left after an extensive process of merging still contain a substantial number of tunnels and voids. More generally, we see that the Betti numbers diminish to small values in sequence: first β_0, then β_1, and finally β_2.

As the matter distribution evolves, the tracks shift in position. The four tracks in the left frame of Figure 9 form a good illustration of this effect. The red dashed track corresponds to an early phase of the kinematic model, at which the particle distribution is still almost uniform, while the orange track represents the final time-step marked by a pattern in which nearly all particles reside in the clusters. In the intermediate stages, most of the particles are located in the walls (blue) and in the filaments (magenta). The decreasing number of tunnels and voids at intermediate scales is a manifestation of the hierarchical formation process, in which small scale structures merge into ever larger ones; see e.g. [49].

6   Topological Analysis of the LCDM Universe

Having discussed the scale-dependent Betti numbers in heuristic Voronoi clustering models, we turn to the analysis of more realistic megaparsec cosmic mass distributions. These are characterized by an intricate multiscale configuration of anisotropic web-like patterns. The crucial question is whether we can exploit the topological information toward determining crucial cosmological parameters, such as the nature of dark energy.

We concentrate on the analysis of computer simulations of cosmic structure formation in the Universe, leaving the analysis of the observed distribution in galaxy redshift surveys for the future. The computer simulations follow the nonlinear evolution of structure as it emerges from the near uniform early Universe. Once the gravitational clustering process has progressed beyond the initial linear growth phase, we see the emergence of intricate patterns in the density field; see Figure 1. The near homogeneous initial conditions evolve into an increasingly pronounced clustering pattern.

6.1   N-body Simulations

In cosmological N-body simulations, the cosmic mass distribution is represented by a large number of particles, which move under the influence of the combined gravitational force of all particles. The initial conditions (location and velocities of the particles) are a realization of the mass distribution expected in the cosmological scenario at hand. Since the majority of the matter in the Universe is non-dissipative dark matter, a major aspect of cosmological N-body simulations concerns itself with dark matter particles that only interact via gravity. State-of-the-art computer simulations, such as the Millennium simulation [57], count in the order of 10^{10} particles and are run on the most powerful supercomputers available to the scientific community.

Figure 1 illustrates the matter distribution in the standard, or "concordance" LCDM cosmological model. It shows the dark matter distribution in a box of 80 h^{-1}Mpc comoving size, based on a 384^3 particle N-body simulation. The LCDM cosmological scenario assumes that most gravitating matter in the Universe consists of as yet undetected and unidentified cold dark matter, accounting for approximately 23% of the energy content of the Universe, i.e. $\Omega_{dm} = 0.23$. The normal, baryonic matter, which consists mostly of protons and neutrons, represents only 4.4% of the energy content of the Universe, i.e. $\Omega_b = 0.044$. Most importantly, the model assumes the presence of a cosmological constant, Λ, or equivalent dark energy component, representing 73% of the density of the Universe, i.e. $\Omega_\Lambda = 0.73$; see Section 7. The Hubble parameter, which specifies the expansion rate of the Universe, is taken to be $H_0 = 70$ km/s/Mpc. The amplitude of the initial fluctuation has been normalized to a level at which the current density fluctuation on a scale of $8\ h^{-1}\mathrm{Mpc}$ is equal to $\sigma_8 = 0.8$.

The most prominent aspect of the cosmic mass distribution is the intriguing network of filaments and high-density clusters, which surround low-density voids. The mass distribution has a distinct multiscale character, marked by clumps over a wide range of scales, reflecting the hierarchical evolution of the distribution.

6.2 Homological Evolution

We follow the developing structure in the LCDM scenario in eleven time-steps that run from an expansion factor $a = 0.25$ ($z = 3.0$) to $a = 1.0$ ($z = 0.0$), i.e. from around 3.4 Gigayears after the Big Bang to the current epoch. At each time-step, we compute the alpha shapes of a randomly sampled subset of the particle distribution for α from 0.0 to 10.0 $h^{-1}\mathrm{Mpc}$, and their Betti numbers, from which we extrapolate the scale-dependent behavior of homology. The resulting curves of the Betti numbers for dimensions $p = 0, 1, 2$ are shown in Figure 10. The curves in the upper left frame show that the value of β_0 decreases for $\alpha < 1.2\ h^{-1}\mathrm{Mpc}$, while it increases for larger values of α. This is a manifestation of the hierarchical structure formation process. It reflects the progressively earlier merging of clumps at small scale and the progressively later merging of massive components at large scale.

The curves of β_1 and β_2 have a slightly different appearance. Both start as highly peaked distributions. Most prominent in the near homogeneous initial conditions, the peaks represent the imprint of a Poisson distribution with a similar particle density. In principle, one should remove this imprint via a persistence procedure; see Section 9. Here we keep to the unfiltered version. As the mass distribution evolves under the influence of gravity, the particles get more and more clustered. This leads to a decrease in the number of holes on small scales. As these merge into ever larger supercluster complexes, their formation goes along with the evacuation of ever larger tunnels and voids enclosed by the higher density structures in the web-like network.

All three curves are marked by a steady shift toward higher values of α. While we find a minor shift of the maximum of β_1, the shift is clear for β_2. It is interesting that a similar shift toward larger α values has has been predicted by Sheth & van de Weygaert [49] in the context of their dynamical model of the evolving void hierarchy. In their two-barrier excursion set formalism, this shift is a manifestation of the merging of smaller voids into ever larger ones, accompanied by the destruction of a large number of small voids inside gravitationally collapsing over-dense regions.

6.3 Alpha Tracks and Multiscale Homology

When assessing the corresponding evolution of alpha tracks in Figure 9, we find a systematic behavior that is similar to what we have seen for the Voronoi clustering models.

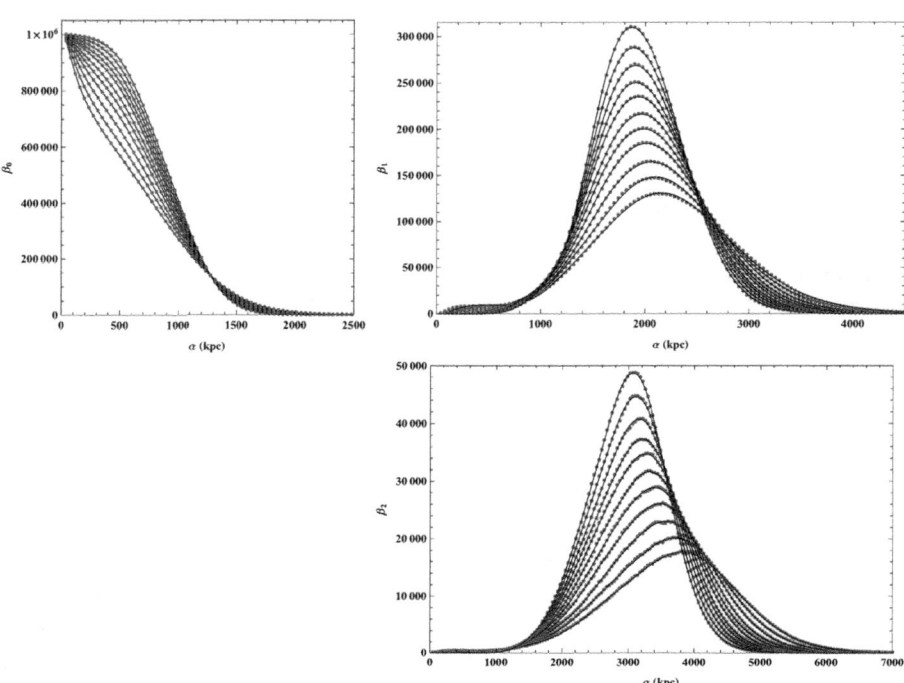

Fig. 10. The evolution of the Betti numbers in an LCDM universe. We show the dependence of β_p on α for dimensions $p = 0, 1, 2$ and for eleven expansion factors running from $a = 0.25$ ($z = 3.0$) to the current epoch $a = 1.0$ ($z = 0.0$). From upper left to lower right: $\beta_0(\alpha)$, $\beta_1(\alpha)$, $\beta_2(\alpha)$. Note that all three curves gradually shift from smaller to larger scales.

As the evolution proceeds, we find fewer tunnels and voids at small scale, while their numbers are still substantial at values of α at which we have only very few remaining components.

However, as the mass distribution evolves, the shifts between the alpha tracks are relatively small compared to those seen in the Voronoi clustering models. The mass distribution retains its multiscale character. The hierarchical evolution of the multiscale mass distribution leads to a more or less self-similar mapping of the LCDM alpha tracks to higher α values. At each phase, we find dominant clusters and conspicuous filamentary features, although their scale shifts as the mass distribution advances from the mildly linear to the quasi-linear phase. This contrasts the evolution of the Voronoi clustering models, which is characterized by clear transitions between distinct topological patterns outlined by the mass distribution.

7 Probing Dark Energy

To assess the utility of our topological methods, we look at the possibility to use the homology of the Cosmic Web towards determining the nature of dark energy in the

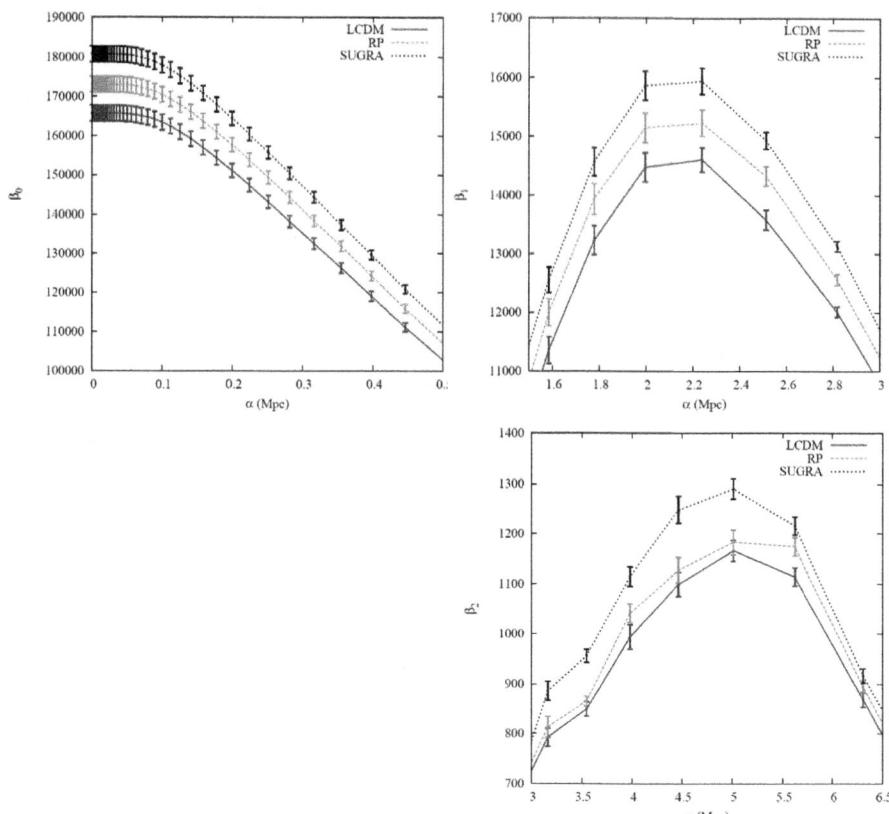

Fig. 11. Homology and dark energy. The Betti number curves for three cosmological models with different dark energy models. The curves are determined from the distribution of collapsed dark matter halos in the corresponding scenarios. Red: the standard LCDM scenario. Green: the Ratra-Peebles quintessence model. Blue: the supergravity SUGRA model. From left to right: $\beta_0(\alpha)$, $\beta_1(\alpha)$, and $\beta_2(\alpha)$. The curves demonstrate that homology is capable of using the web-like matter distribution to discriminate between cosmological scenarios with different dark energy content.

Universe. The parameters for what might now be called "the standard model for cosmology" have been established with remarkable precision.

However, there remains the great mystery of the nature of the so-called *dark energy*, which appears to make up 73% of the total cosmic energy. Dark energy, or the cosmological constant, has been dominating the dynamics of the Universe since the last 7 Gigayears and has pushed it into an accelerated expansion. The simplest model for the dark energy is Einstein's cosmological constant, Λ: it makes a time-independent contribution to the total energy density in the Friedman-Lemaitre equations. Such models are referred to as *LCDM models*. However, there are numerous, possibly more plausible models in which the dark energy evolves as a function of time. These models are generally described in terms of a time- or redshift-dependent function $w(z)$ that describes the history of equation of state of the dark energy component, i.e. the relation between pressure and density:

$$p = w(z)\rho c^2. \tag{18}$$

Different models for dark energy produce different functions $w(z)$, and a number of simple parametrizations can be found in the literature. The classical Einstein cosmological constant corresponds to $w(z) = -1$. In principle, the Cosmic Web should be one of the strongest differentiators between the various dark energy models. As a result of the different growth rates at comparable cosmic epochs, we will see different structures in the matter distribution. To illustrate this idea, we run three N-body simulations with identical initial conditions but different dark energy equation of state: the *standard LCDM model* and two *quintessence models*. The latter assume that the Universe contains an evolving quintessence scalar field, whose energy content manifests itself as dark energy. The two quintessence models are the *Ratra-Peebles (RP) model* and the *SUGRA model*; see [7, 8, 44] for a detailed description.

While the general appearance of the emerging Cosmic Web is structurally similar, there are visible differences in the development. On small to medium scales, the quintessence models are less clustered; see [68] for an illustration. As a result, the alpha shapes contain a larger number of small objects. In Figure 11, we see that this manifests itself in values of β_0 that are systematically higher for the quintessence RP and SUGRA models than for the LCDM model. The two right frames in Figure 11 concentrate on the medium scale range, $1.5\ h^{-1}\mathrm{Mpc} < \alpha < 3.0\ h^{-1}\mathrm{Mpc}$ for β_1 (top righthand panel) and $3.0\ h^{-1}\mathrm{Mpc} < \alpha < 6.5\ h^{-1}\mathrm{Mpc}$ for β_2 (bottom righthand panel). The error bars on the obtained curves are determined on the basis of the obtained values for independent samples. Particularly encouraging is the fact that homology is sensitive to the subtle differences we see in the pattern of the Cosmic Web. The size of the error bars show that it is feasible to find significant and measurable differences between the outcome of the different cosmologies.

8   Betti Numbers of Gaussian Random Fields

To interpret our results for the web-like cosmic matter distribution, we compare them with those for the initial condition out of which our Universe arose. The behavior of the Betti numbers in a Gaussian random field is a reference point for any further assessment of their behavior in the more complex environment of the Cosmic Web.

According to the current paradigm, structure in the Universe grew by gravitational instability out of tiny primordial density perturbations. The evidence provided by the temperature fluctuations in the cosmic microwave background [5, 31, 50, 56] suggests that the character of the perturbation field is that of a homogeneous and isotropic spatial Gaussian process. Such primordial Gaussian perturbations in the gravitational potential are a natural product of an early inflationary phase of our Universe.

Before proceeding to the Betti numbers of a Gaussian random field, we present the necessary nomenclature, focusing on the three-dimensional situation. For fundamentals on Gaussian random fields, we refer to the standard work by Adler and Taylor [2], and for their cosmological application to the seminal study by Bardeen, Bond, Kaiser, and

Szalay [3]. Here we follow the notation from van de Weygaert and Bertschinger [60]. The first papers on the homology and persistence of Gaussian random fields are Adler et al. [1] and Park et al. [38]; see also [43]. Here, we discuss some of the main findings, and we refer to the mentioned papers for more detailed treatments. We alert the reader to the fact that the analysis in this section concerns itself with density fields, which is different from earlier sections in which we studied distance fields defined by particle distributions.

8.1 Gaussian Random Fields

A *random field*, f, on a spatial volume assigns a value, $f(\mathbf{x})$, to each location, $\mathbf{x}$, of that volume. The fields of interest, such as the primordial density or velocity field, are smooth and continuous [11]. The stochastic properties of a random field are defined by its N-*point joint probabilities*, where N can be any arbitrary positive integer. To denote them, we write $\mathbf{X} = (\mathbf{x}_1, \mathbf{x}_2, \cdots, \mathbf{x}_N)$ for a vector of N points and $\mathbf{f} = (f_1, f_2, \ldots, f_N)$ for a vector of N field values. The joint probability is

$$\text{Prob}[f(\mathbf{x}_1) = f_1, \ldots, f(\mathbf{x}_N) = f_N] = \mathcal{P}_{\mathbf{X}}(\mathbf{f}) \, d\mathbf{f} \,,$$

(19)

which is the probability that the field f at the locations $\mathbf{x}_i$ has values in the range f_i to $f_i + df_i$, for each $1 \leq i \leq N$. Here, $\mathcal{P}_{\mathbf{X}}(\mathbf{f})$ is the probability density for the field realization vector $\mathbf{f}$ at the location vector $\mathbf{X} = (\mathbf{x}_1, \mathbf{x}_2, \cdots, \mathbf{x}_N)$. For a *Gaussian random field*, the joint probabilities for $N = 1$ and $N = 2$ determine all others. Specifically, the probability density functions take the simple form

$$\mathcal{P}_{\mathbf{X}}(\mathbf{f}) = C \cdot \exp\left[-\mathbf{f} \mathsf{M}^{-1} \mathbf{f}^T / 2\right] \,,$$

(20)

where $C = 1/[(2\pi)^N (\det \mathsf{M})]^{1/2}$ normalizes the expression, making sure that the integral of $\mathcal{P}_{\mathbf{X}}(\mathbf{f})$, over all $\mathbf{f} \in \mathbb{R}^N$, is equal to 1. Here, we assume that each 1-point distribution is Gaussian with zero mean. The matrix M^{-1} is the inverse of the $N \times N$ covariance matrix with entries

$$M_{ij} = \langle f(\mathbf{x}_i) f(\mathbf{x}_j) \rangle \,,$$

(21)

in which the angle bracket denotes the ensemble average of the product, over the 2-point probability density function. In effect, M is the generalization of the variance of a 1-point normal distribution, and we indeed have $\mathsf{M} = [\sigma_0^2]$ for the case $N = 1$.

Equation (20) shows that a Gaussian random is fully specified by the autocorrelation function, $\xi(r)$, which expresses the correlation between the density values at two points separated by a distance $r = |\mathbf{r}|$,

$$\xi(r) = \xi(|\mathbf{r}|) \equiv \langle f(\mathbf{x}) f(\mathbf{x} + \mathbf{r}) \rangle \,.$$

(22)

---

[11] In this section, the fields $f(\mathbf{x})$ may either be the raw unfiltered field or, without loss of generality, a filtered field $f_s(\mathbf{x})$. A filtered field is a convolution with a filter kernel $W(\mathbf{x}, \mathbf{y})$, $f_s(\mathbf{x}) = \int d\mathbf{y} f(\mathbf{y}) W(\mathbf{x}, \mathbf{y})$.

Here we use the *statistical cosmological principle*, which states that statistical properties of e.g. the cosmic density distribution in the Universe are uniform throughout the
Universe. It means that the distribution functions and moments of fields are the same
in each direction and at each location. The latter implies that ensemble averages depend only on one parameter, namely the distance between the points. In other words,
the entries in the matrix are the values of the *autocorrelation function* for the distance
between the points: $M_{ij} = \xi(r_{ij})$, with $r_{ij} = \|\mathbf{x}_i - \mathbf{x}_j\|$.

An impression of a typical Gaussian random field may be obtained from the three-
dimensional realization shown in the left panel of Figure 12. The field is chosen using
the power spectrum for the standard LCDM cosmology with some Gaussian filtering;
see e.g. [3, 21]. The image illustrates that Gaussian random fields are symmetric, i.e.
negative values are as likely as positive values.

8.2 Power Spectrum

A stochastic random density field is composed of a spectrum of density fluctuations,
each of a different scale. The relative amplitudes of small-scale and large-scale fluctuations is of decisive influence on the outcome of the subsequent gravitational evolution
of the density field and on the emerging patterns in the spatial density distribution.

To describe the multiscale composition of a density field, we write it as a sum of
individual harmonic waves, i.e. in terms of its Fourier sum. Each of the waves is specified by its *wave vector* $\mathbf{k} = (k_x, k_y, k_z) \in \mathbb{R}^3$, describing the direction and spatial
frequency of the wave. The latter is determined by the *magnitude* of the wave vector,

$$k = |\mathbf{k}| = \sqrt{k_x^2 + k_y^2 + k_z^2}, \tag{23}$$

which is the inverse of the wavelength, $\lambda = 2\pi/k$. Subsequently, we write the field
$f(\mathbf{x})$ as the Fourier integral,

$$f(\mathbf{x}) = \int \frac{d\mathbf{k}}{(2\pi)^3} \, \hat{f}(\mathbf{k}) \, e^{-i\mathbf{k}\cdot\mathbf{x}}, \tag{24}$$

where $e^{i\varphi} = \cos\varphi + i\sin\varphi$, as usual, and $\mathbf{k} \cdot \mathbf{x}$ the inner product between the wave
vector, $\mathbf{k}$, and position vector, $\mathbf{x}$. The Fourier components, $\hat{f}(\mathbf{k}) \in \mathbb{C}$, represent the
contributions by the harmonic wave $\exp(i\mathbf{k} \cdot \mathbf{x})$ to the field $f(\mathbf{x})$. They are given by the
inverse Fourier transform,

$$\hat{f}(\mathbf{k}) = \int d\mathbf{x} \, f(\mathbf{x}) \, e^{i\mathbf{k}\cdot\mathbf{x}}. \tag{25}$$

Because the density field is always real, $f(\mathbf{x}) \in \mathbb{R}$, the Fourier components $\hat{f}(\mathbf{k}) \in \mathbb{C}$
obey the symmetry constraint

$$\hat{f}(\mathbf{k}) = \hat{f}^*(-\mathbf{k}). \tag{26}$$

It identifies $\hat{f}(\mathbf{k})$ with the complex conjugate of the Fourier component of the wave
vector $-\mathbf{k}$, i.e. of the wave with the same spatial frequency oriented in the opposite
direction.

The *power spectrum* is formally defined as the mean square of the Fourier components, $\hat{f}(\mathbf{k})$, of the field. It can be computed from the Fourier components using the
Dirac delta function, $\delta_D : \mathbb{R} \to \mathbb{R}$, which is the limit of the normal distribution, whose
variance goes to zero. Most importantly, the integral of the Dirac delta function is assumed to be 1. Now, we have

$$\langle \hat{f}(\mathbf{k})\hat{f}(\mathbf{k}') \rangle = (2\pi)^{3/2} P(k)\,\delta_D(\mathbf{k} - \mathbf{k}'), \tag{27}$$

and we can compute the power spectrum accordingly. We note that because of the sta-
tistical isotropy of the field, we have $P(\mathbf{k}) = P(\mathbf{k}')$, whenever $|\mathbf{k}| = |\mathbf{k}'|$. We can
therefore introduce the 1-dimensional power spectrum, defined by $P(k) = P(\mathbf{k})$ for
every $\mathbf{k}$, which we refer to by the same name, for convenience.

The *power spectrum* is the Fourier transform of the autocorrelation function $\xi(\mathbf{x})$,
which is straightforward to infer from equations (22) and (27),

$$\xi(x) = \int \frac{d\mathbf{k}}{(2\pi)^3} P(k)e^{-i\mathbf{k}\cdot\mathbf{x}}, \tag{28}$$

where $\mathbf{x}$ is a point in the volume with $|\mathbf{x}| = x$.

Because of the (statistical) isotropy of the field $f(\mathbf{x})$, it is straightforward to perform the angular part of the integral over the 2-sphere, to yield the following integral
expression for the autocorrelation function:

$$\xi(x) = \int \frac{k^2\,dk}{2\pi^2} P(k) \frac{\sin kx}{kx}, \tag{29}$$

with $k = |\mathbf{k}|$ and $x = \mathbf{x}..$

From the above we find that the power spectrum is a complete characterization of
a homogeneous and isotropic Gaussian random field. Within its cosmological context,
the power spectrum encapsulates a wealth of information on the parameters and content
of the Universe. Its measurement is therefore considered to be a central key for the
understanding of the origin of the cosmos.

8.3   Fourier Decomposition

The unique nature of Gaussian fluctuations is particularly apparent when considering the
stochastic distribution of the field in terms of its Fourier decomposition. The assumption
of Gaussianity means that the stochastic distribution of the Fourier components

$$\begin{aligned}
\hat{f}(\mathbf{k}) &= \hat{f}_{\mathrm{re}}(\mathbf{k}) + i\hat{f}_{\mathrm{im}}(\mathbf{k}) \\
&= |\hat{f}(\mathbf{k})|\, e^{i\,\theta(\mathbf{k})}
\end{aligned} \tag{30}$$

involves a random phase, $\theta(\mathbf{k})$, in the complex plane. This follows from the mutual
independence of its real and imaginary parts, each of which is a Gaussian variable.
While the phase, $\theta(\mathbf{k})$, has a uniform distribution over the interval $[0, 2\pi]$, the ampli-
tude, $r = |\hat{f}(\mathbf{k})|$, has a Rayleigh distribution:

$$\mathcal{P}(r) = \frac{r}{s}\exp\left[-\frac{r^2}{2s}\right], \tag{31}$$

where $s = P(k)$ is the power spectrum value at spatial frequency k.

Subsequently determining the probability of the field realization $f(\mathbf{x})$ in terms of the probability density function of its Fourier decomposition, one finds the interesting result that it is the product of the individual probability distributions of each of the individual Fourier components $\hat{f}(\mathbf{k})$, each Gaussian distributed with zero mean and variance $\sigma^2(k) = P(k)$. It is most straightforward to appreciate this by assessing the probability of the entire field $f(\mathbf{x})$.

To infer the probability $\mathcal{P}[f]$ of $f(\mathbf{x})$, we take the N-point joint probabilities for the limit of $N \to \infty$ with uniform spatial sampling, the summations appearing in equation (20) may be turned into integrals. The resulting expression for the infinitesimal probability $\mathcal{P}[f]$ with measure $\mathcal{D}[f]$,

$$\mathcal{P}[f] = e^{-S[f]} \mathcal{D}[f],\tag{32}$$

involves the probability density of the field, $\exp\left(-S[f]\right)$. The square brackets in $\mathcal{P}[f]$ and $S[f]$ indicate that these are functionals, i.e., they map the complete function $f(\mathbf{x})$ to one number. The probability density is similar to the quantum-mechanical partition function in path integral form, where S is the action functional (see [60]). For a Gaussian random field the expression for the action S may be inferred from equation (20),

$$S[f] = \frac{1}{2} \int d\mathbf{x}_1 \int d\mathbf{x}_2\, f(\mathbf{x}_1) K(\mathbf{x}_1 - \mathbf{x}_2) f(\mathbf{x}_2),\tag{33}$$

where K is the functional inverse of the correlation function ξ,

$$\int d\mathbf{x}\, K(\mathbf{x}_1 - \mathbf{x}) \xi(\mathbf{x} - \mathbf{x}_2) = \delta_D(\mathbf{x}_1 - \mathbf{x}_2),\tag{34}$$

and δ_D the Dirac delta function. By transforming this expression for the action $S[f]$ to Fourier space, one finds the integral expression (see [60], appendix B),

$$S[f] = \int \frac{d\mathbf{k}}{(2\pi)^3} \frac{|\hat{f}(\mathbf{k})|^2}{2P(k)}.\tag{35}$$

This immediately demonstrates that a Gaussian field has the unique property of its Fourier components being mutually independent. It has the practical virtue of considerably simplifying the construction of Gaussian fields by sampling its Fourier components.

8.4 Betti Numbers

To determine the Betti numbers of a Gaussian field, we consider the density values sampled on a regular grid and assess the topology of the superlevel sets. We adopt a 64^3 grid, and sample the field inside a box of size $100\, h^{-1}\mathrm{Mpc}$. The density values are smoothed on a filter scale $2.0\, h^{-1}\mathrm{Mpc}$. Subsequently, we determine the Betti numbers of superlevel sets, the agglomerate of regions consisting of the voxels whose density is in excess of $\nu\sigma(R_f)$, where $\sigma^2(R_f)$ is the variance of the density field on the filter scale R_f.

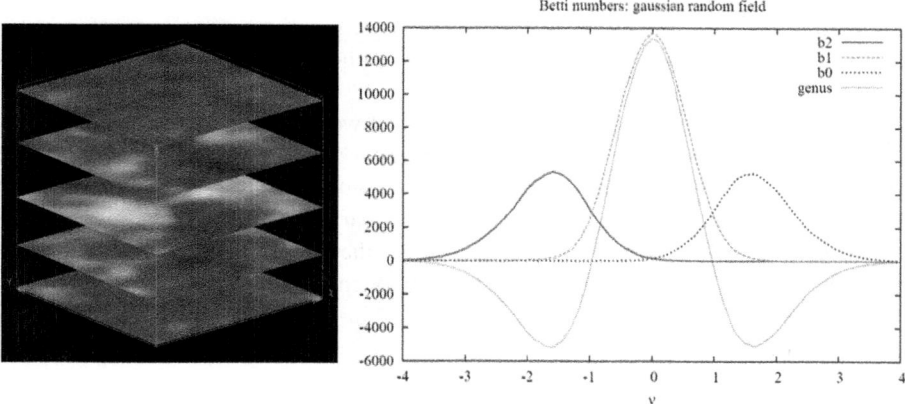

Fig. 12. Left: five slices through a realization of a Gaussian random field, with the color reflecting the density, running from bright yellow (high) to dark red (low). It is an LCDM density field in a box of $100\ h^{-1}$Mpc, Gaussian filtered on a scale of $5\ h^{-1}$Mpc. Right: expected Betti numbers and expected *reduced genus* of the superlevel sets in a Gaussian random field, as functions of the density threshold, ν.

Having outlined the superlevel sets, parameterized by the threshold value ν, we subsequently determine their Betti numbers. To this end, we use the code of Kerber [4], which determines the cycles for the superlevel sets sampled on a grid, and subsequently produces the corresponding homology groups and their ranks, i.e. the Betti numbers $\beta_p(\nu)$. By following this procedure over a threshold level range $\nu \in [-5.0, 5.0]$, we can evaluate the systematic behavior of the Betti numbers as function of ν. Note that a different algorithm was used in [38]. The fact that the results are similar is reassuring.

The right panel of Figure 12 depicts the curves for $\beta_0(\nu)$, $\beta_1(\nu)$, and $\beta_2(\nu)$. The first major impression is the dominant peak of $\beta_1(\nu)$ at $\nu = 0$ (green dashed curve). This relates to the fact that tunnels prevail the topological structure of the outlined regions, which in a Gaussian field are known to possess an intricate sponge-like topology [25]. The region of points with density value at most $\nu = 0$ consists of very few massive structures that percolate through the volume and, by symmetry, the same is true for the region of points with density value at least $\nu = 0$. It is not hard to imagine that this goes along with a large number of tunnels.

Going away from $\nu = 0$ in either direction, we see that β_1 quickly drops to a negligible level. This marks the transition towards a topology marked by individual objects, either clumps for positive ν or voids for negative ν, and this goes along with the disappearance of tunnels. Proceeding to higher levels, we find a quick rise of β_0 (blue dotted curve), the number of over-dense regions in the volume. This is a manifestation of the breaking up into individual high-density regions of the percolating volume at $\nu = 0$. The number of components reaches its maximum for ν approximately equal to $\sqrt{3}$. Beyond that value, the number of individual clumps decreases rapidly, as a result of the decreasing probability of regions having a density that high. It is interesting that the behavior of β_2 (red solid curve) is an almost perfect reflection of the behavior of β_0 through the zero density level. It documents the growing prominence of voids for negative and shrinking density threshold. Here we find a maximum value for ν approximately equal to $-\sqrt{3}$.

We note that we would get perfect symmetry if we replaced the number of components by the number of gaps between them, which is $\beta_0 - 1$. As mentioned, β_2 goes to zero rapidly as ν increases beyond 0, and $\beta_0 - 1$ goes to zero rapidly as ν decreases below 0. However, at $\nu = 0$, both are small but clearly positive. The expected number of components at $\nu = 0$ is thus small but larger than 1, with an expected number of voids that is precisely one less than for the components.

8.5   Gaussian Fields versus the Cosmic Web

It is interesting to compare these results with the homology we find in the web-like configuration of the evolved Universe. Besides the disappearance of symmetry between high-density and low-density regions, reflected in a substantially different behavior of β_0 and β_2, there are a few additional differences as well as similarities.

In evolving LCDM density fields, we find that β_1 is almost always in excess of β_2: the number of tunnels in the Cosmic Web tends to be several factors higher than the number of enclosed voids; see Figure 10. However, the advanced nonlinear mass distribution is marked by a substantially higher number of individual components than found in the Gaussian field. This partially reflects the difference of an alpha shape based analysis and one based on level sets defined for a fixed Gaussian filter radius. In the alpha shape analysis, the substructure of an intrinsically multiscale mass distribution emerging through an hierarchical is not lost. At small values of α one finds the small clumps that in a Gaussian filtered field have been removed.

8.6   Genus and Homology

It is interesting to relate the results on the Betti numbers depending on a threshold value with the analytically known expression for the expected *reduced genus*[12] in a Gaussian random field [25, 26]. This expression is

$$g(\nu) = -\frac{1}{8\pi^2} \left(\frac{\langle k^2 \rangle}{3} \right)^{3/2} (1 - \nu^2)e^{-\nu^2/2}, \qquad (36)$$

where $\langle k^2 \rangle = \langle |\nabla f|^2 \rangle / \langle f^2 \rangle$. Other than its amplitude, the shape does not depend on the power spectrum but only on whether or not the field is Gaussian. Recall from (10) that the genus of an isodensity surface is directly related to the Betti numbers of the enclosed superlevel sets, via the alternating sum

$$g(\nu) = -\beta_0(\nu) + \beta_1(\nu) - \beta_2(\nu). \qquad (37)$$

---

[12] In section 2.1 we remarked that the genus g in cosmological studies is slightly differently defined than the usual definition of the genus G (eqn. 1). The genus, g, in these studies has been defined as the number of holes minus the number of connected regions: $g = G - c$. In this review we distinguish between these definitions by referring to g as the *reduced genus*.

We may therefore compare the outcome of our superlevel set study to the expected distribution for the genus in (36). In Figure 12, the magenta dotted dashed line is the genus $g(\nu)$ computed from the alternating sum (37). It closely matches the predicted relation (36); see [38]. It is most reassuring that the peaks of β_0 and β_2 are reached at the threshold value where, according to (36), $g(\nu)$ has its two minima, at $\nu = -\sqrt{3}$ and $\nu = \sqrt{3}$.

Having established the cosmologically crucial Gaussian basis of the Betti number analysis, a few more interesting findings follow from our analysis in [38]. While the shape of the genus function is independent of the power spectrum, it turns out that the Betti numbers do reflect the power spectrum. In particular, for small absolute values of ν, there are substantial differences between the Betti numbers in Gaussian fields with different power spectra. This makes them potentially strong discriminators between cosmological scenarios. On the other hand, the differences are perfectly symmetric between $\beta_0 - 1$ and β_2, and β_1 is symmetric itself. The Betti numbers may therefore be useful for tracing non-Gaussianities in the primordial Universe, a major point of interest in current cosmological studies.

8.7 Gaussian Betti Correlations

Finally, [38] also assess the correlations in Gaussian fields between the various Betti numbers. They find that the Betti numbers are not independent of one another. For example, β_0 near its maximum ($\nu = 1.7$) is positively correlated with β_1 at low threshold levels, while there are other levels where β_0 and β_1 are anti-correlated.

The important implication is that we need to take into account the mutual dependence of the Betti numbers. This will be largely dependent on the nature of the density field. On the other hand, in general the correlation is not perfect. In other words, the Betti numbers contains complementary information on the topological structure of e.g. the cosmic mass distribution.

9   Persistence

The one outstanding issue we have not yet addressed systematically is the hierarchical substructure of the Cosmic Web data. In standard practice, the multiscale nature of the mass distribution tends to be investigated by means of user-imposed filtering. Persistence rationalizes this approach by considering the range of filters at once, and by making meaningful comparisons. At the same time, it deepens the approach by combining it with topological measurements: the homology groups and their ranks.

Persistence entails the conceptual framework and language for separating scales of a spatial structure or density distribution. It was introduced by Edelsbrunner, Letscher and Zomorodian [19]; see also the recent text on computational topology [17] of which persistent homology forms the core. By separating the scales in a mass distribution, one may analyze and map the topological hierarchy of the cosmos. Within this context, substructures can be separated by determining the range of scales over which they exist. Here, we find a close link to Morse theory, in which the crucial new idea

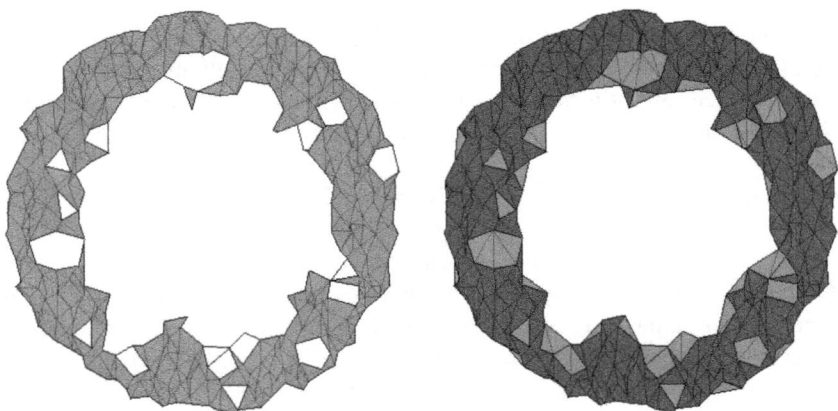

Fig. 13. Illustrating the idea of persistence. The alpha shapes of a set of points randomly chosen in a 2-dimensional annulus. Left: for a small value of α, the alpha shape has 18 holes. Right: for a somewhat larger value of α, the alpha shape has only two holes. One of these holes is new, while the other already existed on the left, albeit in somewhat different form. We say the latter hole *persists* over the entire interval delimited by the two alpha shapes.

contributed by persistence is the pairing of all critical points depending on a global topological criterion. A recent discussions of persistence in the cosmological context can be found in the work of Sousbie et al. [51, 55].

9.1   Persistence of Alpha Shapes

Alpha shapes provide the perfect context to illustrate the essential idea of persistence, which is to follow components, tunnels, and voids over the entire range of the parameter, α. The 2-dimensional illustration in Figure 13 shows how the small scale holes in the left frame disappear when we increase the parameter in the right frame. This is not to say that the value of α chosen on the right is perfect for the data set; indeed, we see a new, larger hole appear that did not yet exist on the left. The key point is that each feature is *born* at some value of α, and *dies* at another value of α. The interval between birth and death sets the position of the feature within the structural hierarchy. In particular, the length of the interval, that is, the absolute difference between the values of α at the birth and at the death, is called the *persistence* of the feature.

The full implementation of persistence in our topological study will be addressed in future publications; e.g. [43]. Here, we briefly address its effect on the Betti number analysis. We assess the matter distribution for three cosmological scenarios: LCDM, and RP and SUGRA quintessence. Each is used in a many-body simulation, generating a discrete point distribution in 3-dimensional space that reflects the intricate multi-scale organization of matter. For each dataset, we compute the entire range of alpha shapes, encoded for efficiency reasons in the Delaunay triangulation of the points, and we follow each cycle through the sequence of alpha shapes, recording when it is born and when it dies. Separating the results for different homological dimensions ($p = 0$ for components, $p = 1$ for tunnels, and $p = 2$ for voids), we show the statistics for $p = 1$

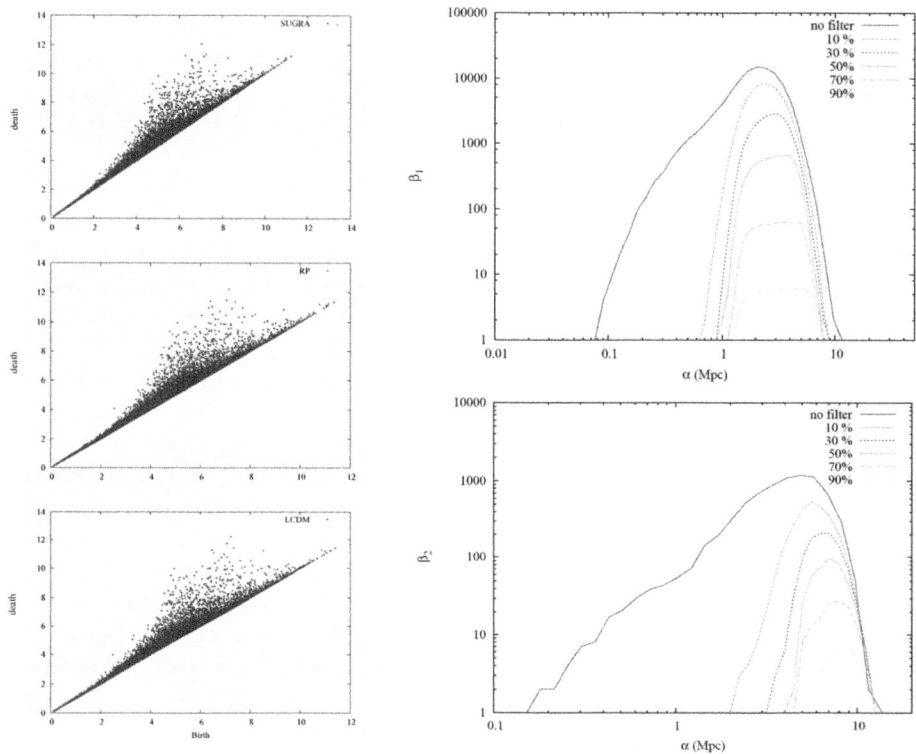

Fig. 14. Persistent homology for the Cosmic Web simulations. Left: the 1-dimensional persistence diagrams of the LCDM simulation and of two quintessence simulations representing the RP and SUGRA dark energy models. Right: the curves describing the evolution of the first and second Betti numbers for growing value of α. In addition to giving the Betti numbers, at every value of α, we show how many of the counted features belong to the top 90%, 70%, 50%, 30%, 10% most persistent features in the entire population.

in Figure 14. On the left, we see the 1-dimensional persistence diagrams of the three datasets. Each dot represents a 1-cycle in the alpha shape filtration. Its horizontal coordinate gives the value of α at which the 1-cycle is born, which happens when its last edge is added to the alpha shape. The vertical coordinate of the dot gives the value of α at which the 1-cycle dies, which happens when the last triangle completes a membrane filling the tunnel formed by the cycle.

In all three diagrams, we see a substantial number of 1-cycles that die shortly after they are born. They correspond to dots close to and right above the diagonal line in the diagram. This is typical for natural data, in which we see features appear and disappear in rapid progression at all times. This "topological noise" creates a deceivingly complicated picture, obscuring the cleaner picture of more persistent features. In our datasets, there are also a good number of 1-cycles that remain for a while after they are born. They correspond to dots that are further up in the persistence diagram, with the vertical distance to the diagonal representing the measured persistence.

9.2 Persistence and Scale

The promise of persistence is that it gives access to data on different scale levels. This is particularly important for the Cosmic Web, where we observe structure on almost every scale. The goal would be to separate scales, or quantify them in such a way that relations between scales become explicit.

To probe this aspect of persistence, we use a *persistence parameter*, λ, that restricts our attention to features of persistence at least λ. In our experiment, we adjusted the parameter so that a specified percentage of the features (dots) have persistence at least λ. All dots with persistence less than λ are removed. The effect on the Betti numbers of the LCDM simulation data is illustrated in the two right frames of Figure 14. The top frame shows six evolutions of β_1 for increasing values of α: for 100%, 90%, 70%, 50%, 30%, and 10% of the most persistent 1-cycles. The bottom frame shows the same information for β_2.

The features with small persistence can be considered noise. They may be reflections of numerical inaccuracies in computations, artefacts of the data representation, measuring errors, or genuine features that are too small for us to distinguish them from noise. In practical astrophysical circumstances, datasets are indeed beset by a variety of artefacts. As a result, the alpha shapes of discrete point distributions, such as the galaxies in our local cosmic environment, will reflect the noise. The induced irregularities in the number of components, tunnels, and voids do not represent any real topological structure but nonetheless influence the Betti numbers we collect. Persistence diagrams offer the real opportunity to remove the noise without introducing side-effects, and they may help in defining the usually ill-specified noise distribution.

10 Conclusions and Prospects

In this overview we have discussed and presented an analysis of the homology of cosmic mass distributions. We have argued that Betti numbers of density level sets and of the corresponding isodensity surfaces provide us with a far more complete description of the topology of web-like patterns in the megaparsec Universe than the well-known genus analysis or the topological and geometric instruments of Minkowski functionals. The genus of isodensity surfaces and the Euler characteristic, the main topological Minkowski functional, are directly related to Betti numbers and may be considered as lower-dimensional projections of the homological information content in terms of Betti numbers.

We have established the promise of alpha shapes for measuring the homology of the megaparsec galaxy distribution in the Cosmic Web. Alpha shapes are well-defined subsets of Delaunay tessellations, selected following a strictly defined scale parameter α. These simplicial complexes constitute a filtration of the Delaunay tessellations, and are homotopy equivalent to the distance function field defined by the point distribution. Alpha shape analysis has the great advantage of being self-consistent and natural, involving shapes and surfaces that are entirely determined by the point distribution, independent of any artificial filtering.

By studying the Betti numbers and several Minkowski functionals of a set of heuristic Voronoi clustering models, as well as dark energy cosmological scenarios (LCDM, and RP and SUGRA quintessence), we have illustrated the potential for exploiting the cosmological information contained in the Cosmic Web. We have analyzed the evolution of Betti numbers in an LCDM simulation, and compared the discriminative power of Betti curves for distinguishing between different dark energy models. In addition, we have addressed the significance of Betti numbers in the case of Gaussian random fields, as these primordial density fields represent the reference point for any further analysis of the subsequent cosmic structure evolution.

Related Work

The mathematical fundamentals of our project, including a full persistence analysis, will be extensively outlined and defined in the upcoming study of Pranav, Edelsbrunner et al. [43]. In addition to an evaluation on the basis of the simplicial alpha complexes, it also discusses and compares the homology analysis on the basis of isodensity surfaces, superlevel and sublevel sets, and their mutual relationship. This includes an evaluation of the results obtained on the basis of a piecewise linear density field reconstructions on Delaunay tessellations, following the DTFE formalism [10, 46, 64].

In related work, [51] and [55] recently published an impressive study on the topological analysis of the Cosmic Web within the context of the skeleton formalism [52–54]. Their study used the DTFE density field reconstructions and incorporated the complexities of a full persistence analysis to trace filamentary features in the Cosmic Web.

In two accompanying letters we will address cosmological applications of homology analysis. In [68] the ability of Betti numbers to discriminate between different dark energy cosmological scenarios will be demonstrated. Crucial for furthering our understanding and appreciation of the significance of measured Betti numbers in a cosmological context, is to know their values in Gaussian random density fields. With cosmic structure emerging from a primordial Gaussian field, the homology of these fields forms a crucial reference point. In the study by Park et al. [38] we have analyzed this aspect in more detail.

Acknowledgement. We thank Dmitriy Morozov for providing us with his code for computing persistence diagrams and Michael Kerber for his homology code for computing Betti numbers of grid represented manifolds. In addition, we thank Bartosz Borucki for his permission to use figure 1. We also acknowledge many interesting and enjoyable discussions with Mathijs Wintraecken. In particular, we wish to express our gratitude to the editor, Mir Abolfazl Mostafevi, for his help and patience at the finishing stage of this project. Part of this project was carried out within the context of the CG Learning project. The project CG Learning acknowledges the financial support of the Future and Emerging Technologies (FET) programme within the Seventh Framework Programme for Research of the European Commission, under FET-Open grant number: 255827.

References

1. Adler, R.J., Bobrowski, O., Borman, M.S., Subag, E., Weinberger, S.: Persistent homology for random fields and complexes. Collections, 1–6 (2010),
   http://arxiv.org/abs/1003.1001
2. Adler, R.J., Taylor, J.E.: Random fields and geometry (2007)

3. Bardeen, J.M., Bond, J.R., Kaiser, N., Szalay, A.S.: The statistics of peaks of Gaussian random fields. Astrophys. J. 304, 15–61 (1986)
4. Bendich, P., Edelsbrunner, H., Kerber, M.: Computing robustness and persistence for images. IEEE Trans. Vis. Comput. Graph. 16(6), 1251–1260 (2010)
5. Bennett, C.L., Halpern, M., Hinshaw, G., Jarosik, N., Kogut, A., Limon, M., Meyer, S.S., Page, L., Spergel, D.N., Tucker, G.S., Wollack, E., Wright, E.L., Barnes, C., Greason, M.R., Hill, R.S., Komatsu, E., Nolta, M.R., Odegard, N., Peiris, H.V., Verde, L., Weiland, J.L.: First-Year Wilkinson Microwave Anisotropy Probe (WMAP) Observations: Preliminary Maps and Basic Results. Astrophys. J. Suppl. 148, 1–27 (2003)
6. Bond, J., Kofman, L., Pogosyan, D.: How filaments are woven into the cosmic web. Nature 380, 603–606 (1996)
7. de Boni, C., Dolag, K., Ettori, S., Moscardini, L., Pettorino, V., Baccigalupi, C.: Hydrodynamical simulations of galaxy clusters in dark energy cosmologies. Mon. Not. R. Astron. Soc. 414, 780 (2011)
8. Bos, P.C. , van de Weygaert, R., Dolag, K., Pettorino, V.: Void shapes as probes of the nature of dark energy. Mon. Not. R. Astron. Soc. (2011)
9. Calvo, M.A.A., Shandarin, S.F., Szalay, A.: Geometry of the cosmic web: Minkowski functionals from the delaunay tessellation. In: International Symposium on Voronoi Diagrams in Science and Engineering, pp. 235–243 (2010)
10. Cautun, M.C., van de Weygaert, R.: The DTFE public software - The Delaunay Tessellation Field Estimator code. ArXiv e-prints (May 2011)
11. Choi, Y.Y., Park, C., Kim, J., Gott, J.R., Weinberg, D.H., Vogeley, M.S., Kim, S.S.: For the SDSS Collaboration: Galaxy Clustering Topology in the Sloan Digital Sky Survey Main Galaxy Sample: A Test for Galaxy Formation Models. Astrophys. J. Suppl. 190, 181–202 (2010)
12. Colless, M.: 2dF consortium: The 2df galaxy redshift survey: Final data release pp. 1–32 (2003); astroph/0306581
13. Delfinado, C.J.A., Edelsbrunner, H.: An incremental algorithm for betti numbers of simplicial complexes. In: Symposium on Computational Geometry, pp. 232–239 (1993)
14. Dey, T., Edelsbrunner, H., Guha, S.: Computational topology. In: Chazelle, B., Goodman, J.E., Pollack, R. (eds.) Advances in Discrete and Computational Geometry, pp. 109–143. American Mathematical Society (1999)
15. Edelsbrunner, H.: Alpha shapes - a survey. In: van de Weygaert, R., Vegter, G., Ritzerveld, J., Icke, V. (eds.) Tessellations in the Sciences; Virtues, Techniques and Applications of Geometric Tilings. Springer, Heidelberg (2010)
16. Edelsbrunner, H., Facello, M., Liang, J.: On the definition and the construction of pockets in macromolecules. Discrete Appl. Math. 88, 83–102 (1998)
17. Edelsbrunner, H., Harer, J.: Computational Topology, An Introduction. American Mathematical Society (2010)
18. Edelsbrunner, H., Kirkpatrick, D., Seidel, R.: On the shape of a set of points in the plane. IEEE Trans. Inform. Theory 29, 551–559 (1983)
19. Edelsbrunner, H., Letscher, D., Zomorodian, A.: Topological persistence and simplification. Discrete and Computational Geometry 28, 511–533 (2002)
20. Edelsbrunner, H., Muecke, E.: Three-dimensional alpha shapes. ACM Trans. Graphics 13, 43–72 (1994)
21. Eisenstein, D.J., Hu, W.: Baryonic Features in the Matter Transfer Function. Astrophys.J. 496, 605–614 (1998)
22. Eldering, B.: Topology of Galaxy Models, MSc thesis, Univ. Groningen (2006)
23. Gott, J.R., Choi, Y.Y., Park, C., Kim, J.: Three-Dimensional Genus Topology of Luminous Red Galaxies. Astrophys. J. Lett. 695, L45–L48 (2009)

24. Gott III, J.R., Miller, J., Thuan, T.X., Schneider, S.E., Weinberg, D.H., Gammie, C., Polk, K., Vogeley, M., Jeffrey, S., Bhavsar, S.P., Melott, A.L., Giovanelli, R., Hayes, M.P., Tully, R.B., Hamilton, A.J.S.: The topology of large-scale structure. III - Analysis of observations. Astrophys. J. 340, 625–646 (1989)
25. Gott, J., Dickinson, M., Melott, A.: The sponge-like topology of large-scale structure in the universe. Astrophys. J. 306, 341–357 (1986)
26. Hamilton, A.J.S., Gott III, J.R., Weinberg, D.: The topology of the large-scale structure of the universe. Astrophys. J. 309, 1–12 (1986)
27. Hoyle, F., Vogeley, M.S., Gott III, J.R., Blanton, M., Tegmark, M., Weinberg, D.H., Bahcall, N., Brinkmann, J., York, D.: Two-dimensional Topology of the Sloan Digital Sky Survey. Astrophys. J. 580, 663–671 (2002)
28. Huchra, J., et al.: The 2mass redshift survey and low galactic latitude large-scale structure. In: Fairall, A.P., Woudt, P.A. (eds.) Nearby Large-Scale Structures and the Zone of Avoidance. ASP Conf. Ser., vol. 239, pp. 135–146. Astron. Soc. Pacific, San Francisco (2005)
29. Icke, V.: Voids and filaments. Mon. Not. R. Astron. Soc. 206, 1P–3P (1984)
30. Kauffmann, G., Colberg, J.M., Diaferio, A., White, S.D.M.: Clustering of galaxies in a hierarchical universe - I. Methods and results at z=0. Mon. Not. R. Astron. Soc. 303, 188–206 (1999)
31. Komatsu, E., Smith, K.M., Dunkley, J., et al.: Seven-Year Wilkinson Microwave Anisotropy Probe (WMAP) Observations: Cosmological Interpretation. eprint arXiv:1001.4538 (January 2010)
32. Liang, J., Edelsbrunner, H., Fu, P., Sudhakar, P., Subramaniam, S.: Analytical shape computation of macromolecules: I. molecular area and volume through alpha shape. Proteins: Structure, Function, and Genetics 33, 1–17 (1998)
33. Liang, J., Edelsbrunner, H., Fu, P., Sudhakar, P., Subramaniam, S.: Analytical shape computation of macromolecules: Ii. inaccessible cavities in proteins. Proteins: Structure, Function, and Genetics 33, 18–29 (1998)
34. Liang, J., Woodward, C., Edelsbrunner, H.: Anatomy of protein pockets and cavities: measurement of binding site geometry and implications for ligand design. Protein Science 7, 1884–1897 (1998)
35. Mecke, K., Buchert, T., Wagner, H.: Robust morphological measures for large-scale structure in the universe. Astron. Astrophys. 288, 697–704 (1994)
36. Moore, B., Frenk, C.S., Weinberg, D.H., Saunders, W., Lawrence, A., Ellis, R.S., Kaiser, N., Efstathiou, G., Rowan-Robinson, M.: The topology of the QDOT IRAS redshift survey. Mon. Not. R. Astron. Soc. 256, 477–499 (1992)
37. Muecke, E.: Shapes and Implementations in three-dimensional geometry, PhD thesis, Univ. Illinois Urbana-Champaign (1993)
38. Park, C., Chingangbam, P., van de Weygaert, R., Vegter, G., Kim, I., Hidding, J., Hellwing, W., Pranav, P.: Betti numbers of gaussian random fields. Astrophys. J. (2011) (to be subm.)
39. Park, C., Gott III, J.R., Melott, A.L., Karachentsev, I.D.: The topology of large-scale structure. VI - Slices of the universe. Astrophys. J. 387, 1–8 (1992)
40. Park, C., Kim, J., Gott III, J.R.: Effects of Gravitational Evolution, Biasing, and Redshift Space Distortion on Topology. Astrophys. J. 633, 1–10 (2005)
41. Park, C., Kim, Y.R.: Large-scale Structure of the Universe as a Cosmic Standard Ruler. Astrophys. J. Lett. 715, L185–L188 (2010)
42. Peebles, P.: The Large Scale Structure of the Universe. Princeton Univ. Press (1980)
43. Pranav, P., Edelsbrunner, H., van de Weygaert, R., Vegter, G.: On the alpha and betti of the universe: Multiscale persistence of the cosmic web. Mon. Not. R. Astron. Soc. (2011) (to be subm.)
44. Ratra, B., Peebles, P.J.E.: Cosmological consequences of a rolling homogeneous scalar field. Phys. Rev. D. 37, 3406–3427 (1988)

45. Sahni, V., Sathyprakash, B.S., Shandarin, S.: Shapefinders: A new shape diagnostic for large-scale structure. Astrophys. J. 507, L109–L112 (1998)
46. Schaap, W.E., van de Weygaert, R.: Continuous fields and discrete samples: reconstruction through Delaunay tessellations. Astron. Astrophys. 32, L29–L32 (2000)
47. Schmalzing, J., Buchert, T.: Beyond Genus Statistics: A Unifying Approach to the Morphology of Cosmic Structure. Astrophys. J. Lett. 482, L1–L4 (1997)
48. Schmalzing, J., Buchert, T., Melott, A., Sahni, V., Sathyaprakash, B., Shandarin, S.: Disentangling the cosmic web. i. morphology of isodensity contours. Astrophys. J. 526, 568–578 (1999)
49. Sheth, R.K., van de Weygaert, R.: A hierarchy of voids: much ado about nothing. Mon. Not. R. Astron. Soc. 350, 517–538 (2004)
50. Smoot, G.F., Bennett, C.L., Kogut, A., Wright, E.L., Aymon, J., Boggess, N.W., Cheng, E.S., de Amici, G., Gulkis, S., Hauser, M.G., Hinshaw, G., Jackson, P.D., Janssen, M., Kaita, E., Kelsall, T., Keegstra, P., Lineweaver, C., Loewenstein, K., Lubin, P., Mather, J., Meyer, S.S., Moseley, S.H., Murdock, T., Rokke, L., Silverberg, R.F., Tenorio, L., Weiss, R., Wilkinson, D.T.: Structure in the COBE differential microwave radiometer first-year maps. Astrophys. J. Lett. 396, L1–L5 (1992)
51. Sousbie, T.: The persistent cosmic web and its filamentary structure - I. Theory and implementation. Mon. Not. R. Astron. Soc., pp. 511–+ (April 2011)
52. Sousbie, T., Colombi, S., Pichon, C.: The fully connected N-dimensional skeleton: probing the evolution of the cosmic web. Mon. Not. R. Astron. Soc. 393, 457–477 (2009)
53. Sousbie, T., Pichon, C., Colombi, S., Novikov, D., Pogosyan, D.: The 3D skeleton: tracing the filamentary structure of the Universe. Mon. Not. R. Astron. Soc. 383, 1655–1670 (2008)
54. Sousbie, T., Pichon, C., Courtois, H., Colombi, S., Novikov, D.: The Three-dimensional Skeleton of the SDSS. Astrophys. J. Lett. 4, L1–L4 (2008)
55. Sousbie, T., Pichon, C., Kawahara, H.: The persistent cosmic web and its filamentary structure - II. Illustrations. Mon. Not. R. Astron. Soc., pp. 530–+ (2011)
56. Spergel, D.N., Bean, R., Doré, O., Nolta, M.R., Bennett, C.L., Dunkley, J., Hinshaw, G., Jarosik, N., Komatsu, E., Page, L., Peiris, H.V., Verde, L., Halpern, M., Hill, R.S., Kogut, A., Limon, M., Meyer, S.S., Odegard, N., Tucker, G.S., Weiland, J.L., Wollack, E., Wright, E.L.: Three-Year Wilkinson Microwave Anisotropy Probe (WMAP) Observations: Implications for Cosmology. Astrophys. J. Suppl. 170, 377–408 (2007)
57. Springel, V., et al.: Simulations of the formation, evolution and clustering of galaxies and quasars. Nature 435, 629–636 (2005)
58. Tegmark, M., et al.: The three-dimensional power spectrum of galaxies from the sloan digital sky survey. Astrophys. J. 606, 702–740 (2004)
59. van de Weygaert, R.: Voids and the geometry of large scale structure, PhD thesis, University of Leiden (1991)
60. van de Weygaert, R., Bertschinger, E.: Peak and gravity constraints in Gaussian primordial density fields: An application of the Hoffman-Ribak method. Mon. Not. R. Astron. Soc. 281, 84–118 (1996)
61. van de Weygaert, R., Bond, J.R.: Clusters and the Theory of the Cosmic Web. In: Plionis, M., López-Cruz, O., Hughes, D. (eds.) A Pan-Chromatic View of Clusters of Galaxies and the Large-Scale Structure. Lecture Notes in Physics, vol. 740, pp. 335–407. Springer, Heidelberg (2008)
62. van de Weygaert, R., Bond, J.R.: Observations and Morphology of the Cosmic Web. In: Plionis, M., López-Cruz, O., Hughes, D. (eds.) A Pan-Chromatic View of Clusters of Galaxies and the Large-Scale Structure. Lecture Notes in Physics, vol. 740, pp. 409–467. Springer, Berlin (2008)
63. van de Weygaert, R., Icke, V.: Fragmenting the universe. II - Voronoi vertices as Abell clusters. Astron. Astrophys. 213, 1–9 (1989)

64. van de Weygaert, R., Schaap, W.: The Cosmic Web: Geometric Analysis. In: Martínez, V.J., Saar, E., Martínez-González, E., Pons-Bordería, M.-J. (eds.) Data Analysis in Cosmology. Lecture Notes in Physics, vol. 665, pp. 291–413. Springer, Berlin (2009)
65. Vegter, G.: Computational topology. In: Goodman, J.E., O'Rourke, J. (eds.) Handbook of Discrete and Computational Geometry, 2nd edn., ch. 32, pp. 719–742. CRC Press LLC, Boca Raton (2004)
66. Vegter, G., van de Weygaert, R., Platen, E., Kruithof, N., Eldering, B.: Alpha shapes and the topology of cosmic large scale structure. Mon. Not. R. Astron. Soc. (2010) (in prep.)
67. Vogeley, M.S., Park, C., Geller, M.J., Huchra, J.P., Gott III, J.R.: Topological analysis of the CfA redshift survey. Astrophys. J. 420, 525–544 (1994)
68. van de Weygaert, R., Pranav, P., Jones, B., Vegter, G., Bos, P., Park, C., Hellwing, W.: Probing dark energy with betti-analysis of simulations. Astrophys. J. Lett. (2011) (to be subm.)
69. van de Weygaert, R., Platen, E., Vegter, G., Eldering, B., Kruithof, N.: Alpha shape topology of the cosmic web. In: International Symposium on Voronoi Diagrams in Science and Engineering, pp. 224–234 (2010)
70. van de Weygaert, R.: Voronoi tessellations and the cosmic web: Spatial patterns and clustering across the universe. In: ISVD 2007: Proceedings of the 4th International Symposium on Voronoi Diagrams in Science and Engineering, pp. 230–239. IEEE Computer Society, Washington, DC (2007)
71. Zhang, Y., Springel, V., Yang, X.: Genus Statistics Using the Delaunay Tessellation Field Estimation Method. I. Tests with the Millennium Simulation and the SDSS DR7. Astrophys. J. 722, 812–824 (2010)
72. Zomorodian, A.: Topology for Computing. Cambr. Mon. Appl. Comp. Math., Cambr. Univ. Press (2005)

Skew Jensen-Bregman Voronoi Diagrams[*]

Frank Nielsen[1] and Richard Nock[2]

[1] École Polytechnique Palaiseau, France
Sony Computer Science Laboratories Inc., Tokyo, Japan
nielsen@lix.polytechnique.fr
[2] CEREGMIA-UAG, University of Antilles-Guyane Martinique, France
rnock@martinique.univ-ag.fr

— Dedicated to the victims of Japan Tohoku earthquake (March 2011).

Abstract. A Jensen-Bregman divergence is a distortion measure defined by a Jensen convexity gap induced by a strictly convex functional generator. Jensen-Bregman divergences unify the squared Euclidean and Mahalanobis distances with the celebrated information-theoretic Jensen-Shannon divergence, and can further be skewed to include Bregman divergences in limit cases. We study the geometric properties and combinatorial complexities of both the Voronoi diagrams and the centroidal Voronoi diagrams induced by such as class of divergences. We show that Jensen-Bregman divergences occur in two contexts: (1) when symmetrizing Bregman divergences, and (2) when computing the Bhattacharyya distances of statistical distributions. Since the Bhattacharyya distance of popular parametric exponential family distributions in statistics can be computed equivalently as Jensen-Bregman divergences, these skew Jensen-Bregman Voronoi diagrams allow one to define a novel family of statistical Voronoi diagrams.

Keywords: Jensen's inequality, Bregman divergences, Jensen-Shannon divergence, Jensen-von Neumann divergence, Bhattacharyya distance, information geometry.

1 Introduction

The Voronoi diagram is one of the most fundamental combinatorial structures studied in computational geometry [2] often used to characterize solutions to geometric problems [3] like the minimum spanning tree, the smallest enclosing ball, motion planning, etc. For a given set of sites, the Voronoi diagram partitions the space into elementary proximity cells denoting portions of space closer to a

---

[*] This journal article revises and extends the conference paper [1] presented at the International Symposium on Voronoi Diagrams (ISVD) 2010. This paper includes novel extensions to matrix-based Jensen-Bregman divergences, and present the general framework of skew Jensen-Bregman Voronoi diagrams that include Bregman Voronoi diagrams as particular cases. Supporting materials available at http://www.informationgeometry.org/JensenBregman/

M.L. Gavrilova et al. (Eds.): Trans. on Comput. Sci. XIV, LNCS 6970, pp. 102–128, 2011.
© Springer-Verlag Berlin Heidelberg 2011

site than to any other one. Voronoi diagrams have been generalized in many ways by considering various types of primitives (points, lines, balls, etc.) and distance functions (L_p Minkowski distances [4], convex distances[1] [5], Bregman divergences [6], etc.) among others.

In this work, we introduce a novel class of information-theoretic distortion measures called *skew Jensen-Bregman divergences* that generalizes the celebrated Jensen-Shannon divergence [7] in information theory [8]. We study both Voronoi diagrams and centroidal Voronoi tessellations [9] with respect to that family of distortion measures. As a by-product, we also show that those skew Jensen-Bregman Voronoi diagrams allow one to characterize statistical Voronoi diagrams induced by the skew Bhattacharyya statistical distance on a given set of probability measures.

Our main contributions are summarized as follows:

– We define the family of Jensen-Bregman divergences extending the concept of Jensen-Shannon divergence, and show that those divergences appear when symmetrizing Bregman divergences,
– By skewing those Jensen-Bregman divergences, we obtain parametric divergences that encapsulate Bregman divergences as limit cases,
– We study the combinatorial complexities of skew Jensen-Bregman Voronoi diagrams (generalizing Bregman Voronoi diagrams [6]),
– We describe an efficient algorithm to arbitrarily finely estimate the Jensen-Bregman centroids, and extend its scope to matrix-valued divergences,
– We show that the statistical Bhattacharyya distance of parametric exponential family distributions amount to compute a Jensen-Bregman divergence on the corresponding parameters.

The paper is organized as follows: Section 2 introduces the class of Jensen-Bregman divergences and described the link with Bregman divergence symmetrization [6]. Section 3 defines the Voronoi diagram with respect to Jensen-Bregman divergences, and bound their complexity by studying the corresponding minimization diagram and investigating properties of the bisectors and level sets. Section 4 presents the Jensen-Bregman centroids, design an efficient iterative estimation algorithm, and provide some experiments on the centroidal Jensen-Bregman Voronoi tessellations. It is followed by Section 5 that extends Jensen-Bregman divergences to matrix-valued data sets. Section 6 introduces a skew factor in the divergence and show how to obtain Bregman Voronoi diagrams [6] as extremal cases. Finally, Section 7 concludes the paper by mentioning the underlying differential geometry.

In order to not overload the paper, Appendix A introduces the class of statistical Bhattacharyya distances, and show how it is equivalent to Jensen-Bregman divergences when distributions belong to the same parametric exponential family.

---

[1] Convex distances may not necessarily be metrics [5]. A metric satisfies both the symmetry and triangular inequality axioms.

2   Jensen-Bregman Divergences

There is a growing interest in studying *classes* of distortion measures instead of merely choosing a single distance. This trend is attested in many fields of computer science including computational geometry, machine learning, computer vision, and operations research. The goal is to study and design *meta-algorithms* that can provably run correct on a class of distances rather than on a single distance at hand. Among such classes of distances, the Bregman distances [6] appear attractive because this family of dissimilarity measures encapsulate both the geometric (squared) Euclidean distance and the information-theoretic relative entropy. A Bregman distance B_F on an open convex space $\mathcal{X} \subseteq \mathbb{R}^d$ is defined for a strictly convex and differentiable function F as

$$B_F(p, q) = F(p) - F(q) - \langle p - q, \nabla F(q) \rangle, \tag{1}$$

where $\langle p, q \rangle = p^T q$ denotes the inner product, and

$$\nabla F(x) = \left[\frac{\partial F}{\partial x_1} \cdots \frac{\partial F}{\partial x_d} \right]^T \tag{2}$$

the partial derivatives. Choosing $F(x) = \sum_{i=1}^{d} x_i^2 = \langle x, x \rangle$ yields the squared Euclidean distance $B_{x^2}(p, q) = \|p - q\|^2$, and choosing $F(x) = \sum_{i=1}^{d} x_i \log x_i = S(x)$ yields the relative entropy, called the Kullback-Leibler divergence [8]. The Kullback-Leibler divergence is defined for normalized d-dimensional "distribution" points (i.e., points falling in the $(d-1)$-dimensional unit simplex denoting discrete distributions) as:

$$I(p, q) = B_S(p, q) = \sum_{i=1}^{d} p_i \log \frac{p_i}{q_i}. \tag{3}$$

Handling Bregman divergences instead of the (squared) Euclidean distance brings the opportunity to enlarge the field of applications of geometric algorithms to other settings like statistical contexts.

The generator function F can be interpreted as a measure of *information* (namely a negative entropy, since entropies are usually concave functions [8]). In information theory, the entropy measures the amount of uncertainty of a random variable. For example, one expects that the entropy is maximized for the uniform distribution. Axiomatizing a few behavior properties [8] of entropy yields the unique concave function

$$H(x) = x \log \frac{1}{x} = -x \log x, \tag{4}$$

called the *Shannon entropy* ($-H(x)$ is the convex Shannon information).

Bregman divergences are *never* metrics, and provably only symmetric for the generalized quadratic distances obtained for generator $F(x) = Qx$ for a positive definite matrix $Q \succ 0$. Thus those distances are preferably technically called divergences instead of distances. Bregman divergences satisfy

$$B_F(p, q) \geq 0, \tag{5}$$

with equality if and only if $p = q$. This is called Gibb's inequality for the particular case of Kullback-Leibler divergence, and can be demonstrated by using a geometric argument as follows: Let $\hat{x} = (x, F(x))$ denote the lifting map of point x to the potential function plot $\mathcal{F} = \{\hat{x} = (x, F(x)) \mid x \in \mathcal{X}\}$. The Bregman divergence measures the vertical distance between two non-vertical hyperplanes: The hyperplane H_q tangent at the potential function $\mathcal{F} = (x, F(x))$ at lifted point $\hat{q}$:

$$H_q(x) = F(q) + \langle x - q, \nabla F(q) \rangle, \tag{6}$$

and its translate H_q' passing through $\hat{p}$:

$$H_q'(x) = F(p) + \langle x - p, \nabla F(q) \rangle. \tag{7}$$

We have

$$B_F(p, q) = H_q'(x) - H_q(x), \tag{8}$$

independent of the position of x. This geometric interpretation is illustrated in Figure 1.

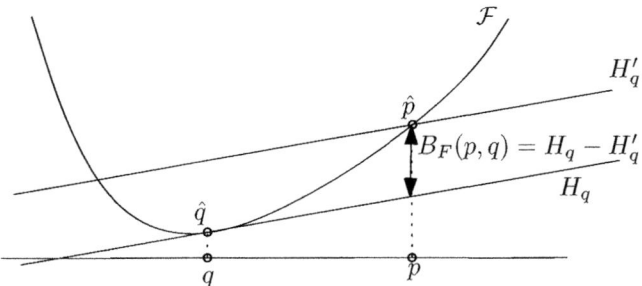

Fig. 1. Interpreting the Bregman divergence as the vertical distance between the tangent plane at q and its translate passing through p (with identical slope $\nabla F(q)$)

Since Bregman divergences are not symmetric, one can naturally symmetrize them as follows:

$$S_F(p, q) = \frac{B_F(p, q) + B_F(q, p)}{2} \tag{9}$$

$$= \frac{1}{2} \langle p - q, \nabla F(p) - \nabla F(q) \rangle. \tag{10}$$

That is indeed what happened historically with Jeffreys [10] considering the J-measure as the sum of sided I measures:

$$J(p,q) = I(p,q) + I(q,p) \tag{11}$$

However, there are two major *drawbacks* for such an information-theoretic divergence:

1. The divergence S_F may be undefined (unbounded): For example, considering the negative Shannon entropy, the symmetrized Kullback-Leibler divergence is undefined if for some coordinate $q_i = 0$ and $p_i \neq 0.$[2]
2. The divergence S_F is not bounded in terms of the variational *metric* distance $V(p,q) = \sum_{i=1}^{d} |p_i - q_i|$ (L_1-metric [4]). For the Kullback-Leibler divergence, it is known that $\mathrm{KL}(p,q) \geq \frac{1}{2} V^2(p,q)$. Such kinds of bounds are called *Pinsker's inequalities* [11].

To overcome those two issues, Lin [7] proposed a new divergence built on the Kullback-Leibler divergence called the *Jensen-Shannon divergence*. The Jensen-Shannon divergence is defined as

$$\mathrm{JS}(p,q) = \mathrm{KL}\left(p, \frac{p+q}{2}\right) + \mathrm{KL}\left(q, \frac{p+q}{2}\right). \tag{12}$$

This divergence is always (1) defined, (2) finite, and furthermore (3) bounded by the variational L_1-metric:

$$V^2(p,q) \leq \mathrm{JS}(p,q) \leq V(p,q) \leq 2 \tag{13}$$

Those two different ways to symmetrize the KL divergence J (S_F for Shannon entropy) and JS are related by the following inequality

$$J(p,q) \geq 4\,\mathrm{JS}(p,q) \geq 0. \tag{14}$$

The Jensen-Shannon divergence can be interpreted as a measure of *diversity* of the source distributions p and q to the *average* distribution $\frac{p+q}{2}$. In the same vein, consider the following Bregman symmetrization [12,13]:

$$J_F(p,q) = \frac{B_F(p, \frac{p+q}{2}) + B_F(q, \frac{p+q}{2})}{2} \tag{15}$$

$$= \frac{F(p) + F(q)}{2} - F\left(\frac{p+q}{2}\right) = J_F(q,p). \tag{16}$$

For d-dimensional multivariate data, we define the corresponding Jensen divergences coordinate-wise as follows:

$$J_F(p,q) = \sum_{i=1}^{d} J_F(p_i, q_i) = \sum_{i=1}^{d} \frac{F(p_i) + F(q_i)}{2} - F\left(\frac{p_i + q_i}{2}\right). \tag{17}$$

---

[2] We may enforce definiteness by assuming the distributions are mutually absolutely continuous to each others [8].

Jensen-Bregman divergences J_F are always finite ($0 \leq J_F < \infty$) on the domain $\mathcal{X}$ (because entropies F measuring uncertainties are finite quantities: $F(x) < \infty$). J_F denote the Bregman divergence of the source distributions to the average distributions. Another way to interpret this family of divergences is to say that the Jensen-Bregman divergence is the average of the (negative) entropies minus the (negative) entropy of the average. For the negative Shannon entropy, we find the celebrated Jensen-Shannon divergence. Those divergences are *not* translation-invariant, and we require F to be *strictly convex* since for linear generators $L(x) = \langle a, x \rangle + b$, one does not discriminate distributions (i.e., $J_L(p, q) = 0 \ \forall p, q$).

This family of divergences can be termed *Jensen-Bregman divergences*.[3] Since F is a strictly convex function, J_F is nonnegative and equal to zero if and only if $p = q$. Figure 2 gives a geometric interpretation of the divergence as the vertical distance between $(\frac{p+q}{2}, F(\frac{p+q}{2}))$ and the midpoint of the segment $[(p, F(p)), (q, F(q))]$. Positive-definiteness follows from the Jensen's inequality.

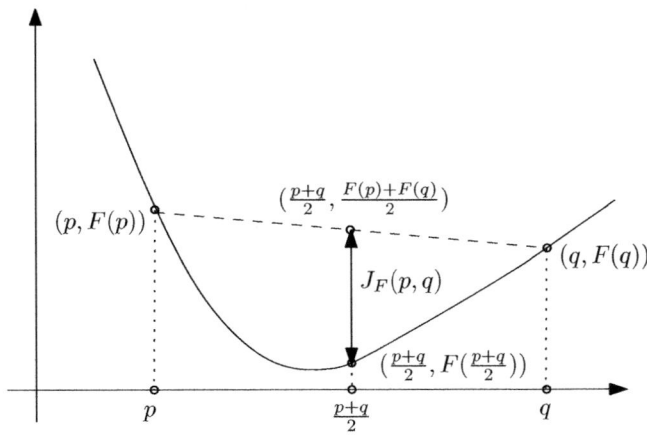

Fig. 2. Interpreting the Jensen-Bregman divergence as the vertical distance between the midpoint of segment $[(p, F(p)), (q, F(q))]$ and the midpoint of the graph plot $\left(\frac{p+q}{2}, F\left(\frac{p+q}{2}\right)\right)$

Note that Jensen-Bregman divergences are defined modulo affine terms $\langle a, x \rangle + b$. (Indeed, let $G(x) = F(x) + \langle a, x \rangle + b$, then one checks that $J_F(p, q) = J_G(p, q)$.) Thus we can choose coefficients $b = -F(0)$ and $a = -\nabla F(0)$ to fix unambiguously the generator. This means that the plot $(x, F(x))$ of the convex function touches the origin at its minimum value.

Jensen-Bregman divergences contain all generalized quadratic distances ($F(x) = \langle Qx, x \rangle$ for a positive definite matrix $Q \succ 0$), well-known in computer

---

[3] Or Burbea-Rao divergences [14]) or Jensen divergences. We prefer the term Jensen-Bregman because as we shall see (1) skew Jensen-Bregman divergences include Bregman divergences in the limit cases, and (2) they are obtained by symmetrizing Bregman divergences *à la* Jensen-Shannon.

vision as the squared Mahalanobis distances (squared Euclidean distance obtained for $Q = I$, the identity matrix).

$$
\begin{aligned}
J_F(p, q) &= \frac{F(p) + F(q)}{2} - F\left(\frac{p+q}{2}\right) \\
&= \frac{2\langle Qp, p\rangle + 2\langle Qq, q\rangle - \langle Q(p+q), p+q\rangle}{4} \\
&= \frac{1}{4}(\langle Qp, p\rangle + \langle Qq, q\rangle - 2\langle Qp, q\rangle) \\
&= \frac{1}{4}\langle Q(p-q), p-q\rangle \\
&= \frac{1}{4}\|p - q\|_Q^2.
\end{aligned}
$$

It is well-known that the square root of the Jensen-Shannon divergence (using Shannon entropy generator $F(x) = -x \log x$) is a metric. However, the square root of Jensen-Bregman divergences are *not* always metric. A Jensen-Bregman divergence is said *separable* if it can be decomposed independently dimension-wise:

$$
F(x) = \sum_{i=1}^{d} f_i(x_i), \tag{18}
$$

with all f_i's strictly convex functions. Usually all f_i's are taken as an identical univariate function. For example, Shannon (convex) information (the negative of Shannon concave entropy) is defined as

$$
I(x) = -\sum_{i=1}^{d} x_i \log x_i, \tag{19}
$$

for x belonging to the $(d-1)$-dimensional simplex $\mathbb{S}_{d-1}$ of discrete probabilities ($\sum_{i=1}^{d} x_i = 1$). Shannon information can be extended to the non-normalized *positive measures* $\mathbb{P}_d$ by taking $I(x) = -\sum_{i=1}^{d} x_i \log x_i - x_i$.

Appendix A shows that those Jensen-Bregman distances encapsulate the class of statistical Bhattacharyya distances for a versatile family of probability measures called the exponential families. Let us now consider Jensen-Bregman Voronoi diagrams.

3   The Voronoi Diagram by Jensen Difference

We have shown that Jensen-Bregman divergences are an important class of distortion measures containing both the squared Euclidean/Mahalanobis distance (non-additive quadratic entropy [8]) and the Jensen-Shannon divergence (additive entropy [8]). Note that one key property of the Euclidean distance is that

$D(\lambda p, \lambda q) = \lambda D(p, q)$. That is, it is a distance function of *homogeneous degree* 1. A distance is said of homogeneous degree α if and only if

$$D(\lambda p, \lambda q) = \lambda^\alpha D(p, q). \tag{20}$$

Usually, Jensen-Bregman divergences are not homogeneous except for the following three *remarkable* generators:

– Burg entropy ($\alpha = 0$)

$$F(x) = -\log x, \quad J_F(p, q) = \log \frac{p + q}{2\sqrt{pq}} \tag{21}$$

  (logarithm of the ratio of the arithmetic mean over the geometric mean),
– Shannon entropy ($\alpha = 1$)

$$F(x) = x \log x, \quad J_F(p, q) = \frac{1}{2} \left(p \log \frac{2p}{p + q} + q \log \frac{2q}{p + q} \right) \tag{22}$$

– Quadratic entropy ($\alpha = 2$)

$$F(x) = x^2, \quad J_F(p, q) = \frac{1}{4}(p - q)^2 \tag{23}$$

Since Jensen-Bregman Voronoi diagrams include the ordinary Euclidean Voronoi diagram [3],[4] the complexity of those diagrams is at least the complexity of Euclidean diagrams [15,6]: namely, $\Theta(n^{\lceil \frac{d}{2} \rceil})$. In general, the complexity of Voronoi diagrams by an arbitrary distance function (under mild conditions) is at most $O(n^{d+\epsilon})$ for any $\epsilon > 0$, see [16,17]. Thus as the dimension increases there is a potential quadratic gap in the combinatorial complexity between the Euclidean and general distance function diagrams.

Let us analyze the class of Jensen-Bregman diagrams by studying the induced minimization diagram and characterizing the bisector structure.

3.1   Voronoi Diagrams as Minimization Diagrams

Given a point set $\mathcal{P} = \{p_1, ..., p_n\}$ of n sites,[5] the Jensen-Bregman Voronoi diagram partitions the space into elementary *Voronoi cells* such that the Voronoi cell $V(p_i)$ associated to site p_i is the loci of points closer to p_i than to any other point of $\mathcal{P}$ with respect to the Jensen-Bregman divergence:

$$V(p_i) = \{p \mid J_F(p, p_i) < J_F(p, p_j) \; \forall j \neq i\}. \tag{24}$$

---

[4] Since the Voronoi diagrams by any strictly monotonous increasing function of a distance coincides with the Voronoi diagrams of that distance, the squared Euclidean Voronoi diagram coincides with the ordinary Voronoi diagram.
[5] Without loss of generality, we assumed points distinct and in general position.

Fig. 3. The 2D Jensen-Burg Voronoi diagram of 4 points from the corresponding lower envelope of corresponding 3D functions

For each Voronoi site p_i, consider the following *anchored distance function* to that site:

$$D_i(x) = J_F(x, p_i) = \frac{F(p_i) + F(x)}{2} - F\left(\frac{p_i + x}{2}\right) \qquad (25)$$

Thus the Voronoi diagram amounts to a minimization diagram. This minimization task can be solved by computing the lower envelope of $(d + 1)$-dimensional functions $(x, D_i(x))$. The projection of the lower envelope (resp. upper envelope) yields the Jensen-Bregman Voronoi diagram (resp. farthest Jensen-Bregman Voronoi diagram). Figure 3 displays the lower envelope of four 3D functions for the Burg entropy generator (homogeneous degree $\alpha = 0$).

In general, besides the ordinary Euclidean case with $F(x) = x^2$, the equation of the bisector can be tricky to manipulate, even in the planar case. For example, consider the Burg entropy $(F(x) = -\log x)$. Using $\sum \log \leftrightarrow \log \prod$, the Burg bisector $B(p,q)$ for the corresponding separable Jensen-Burg distance can be written as:

$$B(p,q) : \prod_{i=1}^{d} \frac{p_i + x_i}{\sqrt{p_i}} = \prod_{i=1}^{d} \frac{q_i + x_i}{\sqrt{q_i}}, \qquad (26)$$

where $p = (p_1, ..., p_d)$ and $q = (q_1, ..., q_d)$ denote the coordinates of p and q, respectively.

We next concentrate on a *concave-convex structural property* of the Jensen-Bregman bisector. But first, we recall some prior work on Voronoi diagrams. The Voronoi diagrams with respect to convex functions has been studied [18]. However, note that Jensen-Bregman divergences are *not* necessarily convex.

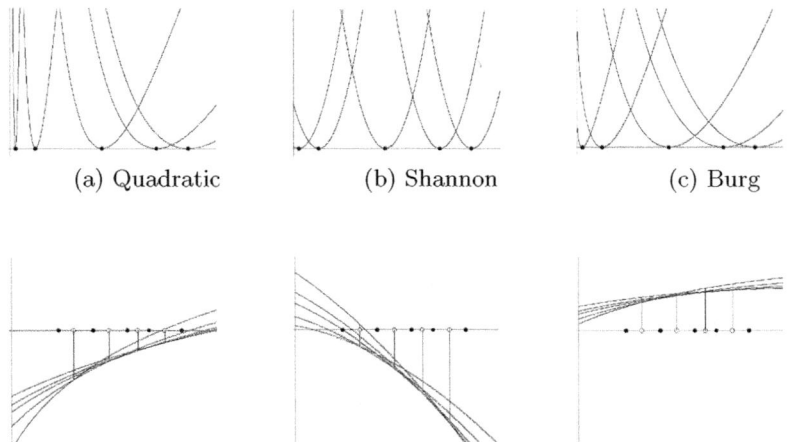

Fig. 4. (Top) 1D Voronoi diagram from the lower envelope of corresponding anchored distance functions for the (a) quadratic, (b) Shannon and (c) Burg entropies. (Bottom) minimum of the functions $D_i'(\cdot)$ (removing the common term $\frac{1}{2}F(x)$).

Indeed, without loss of generality, consider separable Jensen-Bregman divergences, and let us look at the second-order derivatives of a univariate Jensen-Bregman divergence. We have

$$D_p'(x) = J_F'(x, p) = \frac{1}{2}F'(x) - \frac{1}{2}F'\left(\frac{p+x}{2}\right) \tag{27}$$

and

$$D_p''(x) = J_F''(x, p) = \frac{1}{2}F''(x) - \frac{1}{4}F''\left(\frac{p+x}{2}\right). \tag{28}$$

This second-order derivative is *not* necessarily always strictly positive. For example, consider $F(x) = x^3$ on $\mathbb{R}^+$ (with $F''(x) = 6x$). We have $D_p''(x) = 3(x - \frac{p+x}{4})$; This is non-negative for $x \geq \frac{p}{3}$ only. That means that some Jensen-Bregman divergences are neither convex nor concave either. Another typical example is the Jensen-Burg entropy ($F(x) = -\log x$ and $F''(x) = 1/x^2$). Indeed, the anchored distance $J_F(x, p)$ at p is strictly convex if $x > p(1 + \sqrt{2})$ and strictly concave if $x < p(1 + \sqrt{2})$. However, Jensen-Shannon divergence (defined for generator $F(x) = x \log x$ with $F''(x) = 1/x$) is convex for all values on the positive orthant (positive measures $\mathbb{P}_2$): Indeed, $D_F''(x) = \frac{1}{2x} - \frac{1}{4}\frac{2}{p+x} = \frac{1}{2}(\frac{1}{x} - \frac{1}{p+x}) > 0$ for all $p > 0$ ($p \in \mathbb{P}$).

In general, a necessary condition is that F is strictly convex: Indeed, choose $x = p$ (for any arbitrary p), it comes that $D_p''(x) = \frac{1}{4}F''(p)$ that is positive if and only if $F''(p) > 0$. That is, it is required that F be strictly convex.

In the general case, J_F is convex if and only if its Hessian is positive definite:[6] $\nabla^2 J_F(\cdot, p) \succ 0 \; \forall p$:

$$\nabla^2 J_F = \begin{bmatrix} \frac{\partial^2 J_F(x,y)}{\partial x^2} & \frac{\partial^2 J_F(x,y)}{\partial x \partial y} \\ \frac{\partial^2 J_F(x,y)}{\partial x \partial y} & \frac{\partial^2 J_F(x,y)}{\partial y^2} \end{bmatrix} \tag{29}$$

$$= \begin{bmatrix} \frac{F''(x)}{2} - \frac{1}{4}F''(\frac{x+y}{2}) & -\frac{1}{4}F''(\frac{x+y}{2}) \\ -\frac{1}{4}F''(\frac{x+y}{2}) & \frac{F''(y)}{2} - \frac{1}{4}F''(\frac{x+y}{2}) \end{bmatrix} \succ 0 \tag{30}$$

Considering separable divergences J_F, the positive definiteness condition of the Hessian becomes

$$F''(x) > F''(\frac{x+y}{2}) - F''(x) \tag{31}$$

It follows that the Jensen-Shannon (separable) divergence ($F(x) = x \log x - x$, $F''(x) = \frac{1}{x}$) is a strictly convex distance function on the set of positive measures $\mathcal{X} = \mathbb{R}_{++}$ since $\frac{1}{x} > \frac{2}{x+y} - \frac{1}{x}$ for all $x, y > 0$.

Lemma 1. *Jensen-Bregman divergences are not necessarily strictly convex nor strictly concave distortion measures. Jensen-Shannon divergence is a strictly convex function on the set $\mathbb{P}_d$ of positive measures. Separable Jensen-Bregman divergences J_F on domain $\mathcal{X}^d$ are strictly convex distance functions if and only if $F''(x) > F''(\frac{x+y}{2}) - F''(x) > 0$ for $x, y \in \mathcal{X}$.*

Lihong [19,18] studied the Voronoi diagrams in 2D and 3D under a *translation-invariant* convex distance function (e.g., a polyhedral convex distance). Translation-invariant means that a convex object C gives a distance profile, and the distance between two points p and q is the smallest scaling factor so that a homothet of C centered at p touches q. Note that Jensen-Bregman divergences are not invariant under translation.

Recently, Dickerson et al. [20] studied the planar Voronoi diagram for *smoothed* separable convex distances. They show that provided that the functions of the minimization diagrams satisfy the constraint $f'''f' < (f'')^2$, then the 2D Voronoi diagram has linear complexity and can be computed using a randomized algorithm in $\tilde{O}(n \log n)$ time. In fact, in that case, the distance *level sets* $\{D_{p_i}(x) = l\}_l$ (iso-distance level) yield pseudo-circles, and the arrangement of bisectors are pseudo-lines. However, if the condition $f'''f' < (f'')^2$ fails, the 3D minimization diagram (and corresponding 2D Voronoi diagram) may have *quadratic* complexity. For example, choosing $f(x) = e^{x^2}$, and $F(x,y) = e^{x^2} + e^{y^2}$ yields potentially a quadratic complexity diagram.

Consider a d-dimensional finite point set $p_1, ..., p_n$, and let $p_{i,1}, ..., p_{i,d}$ denote the coordinates of point p_i for all $i \in \{1, ..., n\}$. We consider separable Jensen-Bregman divergences. Let $x_1, ..., x_d$ denote the coordinates of point x.

---

[6] A matrix M is said positive definite iff. $x^T M x > 0$ for all $x \neq 0$. A positive definite matrix has all its eigenvalues strictly positive, and hence the trace (sum of eigenvalues) and determinant (product of eigenvalues) are necessarily positive.

Since the term $\frac{F(x)}{2}$ are shared by all D_i's functions, we can remove it equivalently from all anchored distance functions. Therefore the minimization diagram $\min_i D_i(x)$ is equivalent to the minimization diagram of the functions

$$D_i'(x) = \frac{1}{2}F(p_i) - F\left(\frac{p_i + x}{2}\right), \tag{32}$$

or equivalently using separable generator by

$$D_i'(x) = \sum_{k=1}^{d} \frac{1}{2}F(p_{i,k}) - F\left(\frac{p_{i,k} + x_k}{2}\right). \tag{33}$$

This minimization diagram can be viewed as the lower envelope of n concave functions (entropy function $-F$) in dimension $d+1$. The *vertical shift* corresponds to a weight $F(p_i) = \sum_{k=1}^{d} F(p_{i,k})/2$. Let us write the equation of a bisector (p, q):

$$B(p,q) : \frac{F(p)}{2} - F\left(\frac{x+p}{2}\right) = \frac{F(q)}{2} - F\left(\frac{x+q}{2}\right) \tag{34}$$

$$: \sum_{k=1}^{d} \frac{F(p_k)}{2} - F\left(\frac{x_k + p_k}{2}\right) = \sum_{k=1}^{d} \frac{F(q_k)}{2} - F\left(\frac{x_k + q_k}{2}\right). \tag{35}$$

That is, we get

$$B(p,q) : \left(F\left(\frac{x+q}{2}\right) - F\left(\frac{x+p}{2}\right)\right) + \left(\frac{F(p)}{2} - \frac{F(q)}{2}\right) = 0 \tag{36}$$

$$: \sum_{k=1}^{d} \left(F\left(\frac{x_k + q_k}{2}\right) - F\left(\frac{x_k + p_k}{2}\right)\right) + \sum_{k=1}^{d} \left(\frac{F(p_k)}{2} - \frac{F(q_k)}{2}\right) = 0 \tag{37}$$

The bisector is thus interpreted as the sum of a *convex* function

$$\sum_{k=1}^{d} F\left(\frac{x_k + q_k}{2}\right) - \frac{F(p_k)}{2}$$

with a *concave* function

$$\sum_{k=1}^{d} -F\left(\frac{x_k + p_k}{2}\right) - \frac{F(q_k)}{2}.$$

The next section on centroidal Voronoi tessellations show how to handle this concave-convex structural property using a tailored optimization mechanism.

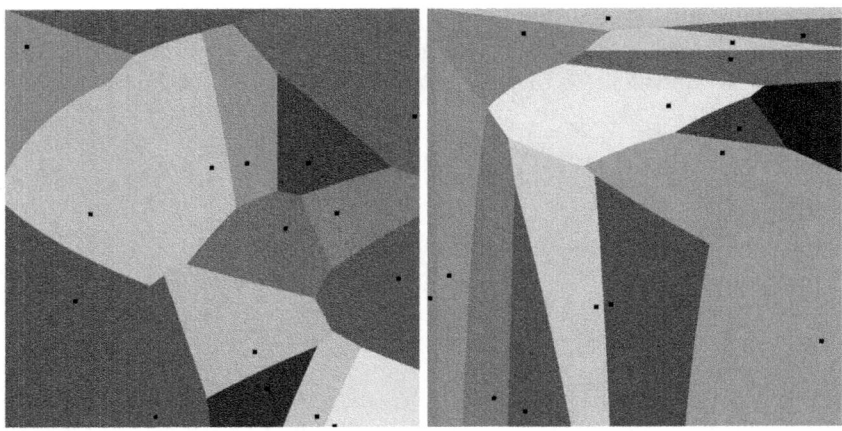

Fig. 5. (Left) The Jensen-Shannon Voronoi diagram for a set of 16 points (positive arrays denoting unnormalized probability distributions). (Right) The Jensen-Burg Voronoi diagram for the Burg entropy $F(x) = -\sum_{i=1}^{d} \log x_i$.

4   Jensen-Bregman Centroidal Voronoi Diagrams

The centroidal Voronoi diagram [9] (or centroidal Voronoi tesselation; CVT for short) is defined as follows: First, we fix a number of generators n, and a compact domain $\mathcal{X}$ (say, a unit square). Then we ask to find the locations of the generators so that the induced Voronoi cells have (approximately) more or less the same area. Figure 6(b) shows a CVT for a set of 16 points. A CVT is computed iteratively by first initializing the generators to arbitrary position (Figure 6(a)), and by iteratively relocating those generators to the center of mass of their Voronoi cell (Lloyd iteration). Proving that such a scheme is always converging is still a difficult open problem of computational geometry [9], although that in practice it is known to converges quickly. Instead of relocating to the center of mass (barycenter) according to a uniform density distribution (i.e., to the centroid or geometric center of the cell), we can relocate those generators to the barycenter of the cell according to an underlying non-uniform density distribution. This is one technique commonly used in non-photorealistic rendering (NPR) called stippling [21], the art of pointillism. Figure 6(c) is a source image representing the underlying grey intensity distribution. Figure 6(d) is the stippling effect produced by computing a CVT with respect to the underlying image density.

To extend the centroidal Voronoi tesselations to Jensen-Bregman divergences, we first need to define centroids (and barycenters) with respect to this dissimilarity measure. Consider a finite set of points $\mathcal{P} = \{p_1, ..., p_n\}$. The *Jensen-Bregman centroid* is defined as the minimizer of the average Jensen-Bregman distance:

$$c^* = \arg\min_c \sum_{i=1}^{n} \frac{1}{n} J_F(p_i, c) \tag{38}$$

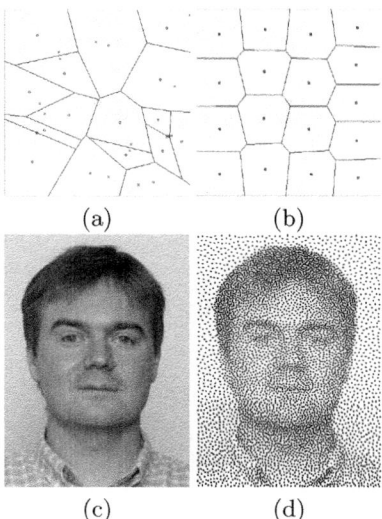

(a)   (b)

(c)   (d)

Fig. 6. (Top) Centroidal Voronoi diagram of 16 sites: (a) initialization, and (b) after a few iterations. (Bottom) Application to image stippling: (c) grey density image and (d) centroidal Voronoi diagram according to the underlying density.

By choosing $F(x) = \langle x, x \rangle$, we minimize the sum of the squared Euclidean distances, and find the usual Euclidean centroid.[7] Similarly, the barycenter is defined with respect to (normalized) weights (interpreted as point multiplicities):

$$c^* = \arg\min_c \sum_{i=1}^{n} w_i J_F(p_i, c) = \arg\min_c L(c) \tag{39}$$

Using the structure of the optimization problem, we can use the Convex-ConCave Procedure [22] (CCCP), a general purpose loss function minimizer. Indeed, we can always decompose an arbitrary (non-convex) function as the sum of a *convex* function and *concave* function (or the difference of two convex functions), provided that the Hessian of the loss function function is bounded:[8]

$$L(c) = L_{\text{convex}}(c) + L_{\text{concave}}(c). \tag{40}$$

For the Jensen-Bregman centroid, this decomposition is given *explicitly* as follows:

$$L_{\text{convex}}(c) = \frac{F(c)}{2} \tag{41}$$

---

[7] If instead of minimizing the squared Euclidean distance, we consider the Euclidean distance, we do not get closed-form solution. This is the so-called Fermat-Weber point.

[8] Always bounded on a compact.

$$L_{\text{concave}}(c) = -\sum_{i=1}^{n} F\left(\frac{p_i + c}{2}\right), \qquad (42)$$

since the sum of concave functions is a concave function. The CCCP approach consists in setting the gradient to zero: $\nabla_x L(x) = 0$. We get

$$\frac{1}{2}\nabla F(x) - \sum_{i=1}^{n} \frac{w_i}{2} \nabla F\left(\frac{x + p_i}{2}\right) = 0. \qquad (43)$$

That is, we need to solve equivalently for

$$\nabla F(x) = \sum_{i=1}^{n} w_i \nabla F\left(\frac{x + p_i}{2}\right) \qquad (44)$$

Since F is *strictly* convex and differentiable, we have ∇F that is *strictly* monotone increasing (because the Hessian is positive definite, i.e. $\nabla^2 F \succ 0$), and the reciprocal gradient ∇F^{-1} is well-defined.

Thus solving Eq. 44 amounts to solve for

$$x = \nabla F^{-1}\left(\sum_{i=1}^{n} w_i \nabla F\left(\frac{x + p_i}{2}\right)\right) \qquad (45)$$

Starting from an arbitrary initial value x_0 of x (say, the Euclidean center of mass), the optimization proceeds iteratively as follows:

$$x_{t+1} = \nabla F^{-1}\left(\sum_{i=1}^{n} w_i \nabla F\left(\frac{x_t + p_i}{2}\right)\right). \qquad (46)$$

For the Jensen-Shannon (separable) divergence defined on positive measures, we thus update the centroid independently on each coordinate by

$$x_{t+1} = \frac{n}{2\sum_{i=1}^{n} \frac{1}{x_t + p_i}}.$$

The CCCP algorithm guarantees monotonicity and convergence to a local minimum or saddle point. For the Jensen-Shannon divergence, this local minimum yields the global minimum since the distance function is strictly convex. Note that for the quadratic entropy $F(x) = \langle x, x \rangle$, we get a closed-form solution (i.e., the center of mass).

However, in general, we do not obtain a closed-form solution, and can only estimate the Jensen-Bregman barycenters up to some arbitrary precision. Thus, we do not have closed-form solutions of computing the Jensen-Bregman centroid of a Voronoi cell. Nevertheless, we can bypass this by finely discretizing the domain, and estimating the centroids using the above generalized mean interations. We implemented and computed the centroidal Jensen-Bregman Voronoi diagrams following such a scheme. Figure 7 presents the Jensen-Bregman centroidal Voronoi tesselations obtained, assuming an underlying uniform density.

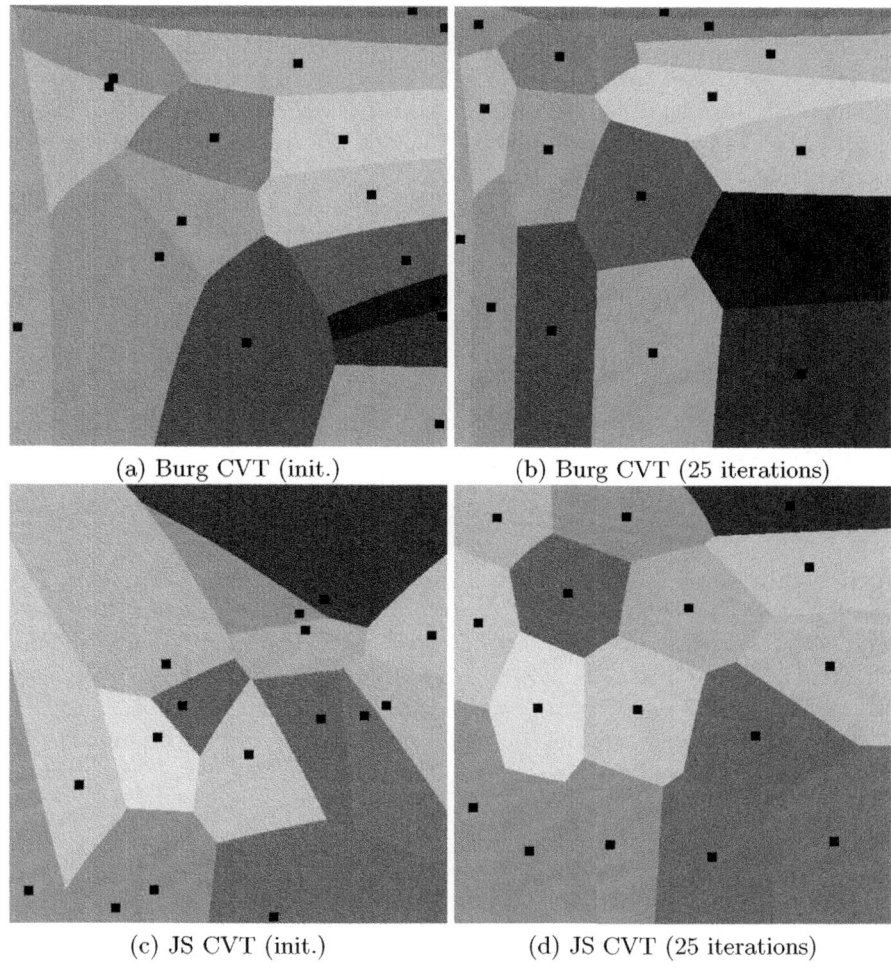

(a) Burg CVT (init.)   (b) Burg CVT (25 iterations)

(c) JS CVT (init.)   (d) JS CVT (25 iterations)

Fig. 7. Centroidal Jensen-Bregman Voronoi diagrams for the Burg and Shannon (JS) entropies. CVTs provide a way to sample uniformly space according to the underlying distance.

The following section shows how to extend those results to matrix-based data sets.

5   Matrix-Based Jensen-Bregman Divergences

A recent trend in data processing is to consider matrix-valued data sets, where each datum is not handled as a scalar or vector but rather as a 2D matrix. Such kind of data sets occurs frequently in many science and engineering application areas where they are termed *tensors*: Gaussian covariance matrices [23] in sound processing, elasticity tensors in mechanical engineering [24], polarimetric synthetic aperture radar [25], diffusion tensor imaging (DTI) [26], kernel-based machine learning [27], etc. Those matrices M are symmetric and positive definite (SPD) $M \succ 0 : \forall x \in \mathbb{R}^d \neq 0, x^T M x > 0$, and can be visualized as ellipsoids: Each matrix M, also called a tensor, is geometrically represented by an ellipsoid $\{x \mid x^T M x = 1\}$. Let us denote by Sym_{++} the open convex cone of symmetric positive definite matrices [28].

We build a matrix-based Jensen-Bregman divergence from a convex generator $F : \mathrm{Sym}_{++} \to \mathbb{R}^+$ as follows:

$$J_F(P, Q) = \frac{F(P) + F(Q)}{2} - F\left(\frac{P + Q}{2}\right) \geq 0, \tag{47}$$

with equality if and only if $P = Q$.

Typical matrix-based convex generators are :

- $F(X) = \mathrm{tr}(X^T X)$: the quadratic matrix entropy,
- $F(X) = -\log \det X$: the matrix Burg entropy, and
- $F(X) = \mathrm{tr}(X \log X - X)$: the von Neumann entropy.

Interestingly, those generators are invariant by a permutation matrix P, ie. $F(PX) = F(P)$. Choosing $F(X) = \mathrm{tr}(X \log X - X)$, we get the *Jensen-von Neumann* divergence, the matrix counterpart of the celebrated Jensen-Shannon divergence. A $d \times d$-dimensional SPD matrix is represented by $D = \frac{d(d+1)}{2}$ matrix entries. Thus 2×2-matrices are encoded by $D = 3$ scalar values.

The matrix-based centroidal Voronoi tesselation requires to compute the SPD centroid (of discretized matrices $M_1, ..., M_n$) using the CCCP iterative optimization technique mentioned in Eq. 45:

$$C_{t+1} = \nabla F^{-1} \left(\sum_{i=1}^{n} \frac{1}{n} \nabla F \left(\frac{M_i + C_t}{2} \right) \right). \tag{48}$$

Table 1 reports the matrix gradients and reciprocal gradients for common matrix-based generators.

We now present a generalization of those Voronoi diagrams and centroidal Voronoi tessellations when skewing the divergences. We shall see that skewing the Jensen-Bregman divergences allows one to generalize Bregman Voronoi diagrams [6].

Table 1. Characteristics of convex matrix-based functional generators

Entropy name	$F(X)$	$\nabla F(X)$	$(\nabla F)^{-1}(X)$
Quadratic	$\frac{1}{2}\mathrm{tr}XX^T$	X	X
log det	$-\log \det X$	$-X^{-1}$	$-X^{-1}$
von Neumann	$\mathrm{tr}(X \log X - X)$	$\log X$	$\exp X$

6   Skew Jensen-Bregman Voronoi Diagrams

Recall that Jensen-Bregman divergences are divergences defined by a Jensen gap built from a convex generator function. Instead of taking the mid-point (for value $\alpha = \frac{1}{2}$), we may consider skewing the divergence by introducing a parameter α as follows:

$$J_F^{(\alpha)} \; : \; \mathcal{X} \times \mathcal{X} \to \mathbb{R}^+$$
$$J_F^{(\alpha)}(p, q) = \alpha F(p) + (1 - \alpha)F(q) - F(\alpha p + (1 - \alpha)q)$$

We consider the open interval $(0, 1)$ since otherwise the divergence has no discriminatory power (indeed, for $\alpha \in \{0, 1\}, J_F^{(\alpha)}(p, q) = 0, \; \forall p, q$). Although skewed divergences are asymmetric $J_F^{(\alpha)}(p, q) \neq J_F^{(\alpha)}(q, p)$, we can swap arguments by replacing α by $1 - \alpha$:

$$\begin{aligned} J_F^{(\alpha)}(p, q) &= \alpha F(p) + (1 - \alpha)F(q) - F(\alpha p + (1 - \alpha)q) \\ &= J_F^{(1-\alpha)}(q, p) \end{aligned} \tag{49}$$

Figure 8 illustrates the divergence as a Jensen gap induced by the convex generator.

Those skew Burbea-Rao divergences are similarly found using a skew Jensen-Bregman counterpart (the gradient terms $\nabla F(\alpha p + (1 - \alpha)q)$ perfectly cancel in the sum of skew Bregman divergences):

$$\alpha B_F(p, \alpha p + (1 - \alpha)q) + (1 - \alpha)B_F(q, \alpha p + (1 - \alpha)q) = J_F^{(\alpha)}(p, q) \tag{50}$$

In the limit cases, $\alpha \to 0$ or $\alpha \to 1$, we have $J_F^{(\alpha)}(p, q) \to 0 \; \forall p, q$. That is, those divergences loose their discriminatory power at extremities. However, we show that those skew Burbea-Rao divergences tend *asymptotically* to Bregman divergences [29]:

$$B_F(p, q) = \lim_{\alpha \to 0} \frac{1}{\alpha} J_F^{(\alpha)}(p, q) \tag{51}$$

$$B_F(q, p) = \lim_{\alpha \to 1} \frac{1}{1 - \alpha} J_F^{(\alpha)}(p, q) \tag{52}$$

Let us consider the Voronoi diagram of a finite point set $p_1, ..., p_n$ with respect to $J_F'^{(\alpha)}$, a *normalized* skew Jensen difference that matches exactly Bregman or reverse Bregman divergences in limit cases:

$$J_F'^{(\alpha)}(p, q) = \frac{1}{\alpha(1 - \alpha)} J_F^{(\alpha)}(p, q)$$

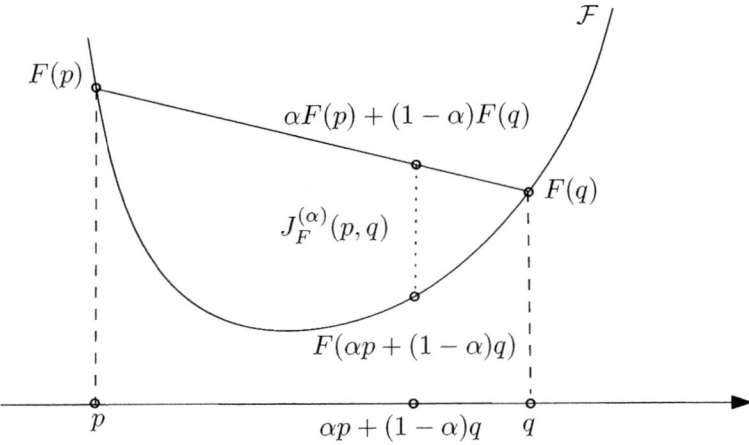

Fig. 8. Skew Jensen-Bregman divergence defined as a Jensen gap induced by a convex generator

The right-sided Voronoi cell associated to site p_i is defined as

$$V_\alpha(p_i) = \{x \mid J'^{(\alpha)}_F(p_i, x) \leq J'^{(\alpha)}_F(p_j, x) \forall j\} \qquad (53)$$

Similarly, the left-sided Voronoi cell

$$V'_\alpha(p_i) = \{x \mid J'^{(\alpha)}_F(x, p_i) \leq J'^{(\alpha)}_F(x, p_j) \forall j\} \qquad (54)$$

is obtained from the right-sided Voronoi cell by changing parameter α to $1 - \alpha$:

$$V'_\alpha(p_i) = V_{1-\alpha}(p_i). \qquad (55)$$

Thus we restrict ourselves to the right-sided Voronoi cells.

The bisector B of points p_i and p_j is defined by the non-linear equation:

$$B : \alpha(F(p_i) - F(p_j)) + F(\alpha p_j + (1 - \alpha)x) - F(\alpha p_i + (1 - \alpha)x) = 0 \qquad (56)$$

Note that for $\alpha \to 0$ or $\alpha \to 1$, using Gâteaux[9] derivatives [29], we find a bisector either linear in x or in its gradient with respect to the generator (i.e, $\nabla F(x)$). Namely, the (normalized) skew Jensen-Bregman Voronoi diagrams become a regular Bregman Voronoi diagram [6].

Figure 9 depicts several skew left-sided/right-sided Jensen-Bregman Voronoi diagrams. Observe that for $\alpha \in \{0, 1\}$, one of the two sided types of diagrams become affine (meaning bisectors are hyperplanes) since they become a sided Bregman Voronoi diagram [6].

---

[9] We assume that $\lim_{\lambda \to 0} \frac{F(x + \lambda p) - F(x)}{\lambda}$ exists and is equal to the Gâteaux derivative: $\langle p, \nabla F(x) \rangle$.

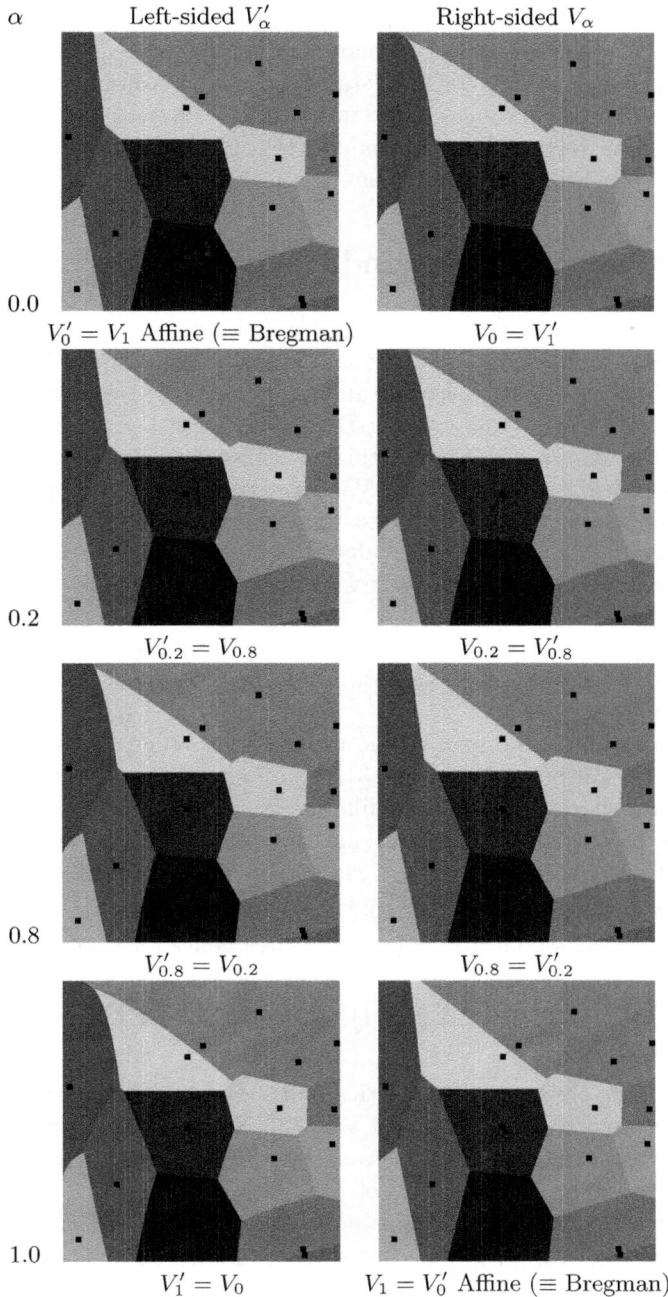

Fig. 9. Skew Jensen-Bregman Voronoi diagrams for various α parameters (Jensen-Shannon divergence). Observe that $V_\alpha = V'_{1-\alpha}$. In the extremal cases $\alpha = 0$ and $\alpha = 1$, the skew Jensen-Bregman diagrams amount to Bregman or reversed Bregman Voronoi diagrams. Note that the left-sided $\alpha = 0$ and right-sided $\alpha = 1$ are affine diagrams since they amount to compute Bregman Voronoi diagrams.

Dealing with skew Jensen-Bregman Voronoi diagrams is interesting for two reasons: (1) it generalizes the Bregman Voronoi diagrams [6] obtained in limit cases, and (2) it allows to consider statistical Voronoi diagrams following [29]. Indeed, consider the parameters of statistical distributions as input, the skew Jensen-Bregman Voronoi diagram amounts to compute equivalently a skew Bhattacharrya Voronoi diagram. Details are left in [29].

7   Concluding Remarks and Discussion

We have introduced a new class of information-theoretic divergences called (skew) Jensen-Bregman divergences that encapsulates both the Jensen-Shannon divergence and the squared Euclidean distance. We showed that those divergences are used when symmetrizing Bregman divergences and computing the Bhattacharyya distance of distributions belonging to the same statistical exponential families (see Appendix). We have studied geometric characteristics of the bisectors. We then introduced the notion of Jensen-Bregman centroid, and described an efficient iterative algorithm to estimate it using the concave-convex optimization framework. This allows one to compute Jensen-Bregman centroidal Voronoi tessellations. We showed how to extend those results to matrix-based Jensen-Bregman divergences, including the Jensen-von Neumann divergence that plays a role in Quantum Information Theory [30] (QIT) dealing with density matrices.

The differential Riemannian geometry induced by such a class of Jensen gaps was studied by Burbea and Rao [14,31] who built quadratic differential metrics on probability spaces using Jensen differences.

The Jensen-Shannon divergence is an instance of a broad class of divergences called the *Csiszár f-divergences*. A f-divergence I_f is a statistical measure of dissimilarity defined by the functional $I_f(p,q) = \int p(x) f(\frac{q(x)}{p(x)}) dx$. It turns out that the Jensen-Shannon divergence is a f-divergence for generator

$$f(x) = \frac{1}{2} \left((x+1) \log \frac{2}{x+1} + x \log x \right). \tag{57}$$

The class of f-divergences preserves the information monotonicity [32], and their differential geometry was studied by Vos [33]. Note that the squared Euclidean distance *does not* belong to the class of f-divergences although it is a Jensen-Bregman divergence.

To conclude, skew Jensen-Bregman Voronoi diagrams extend naturally Bregman Voronoi diagrams [6], but are not anymore affine diagrams in general. Those diagrams allow one to equivalently compute Voronoi diagrams of statistical distributions with respect to (skew) Bhattacharrya distances. This perspective further opens up the field of computational geometry to statistics and decision theory under uncertainty, where Voronoi bisectors denote decision boundaries [34].

Additional material on Jensen-Bregman divergences including videos are available online at:

$$\left| \texttt{www.informationgeometry.org/JensenBregman/} \right|$$

Acknowledgments. The authors would like to thank the anonymous reviewers for their valuable comments and suggestions to improve the quality of the paper. The authors gratefully acknowledge financial support from French funding agency ANR (contract GAIA 07-BLAN-0328-01) and Sony Computer Science Laboratories, Inc.

A   Bhattacharyya Distances as Jensen-Bregman Divergences

This appendix proves that (skew) Jensen-Bregman divergences occurs when computing the (skew) Bhattacharrya distance of statistical parametric distributions belonging to the same probability family, called an exponential family. It follows that statistical Voronoi diagrams of members of the same exponential family with respect to the Bhattacharrya distance amount to compute equivalently Jensen-Bregman Voronoi diagrams on the corresponding measure parameters.

A.1   Statistical Exponential Families

Many usual statistical parametric distributions $p(x; \lambda)$ (e.g., Gaussian, Poisson, Bernoulli/multinomial, Gamma/Beta, etc.) share common properties arising from their common canonical decomposition of probability distribution:

$$p(x; \lambda) = p_F(x; \theta) = \exp\left(\langle t(x), \theta \rangle - F(\theta) + k(x) \right). \tag{58}$$

Those distributions[10] are said to belong to the exponential families (see [35] for a tutorial). An exponential family is characterized by its *log-normalizer* $F(\theta)$, and a distribution in that family is indexed by its *natural parameter* θ belonging to the *natural space* Θ. The log-normalizer F is strictly convex, infinitely differentiable (C^∞), and can also be expressed using the source coordinate system λ using the bijective map $\tau : \Lambda \to \Theta$ that converts parameters from the source coordinate system to the natural coordinate system:

$$F(\theta) = (F \circ \tau)(\lambda) = F_\lambda(\lambda). \tag{59}$$

The vector $t(x)$ denote the *sufficient statistics*, that is the set of linear independent functions that allows to concentrate without any loss all information about the parameter θ carried in the i.i.d. observation sample $x_1, x_2, ..., $. The

---

[10] The distributions can either be discrete or continuous. We do not introduce the framework of probability measures here so as to not to burden the paper.

inner product $\langle p, q \rangle$ is a dot product $\langle p, q \rangle = p^T q$ for vectors. Finally, $k(x)$ represents the carrier measure according to the counting measure or the Lebesgue measure. Decompositions for most common exponential family distributions are given in [35]. To give one simple example, consider the family of Poisson distributions with probability mass function:

$$p(x; \lambda) = \frac{\lambda^x}{x!} \exp(-\lambda), \tag{60}$$

for $x \in \mathbb{N}^*$ a non-negative integer. Poisson distributions are univariate exponential families ($x \in \mathbb{N}$) of order 1 (i.e., a single parameter λ). The canonical decomposition yields

- the sufficient statistic $t(x) = x$,
- $\theta = \tau(\lambda) = \log \lambda$, the natural parameter (and $\tau^{-1}(\theta) = \exp \theta$),
- $F(\theta) = \exp \theta$, the log-normalizer,
- and $k(x) = -\log x!$ the carrier measure (with respect to the counting measure).

A.2   Bhattacharyya Distance

For arbitrary probability distributions $p(x)$ and $q(x)$ (parametric or not), we measure the amount of overlap between those distributions using the Bhattacharyya coefficient [36]:

$$B_c(p, q) = \int \sqrt{p(x)q(x)} \mathrm{d}x, \tag{61}$$

where the integral is understood to be multiple if x is multivariate. Clearly, the Bhattacharyya coefficient measures the *affinity* between distributions [37], and falls in the unit range: $0 \le B_c(p, q) \le 1$. In fact, we may interpret this coefficient geometrically by considering $\sqrt{p(x)}$ and $\sqrt{q(x)}$ as unit vectors (eventually in infinite-dimensional spaces). The Bhattacharyya coefficient is then the dot product, the cosine of the angle made by the two unit vectors. The Bhattacharyya distance $B : \mathcal{X} \times \mathcal{X} \to \mathbb{R}^+$ is derived from its coefficient [36] as

$$B(p, q) = -\ln B_c(p, q). \tag{62}$$

Although the Bhattacharyya distance is symmetric, it is not a metric because it fails the triangle inequality. For distributions belonging to the *same* exponential family, it turns out that the Bhattacharyya distance is always available in closed-form. Namely, the Bhattacharyya distance on probability distributions belonging to the same exponential family is equivalent to a Jensen-Bregman divergence defined for the log-normalizer of the family applied on the natural parameters. This result is not new [38] but seems to have been rediscovered a number of times [39,40]. Let us give a short but insightful proof.

Proof. Consider $p = p_F(x; \theta_p)$ and $q = p_F(x; \theta_q)$ two members of the same exponential families $\mathcal{E}_F$ with natural parameters θ_p and θ_q, respectively. Let us manipulate the Bhattacharyya coefficient $B_c(p, q) = \int \sqrt{p(x)q(x)}dx$:

$$
= \int \exp\left(\left\langle t(x), \frac{\theta_p + \theta_q}{2}\right\rangle - \frac{F(\theta_p) + F(\theta_q)}{2} + k(x)\right) dx
$$

$$
= \int \exp\left(\left\langle t(x), \frac{\theta_p + \theta_q}{2}\right\rangle - F\left(\frac{\theta_p + \theta_q}{2}\right) + k(x) + \right.
$$
$$
\left. F\left(\frac{\theta_p + \theta_q}{2}\right) - \frac{F(\theta_p) + F(\theta_q)}{2}\right) dx
$$

$$
= \exp\left(F\left(\frac{\theta_p + \theta_q}{2}\right) - \frac{F(\theta_p) + F(\theta_q)}{2}\right),
$$

since $\int p_F(x; \frac{\theta_p + \theta_q}{2})dx = 1$. We deduce from $B(p, q) = -\ln B_c(p, q)$ that

$$
B(p_F(x; \theta_p), p_F(x; \theta_q)) = \frac{F(\theta_p) + F(\theta_q)}{2} - F\left(\frac{\theta_p + \theta_q}{2}\right) \tag{63}
$$

It follows that the Bhattacharyya distance for members of the same exponential family is equivalent to a Jensen-Bregman divergence induced by the log-normalizer on the corresponding natural parameters:

$$
B(p_F(x; \theta_p), p_F(x; \theta_q)) = J_F(\theta_p; \theta_q), \tag{64}
$$

with the Jensen-Bregman divergence defined as the following Jensen difference [41]:

$$
J_F(p; q) = \frac{F(p) + F(q)}{2} - F\left(\frac{p + q}{2}\right) \tag{65}
$$

For Poisson distributions, we end up with the following Bhattacharyya distance

$$
\begin{aligned}
B(p_F(x; \theta_p), p_F(x; \theta_q)) &= J_F(\theta_p, \theta_q) \\
&= J_F(\log \lambda_p, \log \lambda_q), \\
&= \frac{\lambda_p + \lambda_q}{2} - \exp\frac{\log \lambda_p + \log \lambda_q}{2}, \\
&= \frac{\lambda_p + \lambda_q}{2} - \sqrt{\lambda_p \lambda_q} \\
&= \frac{1}{2}(\sqrt{\lambda_p} - \sqrt{\lambda_q})^2 \tag{66}
\end{aligned}
$$

Exponential families in statistics are mathematically convenient once again. Indeed, the relative entropy of two distributions belonging to the same exponential family, is equal to the Bregman divergence defined for the log-normalizer on swapped natural parameters [6]: $\mathrm{KL}(p_F(x; \theta_p), p_F(x; \theta_q)) = B_F(\theta_q, \theta_p)$.

For skew divergences, we consider the Chernoff divergences

$$C_\alpha(p, q) = -\ln \int p^\alpha(x) q^{1-\alpha}(x) \mathrm{d}x \tag{67}$$

defined for some α (and generalizing the Bhattacharyya divergence for $\alpha = \frac{1}{2}$). The Chernoff α-divergence amounts to compute a weighted asymmetric Jensen-Bregman divergence:

$$C_\alpha(p_F(x; \theta_p), p_F(x; \theta_q)) = J_F^\alpha(\theta_p, \theta_q) \tag{68}$$
$$= \alpha F(\theta_p) + (1-\alpha)F(\theta_q) - F(\alpha\theta_p + (1-\alpha)\theta_q) \tag{69}$$

References

1. Nielsen, F., Nock, R.: Jensen-Bregman Voronoi diagrams and centroidal tessellations. In: Proceedings of the 2010 International Symposium on Voronoi Diagrams in Science and Engineering (ISVD), pp. 56–65. IEEE Computer Society, Washington, DC (2010)
2. Okabe, A., Boots, B., Sugihara, K., Chiu, S.N.: Spatial tessellations: Concepts and applications of Voronoi diagrams. In: Probability and Statistics, 2nd edn., 671 pages. Wiley, NYC (2000)
3. de Berg, M., Cheong, O., van Kreveld, M., Overmars, M.: Computational Geometry: Algorithms and Applications, 3rd edn. Springer, Heidelberg (2008)
4. Lee, D.T.: Two-dimensional Voronoi diagrams in the L_p-metric. Journal of the ACM 27, 604–618 (1980)
5. Chew, L.P., Dyrsdale III, R.L.S.: Voronoi diagrams based on convex distance functions. In: Proceedings of the First Annual Symposium on Computational Geometry, SCG 1985, pp. 235–244. ACM, New York (1985)
6. Boissonnat, J.D., Nielsen, F., Nock, R.: Bregman Voronoi diagrams. Discrete and Computational Geometry 44(2), 281–307 (2010)
7. Lin, J.: Divergence measures based on the Shannon entropy. IEEE Transactions on Information Theory 37, 145–151 (1991)
8. Cover, T.M., Thomas, J.A.: Elements of information theory. Wiley-Interscience, New York (1991)
9. Du, Q., Faber, V., Gunzburger, M.: Centroidal voronoi tessellations: Applications and algorithms. SIAM Rev. 41, 637–676 (1999)
10. Jeffreys, H.: An invariant form for the prior probability in estimation problems. Proceedings of the Royal Society of London 186, 453–461 (1946)
11. Reid, M.D., Williamson, R.C.: Generalised Pinsker inequalities. CoRR abs/0906.1244 (2009); published at COLT 2009
12. Chen, P., Chen, Y., Rao, M.: Metrics defined by Bregman divergences: Part I. Commun. Math. Sci. 6, 9915–9926 (2008)
13. Chen, P., Chen, Y., Rao, M.: Metrics defined by Bregman divergences: Part II. Commun. Math. Sci. 6, 927–948 (2008)
14. Burbea, J., Rao, C.R.: On the convexity of some divergence measures based on entropy functions. IEEE Transactions on Information Theory 28, 489–495 (1982)

15. Chazelle, B.: An optimal convex hull algorithm in any fixed dimension. Discrete & Computational Geometry 10, 377–409 (1993)
16. Aurenhammer, F.: Voronoi diagrams—a survey of a fundamental geometric data structure. ACM Comput. Surv. 23, 345–405 (1991)
17. Sharir, M., Agarwal, P.K.: Davenport-Schinzel Sequences and their Geometric Applications. Cambridge University Press, New York (2010)
18. Icking, C., Ha, L.: A tight bound for the complexity of Voronoi diagrams under polyhedral convex distance functions in 3d. In: STOC 2001: Proceedings of the Thirty-Third Annual ACM Symposium on Theory of Computing, pp. 316–321. ACM, New York (2001)
19. Ma, L.: Bisectors and Voronoi diagrams for convex distance functions, PhD thesis (2000)
20. Dickerson, M., Eppstein, D., Wortman, K.A.: Dilation, smoothed distance, and minimization diagrams of convex functions, arXiv 0812.0607
21. Balzer, M., Schlömer, T., Deussen, O.: Capacity-constrained point distributions: a variant of Lloyd's method. ACM Trans. Graph. 28 (2009)
22. Yuille, A., Rangarajan, A.: The concave-convex procedure. Neural Computation 15, 915–936 (2003)
23. Arshia Cont, S.D., Assayag, G.: On the information geometry of audio streams with applications to similarity computing. IEEE Transactions on Audio, Speech and Language Processing 19 (2011) (to appear)
24. Cowin, S.C., Yang, G.: Averaging anisotropic elastic constant data. Journal of Elasticity 46, 151–180 (1997), doi:10.1023/A:1007335407097
25. Wang, Y.H., Han, C.Z.: Polsar image segmentation by mean shift clustering in the tensor space. Acta Automatica Sinica 36, 798–806 (2010)
26. Xie, Y., Vemuri, B.C., Ho, J.: Statistical Analysis of Tensor Fields. In: Jiang, T., Navab, N., Pluim, J.P.W., Viergever, M.A. (eds.) MICCAI 2010. LNCS, vol. 6361, pp. 682–689. Springer, Heidelberg (2010)
27. Tsuda, K., Rätsch, G., Warmuth, M.K.: Matrix exponentiated gradient updates for on-line learning and bregman projection. Journal of Machine Learning Research 6, 995–1018 (2005)
28. Bhatia, R., Holbrook, J.: Riemannian geometry and matrix geometric means. Linear Algebra and its Applications 413, 594–618 (2006); Special Issue on the 11th Conference of the International Linear Algebra Society, Coimbra (2004)
29. Nielsen, F., Boltz, S.: The Burbea-Rao and Bhattacharyya centroids. IEEE Transactions on Information Theory (2010)
30. Nielsen, M.A., Chuang, I.L.: Quantum computation and quantum information. Cambridge University Press, New York (2000)
31. Burbea, J., Rao, C.R.: On the convexity of higher order Jensen differences based on entropy functions. IEEE Transactions on Information Theory 28, 961–963 (1982)
32. Csiszár, I.: Information theoretic methods in probability and statistics
33. Vos, P.: Geometry of f-divergence. Annals of the Institute of Statistical Mathematics 43, 515–537 (1991)
34. Hastie, T., Tibshirani, R., Friedman, R.: Elements of Statistical Learning Theory. Springer, Heidelberg (2002)
35. Nielsen, F., Garcia, V.: Statistical exponential families: A digest with flash cards (2009) arXiv.org:0911.4863
36. Bhattacharyya, A.: On a measure of divergence between two statistical populations defined by their probability distributions. Bulletin of Calcutta Mathematical Society 35, 99–110 (1943)

37. Matusita, K.: Decision rules based on the distance, for problems of fit, two samples, and estimation. Annal of Mathematics and Statistics 26, 631–640 (1955)
38. Huzurbazar, V.S.: Exact forms of some invariants for distributions admitting sufficient statistics. Biometrika 42, 533–573 (1955)
39. Kailath, T.: The divergence and Bhattacharyya distance measures in signal selection. IEEE Transactions on Communications [legacy, pre - 1988] 15, 52–60 (1967)
40. Jebara, T., Kondor, R.: Bhattacharyya and expected likelihood kernels. In: 16th Annual Conference on Learning Theory and 7th Kernel Workshop, COLT/Kernel, p. 57 (2003)
41. Sahoo, P.K., Wong, A.K.C.: Generalized Jensen difference based on entropy functions. Kybernetika, 241–250 (1988)

Continuous-Time Moving Network Voronoi Diagram

Chenglin Fan[1,2], Jun Luo[1,*], and Binhai Zhu[3,**]

[1] Shenzhen Institutes of Advanced Technology,
Chinese Academy of Sciences, Shenzhen, China
[2] School of Information Science and Engineering,
Central South University, Changsha, China
[3] Department of Computer Science,
Montana State University, Bozeman, MT59717, USA
{cl.fan,wq.ju,jun.luo}@siat.ac.cn, bhz@cs.montana.edu

Abstract. We study the problem of moving network Voronoi diagram: given a network with n nodes and E edges. Suppose there are m sites (cars, postmen, *etc*) moving along the network edges, we design the algorithms to compute the dynamic network Voronoi diagram as sites move such that we can answer the nearest neighbor query efficiently. Furthermore, we extend it to the k-order dynamic network Voronoi diagram such that we can answer the k nearest neighbor query efficiently. We also study the problem when the query point is allowed to move at a given speed. Moreover, we give the algorithm for the half-online version of moving network Voronoi diagram.

1 Introduction

Voronoi diagram is a fundamental technique in computational geometry and plays important roles in other fields such as GIS and physics [4]. One of the major applications of Voronoi diagram is to answer the *nearest-neighbor* query efficiently. Much has been done about variants of the Voronoi diagrams and the algorithms for computing the Voronoi diagrams in various fields. Many variants of the Voronoi diagrams are based on different definitions of distance in different fields, without limiting to the Euclidean distance [10].

The network Voronoi diagram [2, 6, 8, 11] divides a network (e.g. road network) into Voronoi subnetworks. A network Voronoi diagram is a specialization of a Voronoi diagram in which the locations of objects are restricted to the links that connect the nodes of the network. The distance between objects is defined as the length of the shortest network distance (e.g. shortest path or shortest time), instead of the Euclidean distance. The network Voronoi diagram is useful because in reality, people and cars are moved along road networks. Imagine that somebody gets hurt and we need to drive him

---

* This work is partially supported by Shenzhen Key Laboratory of High Performance Data Mining (grant no. CXB201005250021A) and Shenzhen Fundamental Research Project (grant no. JC201005270342A and grant no. JC201005270334A).
** This research is partially supported by NSF under grant DMS-0901034, by NSF of China under project 60928006, and by the Open Fund of Top Key Discipline of Computer Software and Theory in Zhejiang Provincial Colleges at Zhejiang Normal University.

M.L. Gavriloni et al. (Eds.): Trans. on Comput. Sci. XIV, LNCS 6970, pp. 129–150, 2011.
© Springer-Verlag Berlin Heidelberg 2011

to the nearest hospital. In this scenario, the network Voronoi diagram can answer this kind of query efficiently.

For network Voronoi diagram, any node located in a Voronoi region has a shortest path to its corresponding Voronoi site that is always shorter than that to any other Voronoi site. In this way, the entire graph is partitioned into several subdivisions as shown in Figure 1. We can see that the network Voronoi edges intersect with the network edges in most cases. This means that a network edge may be divided into two parts and placed into two adjacent Voronoi regions.

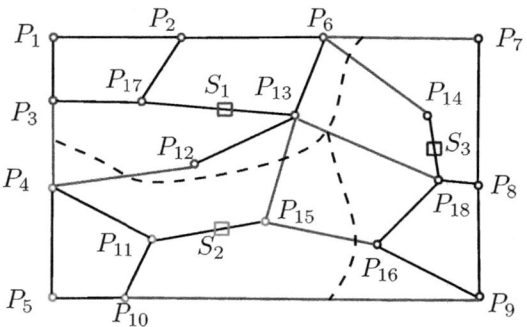

Fig. 1. The network Voronoi diagram. S_1, S_2 and S_3 are Voronoi sites and $P_1 - P_{18}$ are nodes

To construct the network Voronoi diagram, we have to measure network distance between nodes and Voronoi sites. Therefore a shortest-path searching algorithm is required. In [6], Erwig presented a variation of Dijkstra's algorithm, the parallel Dijkstra algorithm, which can be used to compute the multiple source shortest paths, to construct the network Voronoi diagram.

There are two types of dynamic network Voronoi diagrams. The first one is determined by the changes of edge weights. In [13], Ramalingam and Reps presented an incremental search method, the Dynamic SWSF-FP algorithm, to handle "multiple heterogeneous modification" between updates. The input graph is allowed to be reconstructed by an arbitrary mixture of edge insertions, edge deletions, and edge-length changes. The second type of dynamic network Voronoi diagram is determined by moving sites. In paper [5], Devillers and Golin discussed the moving Voronoi diagram based on Euclidean distance in two-dimension. They assumed that each site is moving along a line with constant speed. They gave two types of the "closest", namely closest in static case and kinetic case. In the static case the meaning of "closest" is quite clear. The closest site is the nearest site as to a query point q. In kinetic case, the closest site is the site a customer (who starts from q at time t_0 with speed v) can reach the quickest. In paper[9], Mohammad and Cyrus discuss the continuous K nearest neighbor queries in spatial network, for example, a moving object(e.g. a car) is traveling along a path, the driver in the car wants to find the first 3 closest sites(e.g restaurant) at any given point on the path. In their paper, they assume all sites can not move, only one customer can move on the path. Our problem is different, we assume all sites move at a given speed on their respective path, a customer at any given point on the edge of network want to find the K nearest sites.

In this paper we study the problem of moving network Voronoi diagram when all sites $S_1, S_2, ..., S_m$ move on the edges of network continuously. Every edge of the network is endowed with a positive value which denotes the network length (may not satisfy the triangle inequality) of edge. We can imagine that the set of Voronoi sites is the set of public service platform (bus, postman, policeman, etc). We assume that each site S_i ($1 \leq i \leq m$) moves with constant velocity v_i and their trajectories with time information are also known in advance. The n nodes are denoted as $P_1, P_2, ..., P_n$. Then the kinetic state of site S_i can be described as follows:

$$S_i(t) \in edge[P_{i0}, P_{i1}], t \in [t_0, t_1]$$

$$S_i(t) \in edge[P_{i1}, P_{i2}], t \in [t_1, t_2]$$

$$\vdots$$

$$S_i(t) \in edge[P_{i(w-1)}, P_{iw}], t \in [t_{w-1}, t_w]$$

where $S_i(t)$ is the point where the site S_i is located at time t. We use $length(P_a, P_b)$ to denote the length of edge$[P_a, P_b]$. We assume that all edges are undirected, so $length(P_a, P_b) = length(P_b, P_a)$. When the site S_i moves on edge$[P_a, P_b]$ (from P_a to P_b) during time interval $[t_1, t_2]$, we have:

$$length(S_i(t), P_a) = length(P_a, P_b)(t - t_1)/(t_2 - t_1)$$

We use $d(P_a, P_b)$ to denote the shortest network distance between P_a and P_b. If the length of edge is not satisfied with the triangle inequality, the $d(P_a, P_b)$ may not equal $length(P_a, P_b)$.

In this paper we want to be able to answer the following three different types of queries in continuous-time moving network:

1. Static case query $StaticNearest(q, t', v)$: given a customer at location q (q is on edge of network), find the nearest site at time t'. That is, given q, t', return i such that

$$d(q, S_i(t')) \leq d(q, S_j(t')), j = 1, \ldots, m$$

2. Static case k nearest neighbors query $StaticKNearest(q, t', v)$: given a customer at location q (q is on edge of network), find the k nearest sites at time t'. That is, given q, t', return $i_1, i_2, \ldots, i_k$ such that

$$d(q, S_{i_1}(t')) \leq d(q, S_{i_2}(t')) \leq \ldots$$
$$\leq d(q, S_{i_k}(t')) \leq d(q, S_j(t'))$$
$$where \ j = 1, \ldots, m \wedge j \neq i_1, i_2, \ldots, i_k$$

3. Kinetic case query $KineticNearest(q, t')$: The query inputs are q, t', and $v > 0$. They specify a customer located at q at time t' with walking speed v. The customer

(who can only move on the edges of the network) wants to reach a site as soon as possible. The problem here is to find the site the customer can reach quickest. Set

$$t_j = min\{t \geq t' : (t_j - t')v = d(q, S_j(t))\}, j = 1, \ldots, m$$

be the first time that the customer can catch site S_j starting from q at time t'. Then the query return i such that $t_i = min\{t_1, t_2, \ldots, t_m\}$.

The structure of the paper is as follows. Section 2 gives some definitions that we need in the whole paper. We discuss the algorithms for two static queries $StaticNearest$ (q, t', v) and $KineticKNearest(q, t', v)$ in section 3 and 4 respectively. In section 5, the algorithms for kinetic case query are presented. In section 6, we give the algorithm for the half-online version. Finally we give conclusions in section 7.

2   Preliminaries

In moving network Voronoi, sites move with constant velocity continuously. When a site moves on the edge during some time interval, for example in Figure 2, S_i move from P_a to P_b during time interval $[t_1, t_2]$, for any other node P_j that is connected (we assume the graph is a connected graph) with P_a and P_b,

$$d(P_j, S_i(t)) = min\{d(P_j, P_a) + length(P_a, S_i(t)),$$
$$d(P_j, P_b) + length(P_b, S_i(t))\}$$

where $t \in [t_1, t_2]$. Each site S_i moves on the pre-established path with constant velocity v_i (see Figure 3). For a node P_j, the network distance between P_j and $S_i(t)$ can be described by segmented functions as follows:

$$d(P_j, S_i(t)) = min\{d(P_j, P_{i0}) + v_i(t - t_0),$$
$$d(P_j, P_{i1}) + v_i(t_1 - t)\}, t \in [t_0, t_1]$$
$$\cdots$$
$$d(P_j, S_i(t)) = min\{d(P_j, P_{i(w-1)}) + v_i(t - t_{w-1}),$$
$$d(P_j, P_{iw}) + v_i(t_w - t)\}, t \in [t_{w-1}, t_w]$$

Note that the slope of those line segments is either v_i or $-v_i$ because S_i moves with constant velocity v_i. For simplicity, we only use the absolute value of slope v_i to represent their slope.

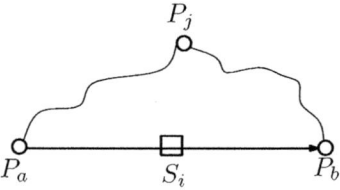

Fig. 2. A site S_i moves from P_a to P_b

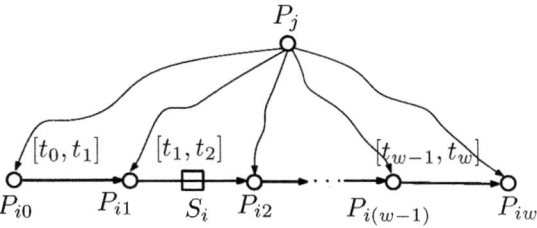

Fig. 3. A site S_i moves from P_{i0} to P_{iw}

Lemma 1. *All segmented functions $d(P_j, S_i(t))$ $(1 \le i \le m, 1 \le j \le n)$ can be computed in $O(n^2 \log n + nE + Y)$ time and $O(E + Y)$ space, where n and E are the number of nodes and edges of the network respectively, and Y is the total number of segmented functions for all nodes.*

Proof. If we already know the shortest path tree rooted at P_j, then each $d(P_j, S_i(t))$ can be computed in constant time. Each shortest path tree rooted at P_j can be computed in $O(n \log n + E)$ time using Dijkstra's shortest path algorithm by implementing the priority queue with a Fibonacci heap. Since we have n nodes, we need $O(n^2 \log n + nE)$ time to compute all n shortest path trees. And we need extra Y time to compute all segmented functions. Therefore the total time is $O(n^2 \log n + nE + Y)$. For the space, we need $O(E)$ space to store the network graph and $O(Y)$ space to store all segmented functions. Therefore, the total space is $O(E + Y)$.

3 Point Location in Static Moving Network

As we described above, for an arbitrary node P_j, the network distance $d(P_j, S_i(t))$ is a piecewise linear function, and it consists of at most w_i positive slope and w_i negative slope line segments, where $i = 1, 2, \ldots, m$. Figure 4 shows an example of the network distance function diagram of $d(P_j, S_1(t))$, $d(P_j, S_2(t))$, $d(P_j, S_3(t))$. P_j's nearest site is S_3 during time intervals $[t_0, t_1]$, $[t_4, t_5]$, $[t_6, t_7]$. During time intervals $[t_2, t_3]$, $[t_8, t_9]$, P_j's nearest site is S_1. Hence we only need to compute the lower envelope of these piecewise linear functions $d(P_j, S_i(t))$, where $i = 1, 2, \ldots, m$.

We use the divide-and-conquer algorithm to compute the lower envelope of those segmented functions. To compute the lower envelope

$$h(t) = \min_{1 \le i \le m} \{d(P_j, S_i(t))\}, t \in [t_0, t_e]$$

we partition

$$\{d(P_j, S_1(t)), d(P_j, S_2(t)), \ldots, d(P_j, S_m(t))\}$$

into two parts,

$$\{d(P_j, S_1(t)), d(P_j, S_2(t)), \ldots, d(P_j, S_{\lceil m/2 \rceil}(t))\}$$

and

$$\{d(P_j, S_{\lceil m/2 \rceil + 1}(t)), d(P_j, S_{\lceil m/2 \rceil + 2}(t)), \ldots, d(P_j, S_m(t))\}$$

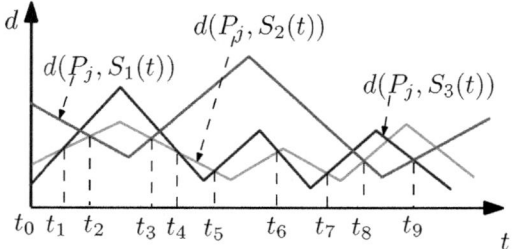

Fig. 4. The network distance function diagram of $d(P_j, S_i(t)), i = 1, 2, 3$

then compute

$$\min\{d(P_j, S_1(t)), d(P_j, S_2(t)), \ldots, d(P_j, S_{\lceil m/2 \rceil}(t))\}$$

and

$$\min\{d(P_j, S_{\lceil m/2 \rceil+1}(t)), d(P_j, S_{\lceil m/2 \rceil+2}(t)),$$
$$\ldots, d(P_j, S_m(t))\}$$

recursively and finally merge them together to obtain $h(t)$.

Now we consider merging two distance functions $d(P_j, S_k(t))$ and $d(P_j, S_l(t))$ (see Figure 5).

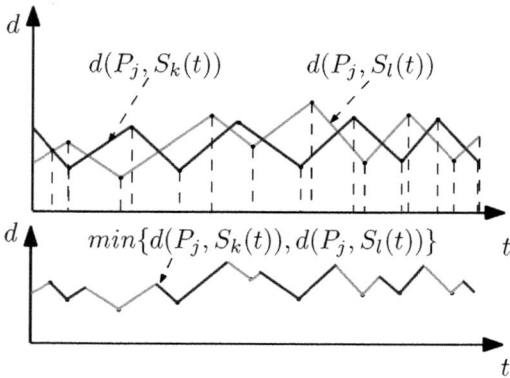

Fig. 5. The example of merging two functions: $d(P_j, S_k(t))$ and $d(P_j, S_l(t))$

Lemma 2. *There are at most* $2(w_k + w_l) - 1$ *intersections between functions* $d(P_j, S_k(t))$ *and* $d(P_j, S_l(t))$.

Proof. $d(P_j, S_k(t))$ and $d(P_j, S_l(t))$ are two continuous and piecewise linear polygonal curves. Suppose that the function $d(P_j, S_k(t))$ is composed of x line segments ($1 \leq x \leq 2w_k$) and the time intervals for those line segments are $[t_{begin}, t_1], [t_1, t_2], \ldots, [t_{x-1}, t_{end}]$. Similarly suppose that the function $d(P_j, S_l(t))$ is composed of y line segments ($1 \leq y \leq 2w_l$) and the time intervals for those line segments are $[t_{begin}, t'_1]$, $(t'_1, t'_2], \ldots, (t'_{y-1}, t'_{end}]$. If we mix those two set of intervals, we obtain $x + y - 1$ new intervals. At each new interval, there is at most one intersection between two functions. Hence the total number of intersections is $x + y - 1 \leq 2(w_k + w_l) - 1$.

It seems that the lower envelope of $d(P_j, S_k(t))$ and $d(P_j, S_l(t))$ has at most $4(w_k + w_l) - 2$ line segments since there are $2(w_k + wl) - 1$ intervals and each interval could have at most two line segments. However, we could prove there are fewer number of line segments for $\min\{d(P_j, S_k(t)), d(P_j, S_l(t))\}$.

We define a *valley* as a convex chain such that if we walk on the convex chain from left to rigt, we always make left turn from one line segment to the next line segment (see Figure 6).

Lemma 3. *The function* $\min\{d(P_j, S_k(t)), d(P_j, S_l(t))\}$ *is composed of at most* $3(w_k + w_l)$ *line segments. The number of valleys does not change after merging*

Proof. The function $d(P_j, S_k(t))$ is composed of x line segments ($1 \leq x \leq 2w_k$) and $d(P_j, S_l(t))$ is composed of y line segments ($1 \leq y \leq 2w_l$). Thus $d(P_j, S_k(t))$ consists of at most $x/2$ valleys and $d(P_j, S_l(t))$ consists of at most $y/2$ valleys. For each valley of $d(P_j, S_k(t))$, it can cut one line segment of $d(P_j, S_l(t))$ into three line segments, of which at most two could be on the lower envelope. It is similar for valleys of $d(P_j, S_l(t))$. Therefore, besides the original line segments of two polygonal lines, $x/2 + y/2$ valleys could add at most $x/2 + y/2$ line segments for the lower envelope. The total number of line segments of the lower envelope is $\frac{3}{2}(x + y) \leq 3(w_k + w_l)$. Note the number of valleys does not change after merging because newly created turns on the lower envelope always go from left to right and make right turns (see Figure 6).

We can decrease furthermore the complexity of the lower envelope because the slope of line segments of $d(P_j, S_k(t))$ is fixed as v_k and the slope of line segments of $d(P_j, S_l(t))$ is fixed as v_l.

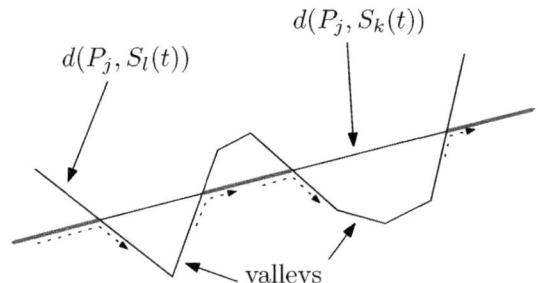

Fig. 6. One line segment of $d(P_j, S_k(t))$ is cut into 5 line segments by two valleys of $d(P_j, S_l(t))$, of which 3 line segments (red) are in the lower envelope of $d(P_j, S_k(t))$ and $d(P_j, S_l(t))$. Dotted lines are newly created turns on the lower envelope.

Lemma 4. *If $v_k < v_l$, the function $\min\{d(P_j, S_k(t)), d(P_j, S_l(t))\}$ is composed of at most $2(w_k + w_l) + w_l$ line segments.*

Proof. The function $d(P_j, S_k(t))$ is composed of x line segments ($1 \leq x \leq 2w_k$). $d(P_j, S_l(t))$ is composed of y line segments ($1 \leq y \leq 2w_l$). Every segments of function $d(P_j, S_k(t))$ has the slope either v_k. $d(P_j, S_k(t))$ consists of at most $x/2$ valleys and $d(P_j, S_l(t))$ consists of at most $y/2$ valleys. For each valley of $d(P_j, S_l(t))$, it can cut one line segment of $d(P_j, S_k(t))$ into three line segments, of which at most two could be on the lower envelope. But it is not true for valleys of $d(P_j, S_k(t))$. Because $v_k < v_l$, the slope of line segments of $d(P_j, S_k(t))$ is less than the slope of line segments of $d(P_j, S_l(t))$ that means no line segments of $d(P_j, S_l(t))$ are cut into more than three pieces by valleys of $d(P_j, S_k(t))$. Therefore, besides the original line segments of two polygonal lines, $x/2 + y/2$ valleys could add at most $y/2$ line segments for the lower envelope. The total number of line segments of the lower envelope is $\frac{3}{2}y + x \leq 2(w_k + w_l) + w_l$. □

We define a valley as *cut-valley* after it cuts a line segment into three pieces. Then the left and right segment adjacent to the cut-valley are from the same line segment and are named as border-edge of the cut-valley (see Figure 7). Two cut-valley may share one common border-edge. A valley that is not cut-valley is called *normal-valley*.

Lemma 5. *Suppose a polygonal line l_1 is cut by a cut-valley of another polygonal line l_2 in the process of merging and the two polygonal lines's start (end) points are on the same vertical line. If the slope of every line segment in l_1 is less than the slope of every segment in l_2, the cut does not add the total number of line segments to the lower envelope.*

Proof. A cut-valley of l_2 may cut a line segment of l_1 into three part, of which at most two could be in the lower envelope. Since the slope of every line segment in l_1 is less than the slope of every segment in l_2, then at least one border-edge of the cut-valley would be shelter by the cut-valley from into a line segment of the lower envelop. That means the total number of line segments of the lower envelope does not change. Furthermore, that cut-valley is still a cut-valley after the cut. □

Lemma 6. *In the process of merging, if the number of cut-valleys decreases by one, then the total number of segments of lower envelope decreased at least by one.*

Proof. There are three cases such that the number of cut-valleys decreases by one:

1. See Figure 8(a). One cut-valley totally disappears because it is above some valleys of the other polygonal line. Then the total number of segments of lower envelope decreased at least by two.
2. See Figure 8(b). One cut-valley disappears and becomes normal-valley because one of its two border-edges is above some valleys of the other polygonal line. The total number of segments of lower envelope decreased by one.
3. See Figure 8(c). One two adjacent cut-valleys becomes one cut-valley because the shared border-edge of two adjacent cut-valleys is above some valleys of the other polygonal line. The total number of segments of lower envelope decreased at least by one.

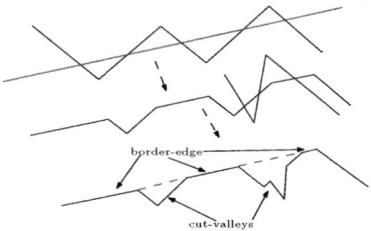

Fig. 7. An example shows how the cut-valley is created

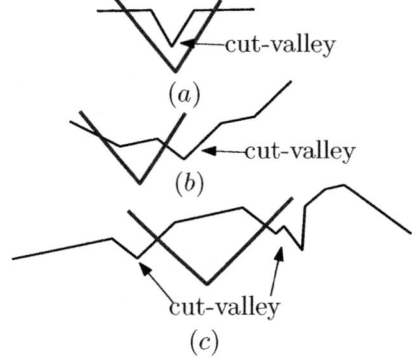

Fig. 8. Three cases such that the number of cut valleys decreases by one

Lemma 7. *The lower envelop* $h(t)=\min\{d(P_j, S_1(t)), d(P_j, S_2(t)), \ldots, d(P_j, S_m(t))\}$ *can be computed in* $O(mW \log m)$ *time and* $O(mW)$ *space where* $W = \max\{2w_1, 2w_2, \ldots, 2w_m\}$ *and it consists of* $O(mW)$ *line segments.*

Proof. The first step of our algorithm is to sort the m polygonal lines according to their slopes in $O(m \log m)$ times. Let m sorted polygonal lines in increasing order of their slopes be $d(P_j, S_1(t)), d(P_j, S_2(t)), \ldots, d(P_j, S_m(t))$. Our next step is to compute $\min\{d(P_j, S_{2a+1}(t)), d(P_j, S_{2a+2}(t))\}$ where $a = 0, \ldots, \frac{m}{2} - 1$. According to Lemma 4, each function $\min\{d(P_j, S_{2a+1}(t)), d(P_j, S_{2a+2}(t))\}$ can be computed in at most $(2 + 1/2)W$ time, which is composed of at most $(2 + 1/2)W$ line segments. The second step is to compute $\min\{\min\{d(P_j, S_{4a+1}(t)), d(P_j, S_{4a+2}(t))\}, \min\{d(P_j, S_{4a+3}(t)), d(P_j, S_{4a+4}(t))\}\}$ where $a = 0, \ldots, \frac{m}{4} - 1$. According to Lemma 3, the number of these valleys does not change as the number of line segments of lower envelope increase. Furthermore, since the slope of segments of all valleys in $\min\{d(P_j, S_{4a+1}(t)), d(P_j, S_{4a+2}(t))\}$ is less than the slope of segments in $\min\{d(P_j, S_{4a+3}(t)), d(P_j, S_{4a+4}(t))\}$, all valleys in $\min\{d(P_j, S_{4a+1}(t)), d(P_j, S_{4a+2}(t))\}$ can not cut the line segments in $\min\{d(P_j, S_{4a+3}(t)), d(P_j, S_{4a+4}(t))\}$ to increase the number of line segment to the lower envelope after merging according to Lemma 4. Suppose there are y cut-valleys in $\min\{d(P_j, S_{4a+3}(t)), d(P_j, S_{4a+4}(t))\}$. Those y cut-valleys are originally normal-valleys in $d(P_j, S_{4a+3}(t))$ and $d(P_j, S_{4a+4}(t))$. Then there are at most $2W + y$ line segments in $\min\{d(P_j, S_{4a+3}(t)), d(P_j, S_{4a+4}(t))\}$ since one normal-valley contributes at most one more line segment in $\min\{d(P_j, S_{4a+3}(t)), d(P_j, S_{4a+4}(t))\}$. There are at most $W - y$ normal-valleys in $\min\{d(P_j, S_{4a+3}(t)), d(P_j, S_{4a+4}(t))\}$ and those $W - y$ normal-valleys contribute at most $W - y$ more line segment in the lower envelope after second step. According to Lemma 5, y cut-valleys in $\min\{d(P_j, S_{4a+3}(t)), d(P_j, S_{4a+4}(t))\}$ does not increase the number of line segment to the lower envelope after second step. Therefore the total line segments of the lower envelope after the second step is $(2 + 1/2)W + W - y + 2W + y = (5 + 1/2)W$, where $(2 + 1/2)W + W - y$ is the number of line segments from $\min\{d(P_j, S_{4a+1}(t)), d(P_j, S_{4a+2}(t))\}$ and $W - y$ is the number of line segments from $\min\{d(P_j, S_{4a+3}(t)), d(P_j, S_{4a+4}(t))\}$. Note that if a cut-valley becomes a normal-valley, it decreases the number of segments of the lower envelope at least by one according to Lemma 6. A common-valley becomes a cut-valley again later on after it cuts a line segment and add one more segment to the lower envelope. Hence the common-valley except those from origin m polygonal lines does not contribute more line segments for the lower envelope. Let $N(i)$ denote the number of line segments of each lower envelope after ith ($1 \leq i \leq \log m$) step. Then we have:

$$N(1) \leq 2W + 1/2W = (2 + 1/2)W = (2 + 1/2)W$$
$$N(2) = (2 + 1/2)W + W - y + 2W + y = (5 + 1/2)W$$
$$N(3) = (5 + 1/2)W + 2W - y' + 4W + y' = (11 + 1/2)W$$

$$\cdots$$

$$N(i) = N(i - 1) + 2^{i-2}W - y'' + 2^{i-1}W + y''$$
$$= (2^i + 2^{i-1} - 1/2)W$$

$$\cdots$$

$$N(\log m) = 2N(\log m - 1) + 1/2W = (3/2m - 1/2)W$$

The computation time for ith step is $\frac{m}{2^i}N(i) = \frac{m}{2^i}(2^i + 2^{i-1} - 1/2)W$. The total computation time is $\sum_{i=1}^{\log m} \frac{m}{2^i}(2^i + 2^{i-1} - 1/2)W = O(mW \log m)$. The space complexity is $O(mW)$ since we need so much space to store $O(mW)$ line segments of lower envelope.

So far we compute the lower envelope for one node. The envelopes for all n nodes can be computed in $O(nmW \log m)$ time. Note that Agarwal and Sharir [1] prove that for mW line segments, the size of the lower envelope is $\Theta(mW\alpha(mW))$ where $\alpha(mW)$ is the inverse of Ackermann's function and the lower envelope can be computed in $O(mW \log(mW))$ time. Our results is better since our inputs are $O(m)$ polygonal lines that are x-monotone and their start (end) points have the same x coordinate instead of $O(mW)$ line segments that are arranged in the plane without any restriction. If the query point q is on some node P_j, we can use binary search for the query time t' on lower envelope associated with P_j to find the nearest site in $O(\log(mW))$ time. If the query point q is on some edge $[P_a, P_b]$, we can just perform the same binary search on both lower envelopes associated with P_a and P_b and then plus the extra length from q to P_a and P_b respectively. The minimum one of those two values and its corresponding site is the answer to the query. Therefore we have following theorem:

Theorem 1. *Given all segmented function $d(P_j, S_i(t))$ $(1 \leq i \leq m, 1 \leq j \leq n)$, for a static query of nearest site in moving network, we can answer it in $\log(mW)$ time with $O(nmW \log m)$ time and $O(nmW)$ space for preprocessing step.*

Sometimes, the number of line segments X of lower envelope could be very small. Here we give an output sensitive algorithm to compute the lower envelope. The general idea of the algorithm is as follows: first, we find function $\min_{1 \leq k \leq m}\{d(P_j, S_k(t_0))\}$, which means the lowest line segment l_e at time t_0. Let l_e correspond to the site S_e. Then we compute the minimum time intersection point of $d(P_j, S_k(t))$ $(1 \leq k \leq m$ and $k \neq e)$ with l_e and get the next line segment $l_{e'}$ of lower envelope. We compute the intersection point from left to right. If the current line segment on $d(P_j, S_k(t))$ has no intersection with l_e then we try the next line segment until there is an intersection point or $d(P_j, S_k(t))$ is out of the time interval of l_e. The line segments of $d(P_j, S_k(t))$ without intersection with l_e will not be tested again later since they are above l_e and can not be on the lower envelope. Let $l_e = l_{e'}$ and we perform above process repeatedly until the algorithm stops. The time complexity of this output sensitive algorithm is $O(mZ + Y) = O(mZ + mW)$, where we need $O(Z)$ space to store the lower envelope. Therefore if $Z \leq W$, then the preprocessing time and space could be reduced to $O(nWm)$ and $O(mW)$ respectively.

4   To Find the k Nearest Site in Static Moving Network

A frequent type of query in road networks is to find the K nearest neighbors (KNN) of a given query point. This is very useful if we want to guarantee some services to be delivered as soon as possible. For example, if there is a car accident and the car driver wants the policeman to arrive as soon as possible, it is not enough to send the nearest policeman to the site scene of the accident since there could be heavy traffic on the road from police to the site scene. So it is a good strategy to send two or more nearest policemen from different locations to the site scene.

Given m x-monotone polygonal lines whose start (end) points have the same x-coordinate, the k nearest sites for some node can be answered easily after we know the k level lower envelopes. Denote kth level lower envelope as L_k.

4.1   Plane Sweep Algorithm to Compute k Level Lower Envelopes

In paper [3], Bentley and Ottmann give a plane sweep algorithm for reporting and counting geometric intersections. In their algorithm, they use a data structure called the *event queue Q* to stores the events such as the end points of line segment and the intersection point of two line segments, and a status structure S, which is a balanced binary search tree. The left-to-right order of the segments along the sweep line corresponds to the left-to-right order of the leaves in S (see an example Figure 9). This algorithm can also be used to compute all level envelopes of polygonal lines. We sweep the m polygonal lines from left to right, instead of from top to bottom. The leftmost leave in status structure is a segment which belongs to L_1 and the kth leftmost leave in status structure is a segment which belongs to the kth level lower envelope. The only difference is that we need to output the changes of k lower envelopes when event happens, not just reporting the intersections. In the analysis of paper [12], The running time of the plane sweep algorithm for a set of n line segments in the plane is $O((n + I) \log n)$ and need $O(n)$ space, where I is the number of intersection points of those segments ($I \leq m(m-1)W/2$). That means the time complexity to compute k level envelopes for m polygonal lines is $(mW + I) \log(mW)$ and space $O(mW + kX)(X = max(X_i)$, X_i denotes the number of segments on the ith envelope.) However, the running time could be reduced.

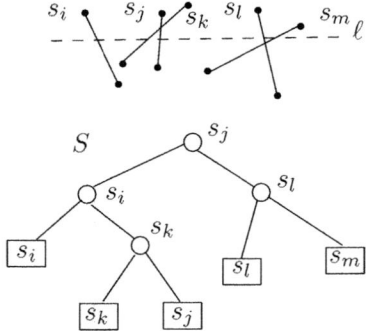

Fig. 9. The left-to-right order of the segments along the sweep line at some time and its status structure in balanced binary search tree

Lemma 8. *Let P be a set of m polygonal lines in the plane, k $(1 \leq k \leq m)$ envelopes can been computed in $O((mW + I) \log m)$ time and $O(mW + kX)$ space using the plane sweep algorithm, where I is the number of intersection points between polygonal lines and X is the maximum number of segments on one of the k envelopes.*

Proof. Since the m polygonal lines are x monotone, the sweep line cross m segments at any time. Hence there are m leaves for the status structure and at most $O(m)$ event in *event queue* at any time, for the event queue only store the intersections between two adjacent segment and segment's endpoints, and these segments must be crossed by the same sweep line. As we know, all segments of a polygonal line are connected from left to right, we do not put all segments' endpoints into *event queue* at a time, but put the leftmost segment's endpoint(including the start point and end point) of every polygonal line into *event queue* at first, when the sweep line meet some polygonal line's segment's endpoint, we put that polygonal line's next segment's endpoint into *event queue*. That means the insertions and deletions and neighbor finding on the status structure S only needs $O(\log m)$ time instead of $O(\log mW)$. Therefore k level $(1 \leq k \leq m)$ envelopes can been computed in $O((mW + I) \log m)$ time and $O(mW + kX)$ space using the plane sweep algorithm.

4.2   Improved Algorithm to Compute k Level Envelopes

The algorithm above actually computes all m level envelopes, but we only need k level envelopes. Therefore the algorithms above can be improved. We do a little modification to the plane sweep algorithm. We only need to know the order of left k segments when m segments intersect with the sweep line. There are $(m - k)$ segments on the right of the sweep line, but we do not need to know their relative order and their intersections with each other (See Figure 10).

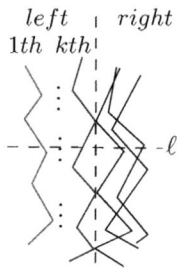

Fig. 10. The k level envelopes of polygonal lines. l is the sweep line.

We use a status structure S (balanced binary search tree) of $O(k)$ size to hold the relative order of that k segments on the left, and a linked list L of $O(m - k)$ size to store the $(m-k)$ segments on the right. But for a segment which belongs to the kth level envelop, we do need to know whether the segments on its right intersect with it. If there are several segments(on its right) intersect with it, only the topmost intersection segment is useful. We also need the *event queue* of $O(m)$ size to store the events. An event occurs when two adjacent segments which belong to the k level envelopes intersect with each other, segments on the right intersect with the segments which belong to the kth envelop, the endpoints of polygonal lines appear on the sweep line. According to Lemma 8, we do not need to put all endpoints into Q at once. When an event happens, we process it as follows:

1. At an endpoint of the segment belonging to k level envelopes, we need to delete one segment and insert adjacent segment from the status structure S, and test intersection between segments that become neighbors after the event.
2. At an endpoint of the segment not belonging to k level envelopes, we need to delete one segment and insert adjacent segment from linked list L, and test if inserted segment intersects with the kth left segment (the rightmost leaves in balanced tree S) along the sweep line. If the inserted segment intersects with the kth segment and the intersection point is above the old intersection point on kth segment, we replace the old intersection event by new intersection event in Q , else we do nothing.
3. At an intersection point, we have to change the order of two segments and do intersection tests between two neighbor segments. If part of a segment becomes a segment of the kth envelope after exchange, we need to examine if it intersects with the segments on its right which are stored in L, then find the topmost intersection event and insert it into Q .

In degenerate cases, where several segments are involved in one event point, the details are a little more tricky, but the method is similar. Hence we do not give details here.

Now we analyze the complexity of the improved algorithm. If we use the Fibonacci heap to implement the event queue, in which the operation for decreasing key only needs $O(1)$ amortized time. That means replacing the old event by a new event only need $O(1)$ time amortized. The total number of events is $O(mW + kX)$. For every segment of the kth envelop, $O(m - k)$ segments (amortized) on its right may intersect with it. Therefore we need $O(m - k)X$ time for the kth envelope. Hence the total running time is $O((mW + kX)\log m + (m - k)X)$ and space is $O(mW + kX)$. Comparing with previous algorithm, the improved one has the same running space but the running time is less because $I = O(mX)$ and usually $k << m$.

Lemma 9. *Let P be set of m polygonal lines in the plane, k $(1 \leq k \leq m)$ envelopes can be computed in $O((mW + kX)\log m + (m - k)X)$ time and $O(mW + kX)$ space, where X is the maximum number of segments in one of the k envelopes.*

Theorem 2. *Given all segmented function $d(P_j, S_i(t))$ $(1 \leq i \leq m, 1 \leq j \leq n)$, for a static k nearest neighbor query in moving network, we can answer it in $O(k \log X)$ time with $O((mW + kX)\log m + (m - k)X)$ time and $O(mW + kX)$ space for preprocessing, where X is the maximum number of segments in one of the k envelopes.*

5   Point Location in Kinetic Moving Network

We now consider kinetic case queries. Imagine that the moving sites are postmen. There is a customer at a query point on the road with walking speed v and searching for a postman that he can reach in minimum time for delivering his package. Both postmen and the customer can only move on the road.

These types of queries differ from the static case query in the previous section. It is possible that the customer might reach a postman further away (that is traveling toward it) quicker than a nearby postman (that is traveling away from it). The answer to the query depends strongly on v. Note that if we let v approach infinity, then the problem becomes the static query case.

We first consider the query $KineticNearest(P_j, t', v)$, which means the customer lies on the node P_j at t', it is clear that customer can reach the point P if $d(P_j, P) = v(t - t')$ at time t. So we add a line $d = v(t - t')$ over $d(P_j, S_i(t))$, $i = 1, 2, \ldots, m$ (see figure 11). If $d = v(t - t')$ intersects with $d(P_x, S_i(t))$, $i = 1, 2, \ldots, m$ at earliest time t'', then S_x which corresponds to the line segment intersecting with $d = v(t - t')$ at t'' would be the site the customer can reach first. Actually it equivalents to find the intersection point of $d = v(t - t')$ with lower envelope of $d(P_j, S_i(t))$, $i = 1, 2, \ldots, m$ (see Figure 12). As we discussed in previous section, for each node P_j, the lower envelope $min\{d(P_j, S_i(t)), i = 1, 2, \ldots, m\}$ can be computed in $O(mW \log m)$ times, which consists of $O(mW)$ line segments. Then we only need to consider the intersection between function $d = v(t - t')$ and function $min\{d(P_j, S_i(t)), i = 1, 2, \ldots, m\}$ (see Figure 12).

If we already compute the function $min\{d(P_j, S_i(t)), i = 1, 2, \ldots, m\}$ in preprocessing step, how to get the first intersection with $d = v(t - t')$. In [7], Guibas *et al.* gave an algorithm: Given a segment e inside a polygon P, preprocess p so that for each query ray r emanating from some point on e into P, the first intersection of r with

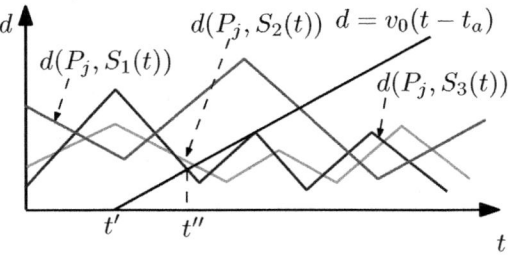

Fig. 11. The diagram of function $d(P_j, S_i(t)), i = 1, 2, 3$ and $d = v(t - t')$

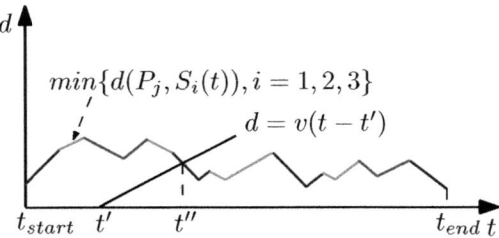

Fig. 12. The diagram of function $min\{d(P_j, S_i(t)), i = 1, 2, 3\}$ and $d = v(t - t')$

the boundary of P can be calculated in $O(\log V)$ time(V denote the number of vertex of polygon P), after given a triangulation of a simple polygon P. Actually the lower envelope $min\{d(P_j, S_i(t)), i = 1, 2, \ldots, m\}$ with two vertical line segments which correspond to $t = t_{start}$ and $t = t_{end}$ (see Figure 12) and the horizontal line segment of time axis can be treated as a simple polygon. The line $d = v(t - t')$ can be treated as a ray emanating from point $(t', 0)$ on horizontal line segment along the time axis. Thus the intersection point can be computed in $O(\log(mW))$ time, which means the $KineticNearest(P_j, t', v)$ can be answered in $O(\log(mW))$ time.

In the above we assume that the customer locates exactly on some node. But the customer may be located in the middle of some edge. So the query becomes $Kinetic$ $Nearest$ $(P_s \in edge[P_a, P_b], t', v)$. Let $length(P_s, P_a) = L_1, length(P_s, P_b) = L_2$, and $t_a = L_1/v, t_b = L_2/v$. If the customer meets some site on some edge other than $edge[P_a, P_b]$, then we know that the customer must walk to P_a or P_b first. Therefore the problem becomes to find the first intersection between function $min\{d(P_a, S_i(t)), i = 1, 2, \ldots, m\}$ and function $d = v(t - t' - t_a)$ where $t \geq t' + t_a$, and the first intersection between function $min\{d(P_b, S_i(t)), i = 1, 2, \ldots, m\}$ and function $d = v(t - t' - t_b)$ where $t \geq t' + t_b$. This is exactly the same as the previous case. If the customer meets some site on edge $[P_a, P_b]$, then we know they must meet during the site travels through the edge $[P_a, P_b]$ at first time (note that one site could travel through the same edge many times). In preprocessing step, we can extract the line segment of $min\{d(P_a, S_i(t)), i = 1, 2, \ldots, m\}$ which corresponds to one site traveling through the edge $[P_a, P_b]$ from P_a to P_b at first time. Then the intersection point between the extracted line segment and the line segment $d = v(t' + t_a - t)$ where $(t' \leq t \leq t' + t_a)$ is the meeting time and position between the customer and that site. Note that there could be no intersection between two line segments. Since any of m sites could meet the customer on edge $[P_a, P_b]$, we just compute all intersection points in $O(m)$ time. Similarly we can compute the intersection point if the customer meet some site when the site travels through the edge $[P_a, P_b]$ from P_b to P_a at first time. Then we have $O(m)$ intersection points on $d = v(t' + t_a - t)$ where $(t' \leq t \leq t' + t_a)$, $O(m)$ intersection points on $d = v(t' + t_b - t)$ where $(t' \leq t \leq t' + t_b)$, one intersection on $d = v(t - t' - t_a)$ where $t \geq t' + t_a$, and the other intersection on $d = v(t - t' - t_b)$ where $t \geq t' + t_b$. The minimum time of all those $O(m)$ intersection points and corresponding site is the answer to the query $KineticNearest(P_s \in edge[P_a, P_b], t', v)$. Then we have the following theorem:

Theorem 3. *Given all segmented function $d(P_j, S_i(t))$ $(1 \leq i \leq m, 1 \leq j \leq n)$, we can answer query $KineticNearest$ $(P_s \in edge[P_a, P_b], t', v)$ in $O(m + \log(mW))$ time with $O(nmW \log m)$ time and $O(nmW)$ space for preprocessing step. If the customer is located at some node, then the query can be answered in $O(\log(mW))$ time.*

6   Half-Online Moving Network Voronoi Diagram

We now consider the online version of moving network, that means we only know the trajectories of m sites up to now. Furthermore, when a site moves on an edge of network, we assume it can only turn around at the endpoint of that edge, and it can not turn around at the midpoint of that edges. This assumption is reasonable for roads in cites since they are either one-way roads, or there are fences in the middle of roads, which separate one road into two-way roads. The car can only make turns at road junctions. Although we know the edge where a site moves on by now, we do not know the next edge it will move on before it arrives the endpoints of that edge. We call this version of moving network *half-online moving network*.

Definition 1. *For a node P_j, suppose its nearest site is S_i at time t. If one of P_j's adjacent nodes' nearest site is not S_i, or there are other sites lie on one of P_j's adjacent edges at t, then we call node p_j a preparation-node at t. Otherwise, p_j is a non-preparation-node. In Figure 13, the node P_2 and P_3 are preparation-nodes, while P_1 is a non-preparation-node. In Figure 14, those preparation-nodes whose nearest site is S_1 are denoted by rectangle, whose nearest site is S_2 are denoted by triangle, whose nearest site is S_3 are denoted by diamond, all non-preparation-nodes are denoted by circle.*

Definition 2. *We define the critical set Set_j for every node P_j as the set of sites on the P_j's adjacent edges and the nearest sites of P_j's adjacent nodes. However P_j's nearest site is not included. For example, in Figure 13, $Set_2 = \{S_2\}$, $Set_3 = \{S_3\}$ and $Set_1 = \{null\}$. If P_j is a preparation-node, then the set_j is not empty, otherwise, set_j is empty.*

For a preparation-node P_j, its nearest site S_j and its set Set_j at time t_0, any site S_i belongs to Set_j satisfies

$$d(P_j, S_j(t_0)) < d(P_j, S_i(t_0))$$

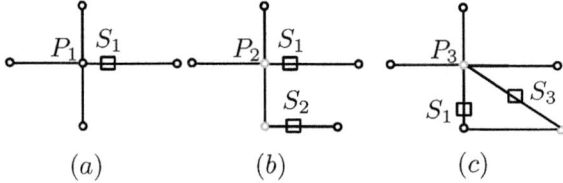

Fig. 13. The example of preparation-node painted by gray and non-preparation-node painted by black

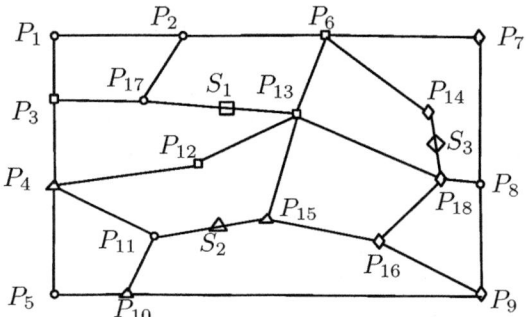

Fig. 14. The network Voronoi diagram, three sites and its preparation-nodes denote by different shapes respectively

We set

$$t_i = min\{t > t_0 : d(P_j, S_j(t)) = d(P_j, S_i(t))$$

$$and \ d(P_j, S_j(t + \triangle t)) > d(P_j, S_i(t + \triangle t))\}$$

be the first time S_i is nearer than S_j next to P_j after time t_0. If there are no solution for t_i, then let $t_i = \infty$.

We set t_{cj} and S_{cj} of P_j to be:

$$t_{cj} = min\{t_i | S_i \in Set_j\}$$

$$S_{cj} = \{S_x | t_x \le t_i, S_i \in Set_j\}$$

If all $t_i = \infty$, we let $t_{cj} = \infty$ and $S_{cj} = \{\}$. We call t_{cj} the critical time of P_j and S_{cj} the critical site of P_j at time t_0.

The main idea of our algorithm that maintains the half-online moving network Voronoi diagram is as follows: All sites move on the edges of network continuously. The network distance between nodes and sites may change continuously. But in some time interval, the nearest site for a node does not change. If the nearest site for a node changes at time t, we call that change as an *output event* at time t. For any non-preparation-node, an output event never happens. The event might only possible happen on the preparation-node. The major step of our algorithm is to only maintain the information of preparation-nodes.

First of all, we need a data structure-the event queue-that stores the events. We denote the event queue by Q (time is the priority). We need an operation that removes the next event that will occur from Q, and returns it so that it can be processed. If two events have the same time priority, they are removed according to the principle of FIFO. The event queue must allow insertion, deletion, because new events or an old events with new early time priority will be computed after an event occurs. Notice two events may have the same time priority.

Besides, we need a two-dimensional array to store all-pair shortest network distance. We also need an adjacency list to store every node and its adjacent nodes (every node's nearest site at present is stored and it needs to be changed after an event happens).

At last we need to store the output result of the algorithm, for the sake of nearest neighbor queries.

Given a preparation-node P_j, and its set Set_j at time t_0, to compute the t_{cj} after t_0 is a difficult work for the half-online problem. Let tt_i denote the time when S_i moves to the endpoint of edge where it is move on currently. But we do not know how S_i will move after tt_i at t_0. If $t_{cj} \leq min\{tt_i, i = 1, ..., x\}$ at t_0, then we can determine the t_{cj} at t_0. Otherwise t_{cj} can not be determine at t_0. When a site S_i arrives at the endpoints of an edge at tt_i, then we know the next edge it will move on. We call it an *input event* at tt_i. If $t_{cj} > tt_i$, then we need to recompute t_{cj} based that input event. We create a link list L_i for the S_i above, that stores those nodes like P_j, such that S_i belong to Set_j and $tt_i < t_{cj}$. When an input event happens on S_i (S_i moves to the endpoint of another edge), then we recompute the t_{cj} for the node P_j in L_i (only consider the site S_i and S_j by the time when event happens, no need to consider other sites in Set_j). Notice that t_{cj} cannot be determined by S_j itself. Then we need to add a node into L_j and recompute the t_{cj} when S_j moves to the endpoints of edge where it moves on currently.

How would the network Voronoi diagram look like after an input event happens? There are three cases:

1. The node in L_i is invalid when input event happens. For a node which stores the information (P_j, S_i, S_{cj}) in L_i, and S_{cj} which is P_j's nearest site when the node is created but is not P_j's nearest site when event happens, we only delete the node in L_i and do nothing.
2. The node (P_j, S_i, S_{cj}) in L_i is valid when input event happens and we still can not determine t_{cj}. Then we move that node to the tail of L_i.
3. The node (P_j, S_i, S_{cj}) in L_i is valid when input event happens, and new t_{cj} can be computed. If new t_{cj} is earlier than old t_{cj} in Queue Q, we update it with earlier time event and delete that node.

How would the network Voronoi diagram looks like just after an output event happens? We assume the next event at t_{cm} occur on the node P_m and its nearest site be S_{cm} just after the event. We first need to find the Set_m of P_m at time t_s, then we compute the critical time and critical site of P_m at t_s, and insert the new critical time into the queue Q. We also need to consider the effects of event to P_m's adjacent nodes. There are three types of changes need to be processed (see Figure 15):

1. Some P_m's adjacent nodes change from non-preparation-node to preparation-node. For example in Figure 15(a), P_d is a non-preparation-node at t_0. Its nearest site

is identical with P_m's nearest site, namely S_1. After the event, P_d becomes a preparation-node and Set_d changes from $\{\}$ to S_2 (see Figure 15(b)). Hence we need to compute the new critical time t_z for this kinds of vertices, and insert P_z's critical time t_z into queue Q.

2. Some P_m's adjacent nodes change from preparation-node to non-preparation-node. For example P_b is a preparation-node at time t_0, but it becomes a non-preparation-node at t_{cm}. For adjacent nodes of this type, we need to do nothing.

3. The remaining adjacent nodes are preparation-node back and forth. There are changes need to deal with for these nodes (we just name them P_z and its nearest site S_z). The P_z's set Set_z may change after event. If S_{cm} belongs to Set_z, we need to do nothing; else we need to add S_{cm} into Set_z. Furthermore we need to compute tp at time t_{cm}:

$$tp = min\{t > t_{cm} : d(P_z, S_z(t)) = d(P_z, S_{cm}(t))$$

$$and \ d(P_z, S_z(t + \triangle t)) > d(P_z, S_{cm}(t + \triangle t))\}$$

If $tp \geq t_{cz}$, then we do nothing either, else we need to update t_{cz} with tp, S_{cz} with S_{cm}, and decrease the t_{cz}'s value in the queue. If t_{cz} can not be determined by now because of S_{cm}, add a node (P_z, S_{cm}) for site S_{cm} into link list L_{cm}.

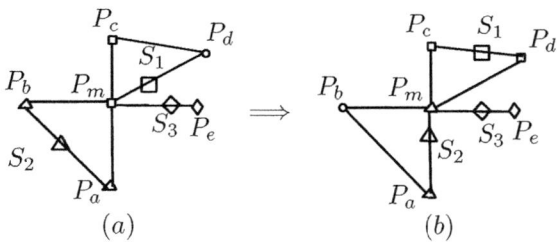

Fig. 15. The changes of network Voronoi diagram caused by an event occurs on the node P_m

The data structure we need in the algorithm has been described in details above. The static network Voronoi diagram at time t_{begin} can be determined by the way we introduced in session 3. Then we can get initial network Voronoi diagram at time t_{begin}. We set all non-preparation-node's critical time as ∞ and its critical site as empty. The pseudo code of this algorithm is described in Algorithm 1.

Now we analyze the time complexity of our algorithm. Let $B(B < V)$ be the maximum number of prepare-nodes at any time, D be the maximum degree of any vertex in G. V is the total number of vertices of G. n_e is the number of output events happening in time interval $[t_{begin}, t_{end}]$, and w_i is the number of edges S_i passes in $[t_{begin}, t_{end}]$.

To compute prepare-nodes in G takes $O(V + E)$ time (line [1.3]). Lines [1.7-1.8] take $O(V)$ times. There are at most D elements in Set_j without the sites like S_3 in Figure 13(c), which lie in the middle of an edge but do not control (nearest) any vertices in G. To compute t_{cj} for P_j take $O(D)$ times without considering the sites like S_3 in

---

input : The Moving Network Voronoi Diagram G at time t_{begin}, and the every site's motion state

output: All events in chronological order

1.1 **begin**

1.2   Initialize an empty event queue Q;

1.3   Compute all prepare-node in G at t_{begin};

1.4   **for** *every prepare-node $P_j \in G$* **do**

1.5   Compute set Set_j, critical time t_{cj} and site S_{cj} at t_{begin};

1.6   Insert the event(t_{cj}, S_{cj}, P_j) into Q;

1.7   **for** *every non-prepare-node $P_j \in G$* **do**

1.8   Set critical time t_{cj} be ∞ and site S_{cj} at empty ;

1.9   **while** *Q is not empty* **do**

1.10   **if** *an input event happened for site S_i* **then**

1.11   **for** *all nodes $n_d(P_j, S_i, S_{cj})$ in L_i* **do**

1.12   **if** *n_d became invalid when input event happened* **then**

1.13   Just delete it and do nothing;

1.14   **else if** *The node n_d (P_j, S_i, S_{cj}) in L_i whose t_{cj} still can not be determined* **then**

1.15   Just move it to the tail of L_i;

1.16   **else**

1.17   Recompute P_j's critical time t_{cj},update p_j's t_{cj} and new nearest site S_i ;

1.18   **if** *There is old critical time t'_{cj} for vertex P_j in Q and $t_{cj} < t'_{cj}$* **then**

1.19   Decrease the P_j's critical time in Q to t_{cj} ;

1.20   **else if** *No old critical time t'_{cj} for vertex P_j in Q* **then**

1.21   Insert P_j's critical time t_{cj} into Q;

1.22   **else**

1.23   Extract and delete earliest event(just suppose vertex P_m at time t'_{cm}) in Q ;

1.24   Report an event happened and update P_m's nearest site with S_{cm};

1.25   Compute new critical time t_{cm}, Set_m, S_{cm} just after t'_{cm};

1.26   **if** *t_{cm} can not be determined at t'_{cm}* **then**

1.27   For those site S_i in Set_m and satisfy $t_{cm} > tt_i$, add a node for site S_i into L_i;

1.28   **else**

1.29   Insert the new event(t_{cm}, S_{cm}, P_m) into Q;

1.30   **for** *every P_m's adjacent vertex P_z* **do**

1.31   **if** *$S_{cm} \notin Set_{cz}$ and $S_{cm} \neq S_z$(P_z's nearest site at moment)* **then**

1.32   Add element S_{cm} to Set_{cz};

1.33   Compute the P_z's new t_{cz} at time S_{cm};

1.34   **if** *t_{cz} can not be determined by now as S_{cm}* **then**

1.35   Add a node(P_z, S_{cm}, S_{cz}) into linklist L_{cm};

1.36   **else**

1.37   Recompute new t_{cz} for node P_z, update p_z's t_{cz} and new nearest site S_{cm};

1.38   **if** *old time $t'_{cz} \neq \infty$ and $t'_{cz} > t_{cz}$* **then**

1.39   Decrease the P_z's critical time in Q to t_{cz} ;

1.40   **else if** *$t'_{cz} = \infty$* **then**

1.41   Insert new time t_{cz} into Q;

1.42 **end**

Algorithm 1. Half-online moving Voronoi diagram algorithm

Figure 13(c). The sites like S_3 in Figure 15(c) appear at most $2m$ times in the Set_j of all preparation-nodes at any time, for prepare-nodes in G, to compute Set_j, critical time t_{cj} and critical site S_{cj}, which take $O(BD + m)$ time totally (lines [1.4-1.6]), and line [1.17] takes $O(BD + m)$ time for all valid nodes when line [1.10] for one loop. There are $\sum w_i$ input events for line[1.10-1.21], which sum up $B\sum w_i$ nodes. Hence line [1.17] costs $\sum w_i(BD+m)$ time for the whole algorithm. Line [1.15] puts at most $O(B)$ valid nodes to the tail of L_i each time because there at most $O(B)$ valid nodes for any site. Lines [1.27,1.35] add at most $m + D$ nodes when every output event happens, which is sum up $n_e(m+D)$. The total number of nodes is at most $n_e(m+D)+\sum w_iB$. Therefore lines [1.11-1.15] cost $O(n_e(m+D)+\sum w_iB)$ time. For the whole algorithm, lines [1.21,1.29,1.41] inserting element into Q cost $O(n_e \log B)$ time because every insertion event would happen soon or later. Lines [1.19,1.39] which decrease values in Q take $O(1)$ time using Fibonacci heap, and in total $O(\sum w_i + Dn_e)$ time. Lines [1.23-1.26,1.31-1.34] take $O((\log B+m+D)n_e)$ time (Line [1.33] only takes $O(1)$ time but not $O(m)$ time). Hence the total time complexity for whole algorithm is $O(n_e(\log B + D + m) + (\sum w_i(BD + m)))$. For the space complexity, we need $O(n_e)$ to store all events, and $O(n_e(m + D) + B\sum w_i)$ to store all nodes. Therefore space complexity is $O(n_e(m + D) + B\sum w_i)$.

Theorem 4. *The algorithm of maintaining half-online moving network Voronoi diagram costs $O(n_e(\log B + D + m) + (\sum w_i(BD + m)))$ time and $O(n_e(m + D) + B\sum w_i)$ space.*

7 Conclusions

In this paper, we consider three variants of the point location problem in moving network. First we computed all segmented functions $d(P_j, S_i(t))$ $(1 \leq i \leq m, 1 \leq j \leq n)$ in $O(n^2 \log n + nE + Y)$ time and $O(E + Y)$ space, where n and E are the number of nodes and edges of the network respectively, and Y is the total number of segmented functions for all nodes. After that, both the static case query and kinetic case query (if customer is located at some node) can be answered in $O(\log(mW))$ time with $O(nmW \log m)$ time and $O(nmW)$ space for preprocessing step. For static k nearest neighbor query, we can answer it in in $O(k \log X)$ time with $O((mW + kX) \log m + (m - k)X)$ time and $O(mW + kX)$ space for preprocessing, where X is the maximum number of segments in one of the k envelopes. For all queries, we assume that the trajectories of m sites are known in advance. For maintaining half-online moving network Voronoi diagram, we give $O(n_e(\log B + D + m) + (\sum w_i(BD + m)))$ time and $O(n_e(m + D) + B\sum w_i)$ space algorithm, where B is the maximum number of prepare-nodes at any time, D is the maximum degree of any vertex in G and n_e is the number of all output events happening. In the future work, it will be interesting to study the complete online version of this problem: we only know the trajectories of m sites up to now. How to maintain the Voronoi diagram such that we can answer the static case query efficiently? Note that we do not have kinetic case query for online problem since we do not know the trajectories of m sites in advance.

References

1. Agarwal, K.P., Sharir, M.: Davenport–Schinzel Sequences and Their Geometric Applications. Cambridge University Press, Cambridge (1995)
2. Bae, S.W., Kim, J.-H., Chwa, K.-Y.: Optimal Construction of the City Voronoi Diagram. In: Asano, T. (ed.) ISAAC 2006. LNCS, vol. 4288, pp. 183–192. Springer, Heidelberg (2006)
3. Bentley, J.L., Ottmann, T.A.: Algorithms for reporting and counting geometric intersections. IEEE Trans. Comput. 28(9), 643–647 (1979)
4. de Berg, M., van Kreveld, M., Overmars, M., Schwarzkopf, O.: Computational geometry algorithms and applications. Springer, Heidelberg (1997)
5. Devillers, O., Golin, M.J.: Dog Bites Postman: Point Location in the Moving Voronoi Diagram and Related Problems. In: Lengauer, T. (ed.) ESA 1993. LNCS, vol. 726, pp. 133–144. Springer, Heidelberg (1993)
6. Erwig, M.: The graph voronoi diagram with application. Networks 36, 156–163 (2000)
7. Guibas, L., Hershberger, J., Leven, D., Sharir, M., Tarjan, R.: Linear time algorithms for visibility and shortest path problems inside simple polygons. In: SCG 1986: Proceedings of the Second Annual Symposium on Computational Geometry, pp. 1–13. ACM, New York (1986)
8. Hakimi, S., Labbe, M., Schmeiche, E.: The voronoi partition of a network and its implications in location theory. INFIRMS Journal on Computing 4, 412–417 (1992)
9. Kolahdouzan, M.R., Shahabi, C.: Alternative solutions for continuous k nearest neighbor queries in spatial network databases. GeoInformatica 9(4), 321–341 (2005)
10. Lee, D.T.: Two-dimensional voronoi diagrams in the lp-metric. J. ACM 27(4), 604–618 (1980)
11. Okabe, A., Satoh, T., Furuta, T., Suzuki, A., Okano, K.: Generalized network voronoi diagrams: Concepts, computational methods, and applications. Int. J. Geogr. Inf. Sci. 22(9), 965–994 (2008)
12. Pach, J., Sharir, M.: On vertical visibility in arrangements of segments and the queue size in the bentley-ottmann line sweeping algorithm. SIAM J. Comput. 20(3), 460–470 (1991)
13. Ramalingam, G., Reps, T.: An incremental algorithm for a generalization of the shortest-path problem. J. Algorithms 21(2), 267–305 (1996)

A GIS Based Wireless Sensor Network Coverage Estimation and Optimization: A Voronoi Approach

Meysam Argany[1], Mir Abolfazl Mostafavi[1], Farid Karimipour[2], and Christian Gagné[3]

[1] Center for Research in Geomatics, Laval University, Quebec, Canada
meysam.argany.1@ulaval.ca
mir-abolfazl.mostafavi@scg.ulaval.ca
[2] Department of Surveying and Geomatics, College of Engineering,
University of Tehran, Tehran, Iran
fkarimipr@ut.ac.ir
[3] Department of Electrical Engineering and Computer Engineering, Laval University
Quebec, Canada
christian.gagne@gel.ulaval.ca

Abstract. Recent advances in sensor technology have resulted in the design and development of more efficient and low cast sensor networks for environmental monitoring, object surveillance, tracking and controlling of moving objects, etc. The deployment of a sensor network in a real environment presents several challenging issues that are often oversimplified in the existing solutions. Different approaches have been proposed in the literatures to solve this problem. Many of these approaches use Voronoi diagram and Delaunay triangulation to identify sensing holes in the network and create an optimal arrangement of the sensors to eliminate the holes. However, most of these methods do not consider the reality of the environment in which the sensor network is deployed. This paper presents a survey of the existing solutions for geosensor network optimization that use Voronoi diagram and Delaunay triangulation and identifies their limitations in a real world application. Next, it proposes a more realistic approach by integrating spatial information in the optimization process based on Voronoi diagram. Finally the results of two cases studies based on the proposed approach in natural area and urban environment are presented and discussed.

Keywords: geosensor networks deployment, coverage problem, Voronoi diagram, Delaunay triangulation, GIS.

1 Introduction

Recent advances in electomechanical and communication technologies have resulted in the development of more efficient, low cost and multi-function sensors. These tiny and ingenious devices are usually deployed in a wireless network to monitor and collect physical and environmental information such as motion, temperature, humidity, pollutants, traffic flow, etc [7]. The information is then communicated to a process center where they are integrated and analyzed for different application. Deploying sensor networks allows inaccessible areas to be covered by minimizing the

M.L. Gavrilova et al. (Eds.): Trans. on Comput. Sci. XIV, LNCS 6970, pp. 151–172, 2011.

sensing costs compared to the use of separate sensors to completely cover the same area. Sensors may be spread with various densities depending on the area of application and the details and the quality of the information required.

Despite the advances in the sensor network technology, the efficiency of a sensor network for collection and communication of the information may be constrained by the limitations of sensors deployed in the network nodes. These restrictions may include sensing range, battery power, connection ability, memory, and limited computation capabilities. These limitations have been addressed by many researchers in recent years from various disciplines in order to design and deploy more efficient sensor networks [28].

Efficient sensor network deployment is one of the most important issues in sensor network filed that affects the coverage and communication between sensors in the network. Nodes use their sensing modules to detect events occurring in the region of interest. Each sensor is assumed to have a sensing range, which may be constrained by the phenomenon being sensed and the environment conditions. Hence, obstacles and environmental conditions affect network coverage and may result in holes in the sensing area. Communication between nodes is also important. Information collected from the region should be transferred to a processing center, directly or via its adjacent sensor. In the later case, each sensor needs to be aware of the position of other adjacent sensors in their proximity.

Several approaches have been proposed to detect and eliminate holes and hence increase sensor networks coverage through optimization methods [9, 23, 24, 26, 39, 40, 41]. Many of these approaches use Voronoi diagram and Delaunay triangulation to identify sensing holes in the network and create an optimal arrangement of the sensors to eliminate the holes. However, most of these methods over simplify the environment in which the sensor networks are deployed reducing the quality of spatial coverage estimation and optimization. This paper makes a critical overview of the existing solutions based on Voronoi diagrams and Delaunay triangulation for geosensor network coverage estimation and optimization. Next, it proposes a novel sensor network deployment approach by integrating spatial information in the optimization process based on Voronoi diagram.

The remainder of this paper is as follows. Section 2 presents a state of the art on the geosensor networks and their related issues. Section 3, describes the coverage problem in geosensor networks and different solutions found in the literature for its estimation and optimization. Section 4 presents the coverage determination and optimization solutions based on Voronoi and Delaunay triangulation and their limitations. Section 5 proposes a novel sensor network deployment approach by integrating spatial information in the optimization process based on Voronoi diagram. In section 6, we present the results of the two experimentations based on the proposed approach both in natural and urban areas. Finally, section 7 concludes the paper and proposes new avenues for the future works.

2   State of the Art on Geosensor Networks and Their Applications

Sensor networks were announced as one of the most important technologies for the 21st century in 1999 by *Business Week* [46]. These networks are usually composed of a set of small, smart and low-cost sensors with limited on-board processing capabilities, storage and short-range wireless communication links based on radio technology. Previously, sensor networks consisted of small number of sensor nodes that were usually wired to a central processing station. However, nowadays, the focus is more on wireless, distributed, sensing nodes [6, 35, 42]. A sensor node is characterized by its sensing field, memory and battery power as well as its computation and communication capabilities. A sensor can only cover a small area. However, collaboration of a group of sensors with each other can cover a more significant sensing field and hence accomplishing much larger tasks. Each element of a group of sensors can sense and collect data from the environment, apply local processing, communicate it to other sensors and perform aggregations on the observed information [31].

Sensor networks are also referred to as *Geo*sensor networks as they are intensively used to acquire spatial information [28]. Hereafter, we will use both of the terms "sensors" and "geosensors" interchangeably. Geosensors can be deployed on the ground, in the air, under water, on bodies, in vehicles, and inside buildings.

Sensor networks have several applications including environmental monitoring, change detection, traffic monitoring, border security, and public security, etc. They are used for collecting the information needed by smart environments quickly and easily, whether in buildings, utilities, industries, home, shipboard, transportation systems automation, or elsewhere. Sensor networks are useful in vehicle traffic monitoring and control. Most traffic intersections have either overhead or buried sensors to detect vehicles and control traffic lights. Furthermore, video cameras are frequently used to monitor road segments with heavy traffic, with the video sent to human operators at central locations [7]. Sensor networks can be used for infrastructure security in critical buildings and facilities, such as power plants and communication centers. Networks of video, acoustic, and other sensors provide early detection of possible threats [34]. Commercial industries has long been interested in sensing as a means of lowering cost and improving machine (and perhaps user) performance and maintainability. Monitoring machine "health" through determination of vibration or wear and lubrication levels, and the insertion of sensors into regions inaccessible by humans, are just two examples of industrial applications of sensors [7]. A broad classification of geosensor network applications is monitoring continuous phenomena (e.g., to assess plant health and growth circumstances, or to observe and measure geophysical processes), detecting real time events (e.g., flood and volcano), and tracking objects (e.g., animal monitoring) [28, 35, 42].

Sensor networks have some limitations when it comes to the modeling, monitoring and detecting environmental processes. Monitoring and analyzing dynamic objects in real time are also difficult. Examples of such processes include the observations of dynamic phenomena, (e.g., air pollution) or monitoring of mobile objects (e.g., animals in a habitat). It is necessary to know how to use this technology to detect and monitor those phenomena appropriately and efficiently. For this purpose, one needs to identify the relevant mix of hardware platforms for the phenomena type, the accessibility or inaccessibility of the observation area, hazardous environmental conditions, and power availability, etc. Today wireless sensor network technology are more effectively used for detecting and monitoring time-limited events (e.g., earthquake tremors), instead of continuous sampling in remote areas due to the battery constraints of geosensor platforms. [28].

3   Coverage Problem in Geosensor Networks

An important issue to deploy a sensor network is finding the best sensor location to cover the region of interest. Definition of coverage differs from an application to another. The so-called art gallery problem, for example, aims to determine the minimum number of required observers (cameras) to cover an art gallery room such that every point is seen by at least one observer [5]. Hence, here, the coverage is defined based on a direct visibility between the observer and the target point. In sensor networks, however, the coverage of a point means that the point is located in the sensing range of a sensor node, which is usually assumed to be uniform in all directions. In this case, the sensing range is represented by a disk around the sensor [3]. Failing this condition for some points in the region of interest will result in coverage holes (Fig. 1).

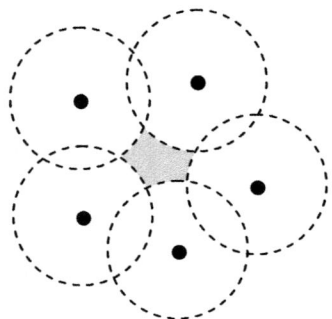

Fig. 1. Coverage hole (shaded region) in a sensor network with disk model sensing range

Regarding this definition of coverage in sensor network, the coverage problem basically means placing minimum number of nodes in an environment, such that every point in the sensing field is optimally covered [1, 11]. Nodes can either be placed manually at predetermined locations or dropped randomly in the environment. It is difficult to find a random scattering solution that satisfies all the coverage and connectivity conditions. Thus, the term of area coverage plays an important role in sensor networks and their connectivity.

The existing solutions to determine and optimize the coverage in sensor networks can be classified in two main categories of "exposure based" and "mobility based" approaches [11]. Exposure based solutions evaluate unauthorized intrusions in the networks. Mobility based solutions, however, exploit moving properties of nodes to get better coverage conditions and try to relocate sensor nodes to optimal locations that serve maximum coverage.

3.1   Coverage Based on Exposure

The estimation of coverage can be defined as a measure of the ability to detect objects within a sensor filed. The notion of *exposure* can represent such a measurement. It is described as the expected average ability of observing a target moving in a sensor field. It is related to coverage in the sense that "it is an integral measure of how well the sensor network can observe an object [exists in the field or] moving on an arbitrary path, over a period of time" [25].

A very simple, but nontrivial example of exposure problem is illustrated in Fig. 2. An object moves from point *A* to point *B* and there is only one sensor node *S* in the field. Obviously, the path 2 has the maximum exposure, because it is the shortest path from *A* to *B* and it passes through the sensor node *S*. Thus, the object moves along this path is certainly tracked by *S*. However, finding the path with the minimum exposure is tricky: although path 1 is the farthest path from the sensor node *S* and so intuitively seems to have the lowest exposure, it is also the longest path. Therefore, travelling along this path takes longer time and the sensor has longer time to track the moving object. It is shown that the minimum exposure path is 3, which is a trade-off between distance from the sensor and travelling time [18].

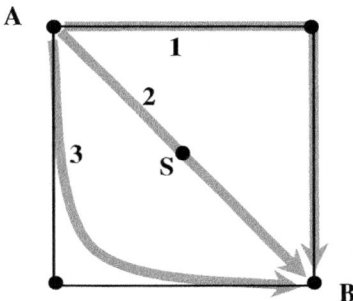

Fig. 2. Minimum and maximum exposure paths in a simple sensor network [18]

The so-called *worst* case and *best* case coverage are examples of methods for exposure evaluation [23, 26]. Worst-case coverage is the regions of lower observability from sensor nodes, so objects move along this path has the minimum probability to be detected. Best-case coverage, however, is the regions of higher observability from sensors, thus probability of detecting an object moving along this path is maximum [11]. These two parameters together give an insight of the coverage quality of the network and can help to decide if additional sensors must be deployed. Different approaches have been proposed in the literatures for the worst- and best-case coverage problems [19, 25, 27, 30, 37]. A Voronoi based solution for this problem is presented in section 4.

3.2 Coverage Based on Mobility

In some sensor placements approaches, where there is no information available about terrain surface and its morphology, random sensor placement is used. This method does not guarantee the optimized coverage of the sensing region. Thus, some deployment strategies take advantage of mobility options and try to relocate sensors from their initial places to optimize the network coverage. Potential field-based, virtual force-based and incremental self-deployment methods [14, 15, 44] are examples of such approach and are introduced, here. Other methods such as VEC, VOR and MiniMax, which are mobility based methods that use Voronoi diagram in their approach, are explained in the next section.

The idea of potential field is that every node is exposed to two forces: (i) a repulsive force that causes the nodes to repel each other, and (ii) the attractive force that makes nodes moving toward each other when they are on the verge of being disconnected [11, 15]. These forces have inverse proportion with the square of distance between nodes. Each node repels all its neighbors. This action decreased the repulsive force, but at the same time, it stimulates the attractive force. Eventually, it ends up in an arrangement in which all the nodes reach an equilibrium situation and uniformly cover the sensing field.

Virtual force-based method is very similar to potential-based, but here each node is exposed to three types of forces: (i) a repulsive force exerted by obstacles, (ii) an attractive force exerted by areas where the high degree of coverage is required, and (iii) attractive or repulsive force by another point based on its location and orientation [44, 45].

In incremental self-deployment algorithm each node finds its optimal location through previous deployed nodes information in four steps [12, 13, 14]: (i) *initialization* that classifies the nodes to three groups: waiting, active and deployed; (ii) *goal selection* that selects the best destination for the node to be deployed based on previous node deployment; (iii) *goal resolution* that assigns this new location to a waiting node and the plan for moving to this location is specified; (iv) Finally, *execution* that deploys the active nodes in their place.

As it is realized in the above algorithms, spatial coverage of sensor networks is much related to the spatial distribution of the sensors in the environment. In other words, the described algorithms try to distribute the sensors in the field so that the much possible coverage is obtained. Voronoi diagram and Delaunay triangulation are the data structures that directly satisfy the required distribution. They have been used for developing algorithms for both exposure and mobility based approaches.

4   Role of Voronoi Diagram and Delaunay Triangulation

This section presents the solutions for sensor network coverage optimisation that use Voronoi diagram and Delaunay triangulation for coverage determination and optimization in sensor networks. The solutions are categorized as coverage hole detection, healing the holes, and node scheduling. Some other challenges are introduced at the end of this section.

4.1   Coverage Hole Detection

In a simple sensor network – where the sensing regions of all sensors are identical circles – if a point is not covered by its closest sensor node, obviously it is not covered by any other sensor node. This property is the basis to use Voronoi diagram in sensor coverage problem: in a Voronoi diagram, all the points within a Voronoi cell are closest to the generating node that lies within this cell. Thus, having constructed the Voronoi diagram of the sensor nodes and overlaid the sensing regions on it (Fig. 3), if a point of a Voronoi cell is not covered by its generating node, this point is not covered by any other sensors [3, 9, 38, 39]. Although computing the area of a Voronoi cell is straightforward, computing the area of the uncovered region in a Voronoi cell is a complicated task, because the sensing regions may protrude the Voronoi cells and overlay each other. Strategies for this computation can be found in [9, 39].

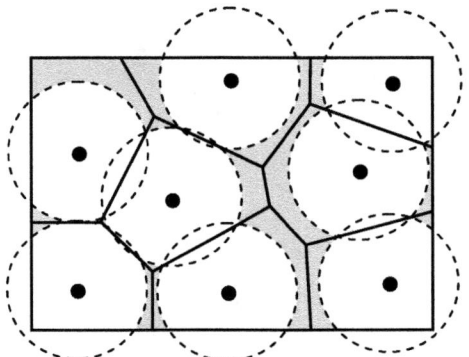

Fig. 3. Using Voronoi diagram to detect the coverage holes (shaded regions) in a sensor network

Another Voronoi-based approach to evaluate the coverage of a sensor network is based on the notion of *exposure*, which was discussed earlier in section III. To solve the worst-case coverage problem, a very similar concept, i.e., *maximal breach path* is used. It is the path through a sensing field between two points such that the distance from any point on the path to the closest sensor is maximized. Since the line segments of the Voronoi diagram has the maximum distance from the closest sites, the maximal breach path must lie on the line segments of the Voronoi diagram corresponding to the sensor nodes (Fig. 4). The Voronoi diagram of the sensor nodes is first constructed. This diagram is then considered as a weighted graph, where the weight of each edge is the minimum distance from the closest sensor. Finally, an algorithm uses breadth first and binary searches to find the maximal breach path [23, 26].

The best-case coverage problem is solved through the similar concept of *maximal support path*. This is the path through a sensing field between two points for which the distance from any point on it to the closest sensor is minimized. Intuitively, this is traveling along straight lines connecting sensor nodes. Since the Delaunay triangulation produces triangles that have minimal edge lengths among all possible triangulations, maximal support path must lie on the lines of the Delaunay triangulation of the sensors (Fig. 5). Delaunay triangulation of the sensor nodes is constructed and considered as a weighted graph, where the weight of each edge is the length of that edge. The maximal breach path is found through an algorithm that uses breadth first and binary searches [23, 26].

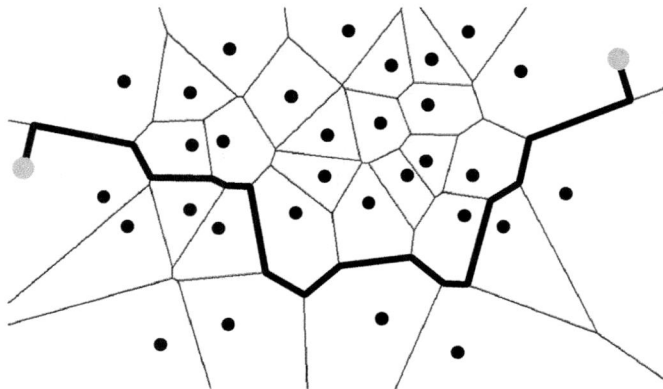

Fig. 4. Maximum breach path in a sensor network and its connection to Voronoi diagram

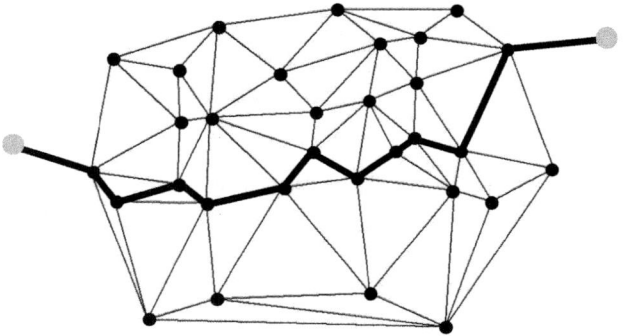

Fig. 5. Maximum support path in a sensor network and its connection to Delaunay triangulation

4.2 Healing the Holes

Having detected the coverage holes, the sensors must be relocated in order to heal the holes. For this, we classify the Voronoi-based solutions based on the sensor types used in the network: (1) Static sensor networks, (2) mobile sensor networks, and (3) hybrid sensor networks, where a combination of static and mobile sensors is deployed. For static sensor networks, new sensors are added. For mobile and hybrid networks, however, existing sensors moves to heal the holes.

Static Sensor Networks

To the best of our knowledge, there are two suggestions to deploy an additional sensor to heal the holes in a static sensor network. Gosh [9] proposes that for each Voronoi vertex, one node should be added to heal the coverage hole around this Voronoi vertex. As Fig. 6 shows, to heal the hole around Voronoi vertex v_2, the target location p_1 lies on the bisector of the angle $v_1v_2v_3$ and $d(s, p_1) = \min \{2R, d(s,v_2)\}$, where d is the

Euclidean distance and R is the sensing radius of the sensors. Wang *et al.* [39], however, deploy only one mobile node to heal the coverage hole of a Voronoi cell. As illustrated in Fig. 6, the target location p_2 lies on the line connecting the sensor node and its furthest Voronoi vertex (v_4 here) and $d(s, p_2) = \max \{ \sqrt{3} R, d(s, v_4) \}$.

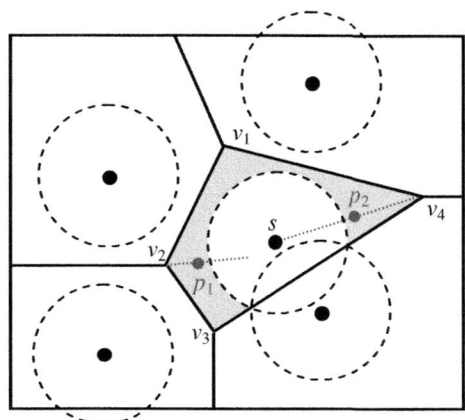

Fig. 6. Deploying an additional sensor to heal the hole in a static sensor network

Mobile Sensor Networks
In mobile sensor networks, all sensors have the ability to move in order to heal the holes. Wang *et al.* [40] proposes three Voronoi-based strategies for this movement: Vector-based (VEC), Voronoi-based (VOR), and Minimax. They all are iterative approaches and gradually improve the coverage of the sensor network.

VECtor-based Algorithm (VEC)
VEC pushes sensors away from a densely covered area. It imitates the electromagnetic force that exists between two particles: if two sensors are too close to each other, they exert a repulsive force. By knowing the target area and the number of sensors, an average distance between the sensors, d_{avg} can be calculated beforehand. If the distance between two sensors s_i and s_j is smaller than d_{avg} and none of their Voronoi cells is completely covered, the virtual force pushes them to move ($d_{avg} - d(s_i, s_j)$)/2 away from each other. However, if one of the sensors completely covers its Voronoi cell, and so it should not move, then the other sensor pushes ($d_{avg} - d(s_i, s_j)$) away.

In addition to the repulsive forces between sensors, the boundaries also exert forces to push sensors that are too close to the boundary inside. If the distance of the sensor i, i.e., $d_b(s_i)$, from its closets boundary is smaller than $d_{avg}/2$, then it moves ($d_{avg}/2 - d_b(s_i)$) toward the inside of the network.

Note that movements of the sensors change the shape of the Voronoi cells, which may result in decreasing the coverage in the new configuration. Thus, the sensors move to the target position only if their movement increase the local coverage within their Voronoi cell. Otherwise, they take the midpoint position between its current and target positions, as the new target position, and again check the improvement, and so on. This process is called movement adjustment). Fig. 7 shows an example of using VEC algorithm.

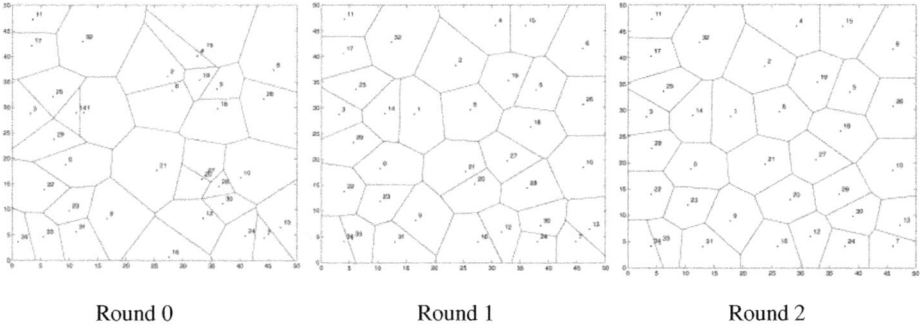

Round 0   Round 1   Round 2

Fig. 7. An example of using VEC algorithm to move the sensors [40]

VORonoi-based Algorithm (VOR)
Unlike VEC algorithm, VOR is a pulling strategy so that sensors cover their local maximum coverage holes. In this algorithm, each sensor moves toward its furthest Voronoi vertex till this vertex is covered (Fig. 8). The movement adjustment mentioned for VEC is also applied here. Furthermore, VOR is a greedy algorithm that heals the largest hole. However, after moving a sensor, a new hole may be created that is healed by a reverse movement in the next iteration, so it results in an oscillation moving. An *oscillation control* is added to overcome this problem. This control does not allow sensors to move backward immediately: Before a sensor moves, it first checks if the direction of this moving is opposite to that in the previous round. If so, it stops for one round to see if the hole is healed by the movement of a neighbouring sensor. Fig. 9 shows an example that moves the sensors based on VOR algorithm.

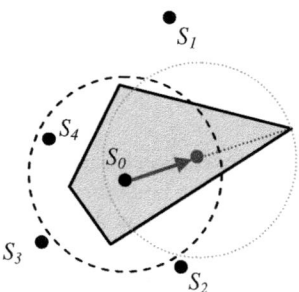

Fig. 8. Movement of a sensor in VOR algorithm

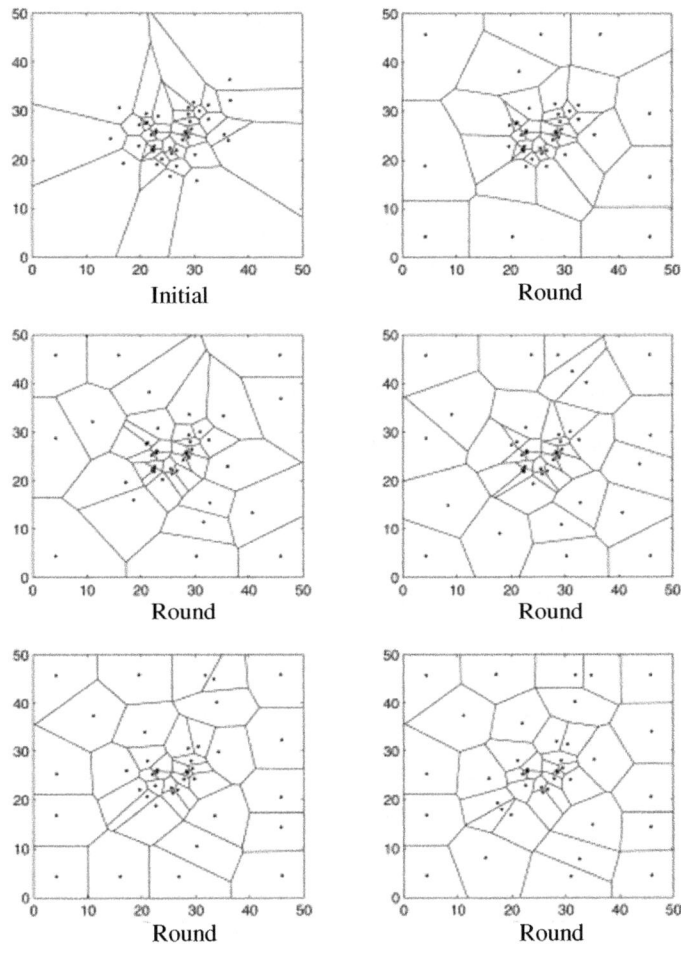

Fig. 9. An example of using VOR algorithm to move the sensors [40]

Minimax Algorithm
This algorithm is based on the fact that when the sensors are evenly distributed, a sensor should not be too far away from any of its Voronoi vertices. In other words, the disadvantage of VOR algorithm is that it may result in a case where a vertex that was originally close becomes a new farthest vertex. The MiniMax algorithm solves this by choosing the target location as the point inside the Voronoi cell whose distance to the farthest Voronoi vertex is minimized. This point, which is called *Minimax* point, is the center of the smallest enclosing circle of the Voronoi vertices and can be calculated by the algorithms described in [24, 32, 41]. Minimax algorithm has some advantages. Firstly, it can reduce the variance of the distances to the Voronoi Vertices, resulting in more regular shaped Voronoi cells, which better utilizes the sensor's sensing circle. Secondly, Minimax considers more information

than VOR, and it is more conservative. Thirdly, Minimax is more "reactive" than VEC, i.e., it heals the hole more directly by moving toward the farthest Voronoi vertex.

Hybrid Sensor Networks
In a hybrid sensor network, having detected a hole around a static sensor, a mobile sensor moves in order to heal this hole. The location to which the mobile sensor should move is computed similar to the solutions proposed for the static networks in section IV.B.1. Then, the static sensor requests the neighbouring mobile sensors to move to the calculated destination. Each of the mobile sensors that have received this request calculates the coverage holes formed at its original location due to its movement. It decides to move if the new hole is smaller than the hole size of the requesting static sensor. It is noted that since movements of the mobile sensors may create new (but smaller) holes, this solution is an iterative procedure. More discussion on this movement and its technical considerations (e.g., bidding protocols) can be found at [9, 39].

4.3   Node Scheduling

As it was mentioned earlier, energy is an important issue in sensor networks. Thus, strategies to save energy are of most interest in this regards. A relevant case to save the energy is turning temporarily some sensor nodes to sleep mode in the multi-covered areas. This is also important to avoid other problems (e.g., the intersection of sensing area, redundant data, and communication interference), in areas with a high density of sensor nodes [21]. Different methods have been proposed for this problem [29, 36].

Augusto *et al.* [21] proposed a Voronoi-based algorithm to find the nodes to be turned on or off. The Voronoi diagram of the sensor nodes is constructed. Each Voronoi cell represents the area that the corresponding node is responsible for. The sensors whose responsible areas are smaller than a predefined threshold are turned off. By updating the Voronoi diagram, the neighbours of that sensor become responsible for that area. This process continues until there is no node responsible for an area smaller than the given threshold.

4.4   Other Challenges

This section shortly introduced more complicated issues in sensor coverage problem that can be dealt using Voronoi diagram and Delaunay triangulation.

K-Coverage Sensor Networks
In some applications, such as military or security control, it is required that each point of the region is covered by at least k $(k>1)$ sensors. Among different solution proposed in the literatures [43], So and Ye [33] has developed an algorithm based on the concept of *Voronoi regions*. Suppose that $P=\{p_1, p_2, ..., p_n\}$ is a set of n point in $\mathbf{R}^n$. For any subset U of P, the Voronoi region of U is set of points in $\mathbf{R}^n$ closer to all points in U than to any point in $P-U$. The proposed algorithm can check the k-coverage for the area, but developing the algorithms to heal the holes is still an open question.

Sensor Networks with Various Sensing Ranges

So far, we have assumed that all sensors are identical. In reality, however, a sensor network could be composed of multiple types of sensors with different specifications, including their sensing range and sensing model (e.g., circular, ellipsoidal or irregular sensing model [3, 33]). Weighted Voronoi diagram is a solution in such cases to examine the coverage quality of the network (Fig. 10) [33]. However, to the best of our knowledge, the movement strategies have not been researched deeply for such heterogeneous sensor networks.

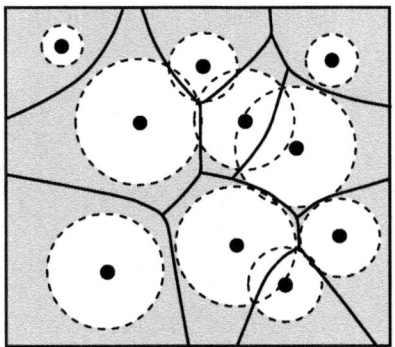

Fig. 10. Using weighted Voronoi diagram to examine the coverage quality of a sensor network with various sensing ranges

Directional Sensor Networks

Coverage determination for directional sensor networks (i.e., networks composed of sensor with limited field of views) is a practical area of research. Adriaens *et al.* [2] has extended the research done in [23, 26] and developed a Voronoi-based algorithm to detect the worst-case coverage (maximal breach path) in such networks.

Sensor Networks in a 3D Environment

The approaches mentioned in this paper assume that a sensor network is deployed in a 2D flat environment (i.e., a 2D Euclidean plane). However, this assumption oversimplifies sensor network reality. The real world environment is mostly 3D heterogeneous filed and contains obstacles (Fig. 11). Hence, 3D sensor networks have considerable interest in diverse applications including structural monitoring networks and underwater networks [17]. In addition to the form and the relief of the sensor network area, various obstacles may prevent the sensors from covering an invisible region or communicating data between each other.

Several algorithms have been proposed for the coverage problem of 3D sensor networks [4, 7, 17]. The algorithms presented here can be extended to use 3D Delaunay triangulation and Voronoi diagram for coverage determination and optimization of such sensor networks [10, 20, 22]. There are also suggestions to use Delaunay triangulation and Voronoi diagram when the environment contains obstacles [16]. Although these extensions are interesting in some applications, they may have deficiencies for the geographical fields, because they consider 3D

Euclidean filed and man-made obstacles, e.g., walls. Real world environment, however, is a 3D heterogeneous filed full of man-made and natural obstacles. Even, the terrain could play the role of an obstacle in this case. Using capabilities of geographical information systems (GIS) seems a promising solution in this regard, which has not been investigated. It can provide the information (e.g., digital terrain models) or spatial analyses (e.g., visibility analysis) required to evaluate and optimize the sensor networks installed in the nature environment. Hence, 3D Delaunay triangulation and Voronoi diagrams present interesting solutions for the sensor network modeling and optimization in 3D environment. However, their application is not straightforward and several challenging conceptual and implementation problems should be addressed.

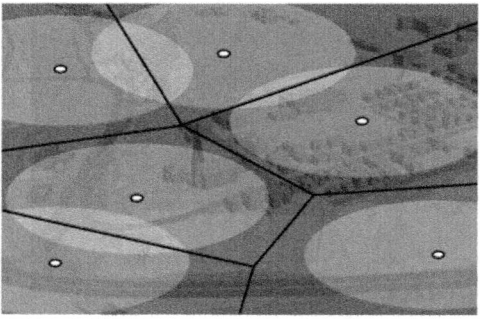

Fig. 11. A sensor network in a 3D environment with various obstacles. The superimposed 2D Voronoi diagram cannot determine the network coverage

5   Proposed Approach for a Realistic Sensor Network Deployment

Although efficient sensor deployment for maximum network coverage has been extensively addressed in the literatures (sections 3 and 4), they are not adequately adopted to consider the reality of the terrain and the environment where the sensor networks are deployed. The main reasons are:

- Most of the existing solutions suffer from the lack of integrating environmental information with sensor network deployment algorithms. They do not consider the form and the topography of the area covered by the sensor network as well as various existing obstacles that may prevent the sensors from covering the whole area or allowing data communication between sensors. To carry out a realistic sensor placement scheme, it is necessary to involve the environmental information that affects sensor performance and network coverage.
- The sensor network region of interest may change over the sensing experiments. For instance, in a battlefield all parameters of the study area may rapidly change. In urban areas, new constructions may happen, urban facilities may be added or removed or changes may occur in land cover and land use information. These changes may significantly affect the sensor network coverage. Furthermore, characteristics of sensor platforms may change during the sensing steps. For example, fluctuation of the battery power for each platform decreases the sensing

range of nodes, so the network arrangement must be modified to stay in good network performance. These changes must be considered by the network and the development methods must be adopted to deal with them.

For establishing a realistic sensor network, we propose an innovative sensor placement method using Voronoi-based optimization methods integrated with terrain information and realistic sensors models. For that purpose, an optimization process is coupled to a Geographical Information System (GIS) for integrating spatial information, including man-made (buildings, bridges, etc) or natural objects. Moreover, the functions and capabilities available in GIS serve more facilities in sensor network deployment. Visibility, line of sight and viewshed analysis are examples of GIS operations that will be used in this regard. Finally, we deploy a dynamic geometric data structure based on Voronoi diagram in order to consider the topology of the sensor network and its dynamics (e.g. inserts, move, delete). In short, our approach focuses on definition and implementation of a framework that integrates environmental information for optimal deployment of sensor nodes based on a geometric data structure (e.g., Voronoi diagram) and optimization algorithms.

A GIS aided simulation platform based on a geometric data structure is used to reduce the coverage holes and to make an optimum sensor network deployment. This is done by using the functionality of a GIS to locate environmental objects such as buildings, vegetations, and sensor nodes in their accurate positions. It also uses other environmental information such as Digital Terrain Models (DTM) to get more reliable results. DTMs are very important issues to be included in the realistic modeling of sensor placement, which have not been considered in most of the previous works. Using GIS helps the deployment process in terms of analyzing the visibility between the sensors (viewshed) and line of sight for sensing area of each sensor in the network.

The proposed framework consists of three major parts including a spatial database (GIS), a knowledgebase and a simulation engine, based on Voronoi diagram (Fig. 12). The spatial database is implemented using a GIS, where different environmental elements organized as different layers, such as man-made and natural obstacles (e.g., streets, building blocks, trees, poles and terrain topography). Another layer will contain the coverage, which is calculated in different steps of the sensor network placement process. An extra layer is defined to keep the sensors positions. These various GIS layers may be updated during the sensing mission considering the fact that the coverage layer may change following the changes in the environmental information layers or sensor nodes positions. All attributes are defined in this database and all metric and topologic operations are exported based on the analyses that are carried out in this database. Other GIS processes such as visibility and viewshed analyses are done in this database.

The second component is the knowledgebase. All environmental and network parameters are used to define basic rules and facts that are stored in this knowledgebase. The knowledgebase is used by a simulation platform for sensor network deployment. The simulation engine consists of a local optimization algorithm based on Voronoi diagram. A reasoning engine will help to extract the appropriate commands to move or delete existing sensors or add new sensors in the network to satisfy the optimum

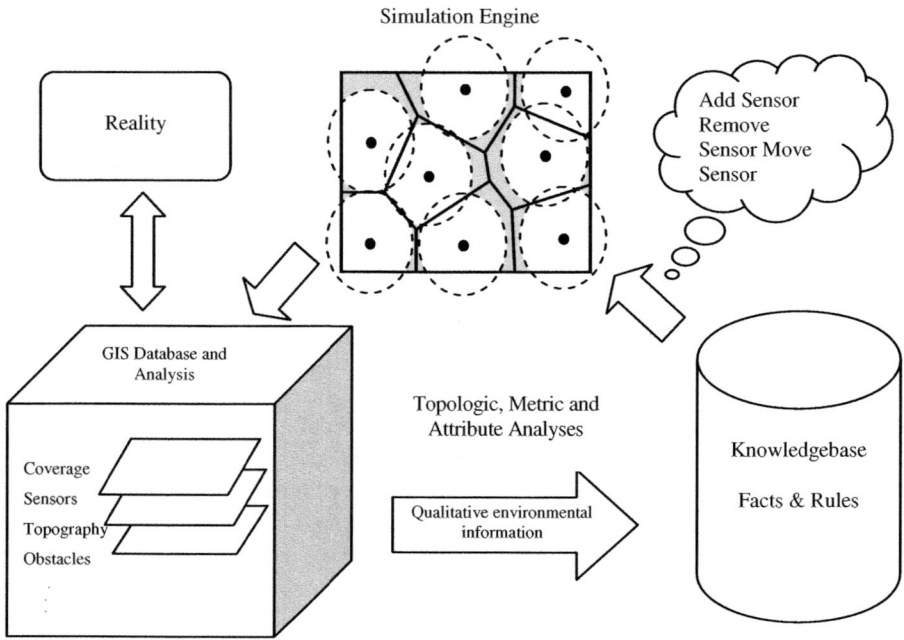

Fig. 12. The proposed framework

coverage. In fact, the optimization algorithm tries to relocate the sensors based on the defined rules in this knowledgebase. Both the database and the knowledgebase components are in relation with the simulation engine as shown in the figure 12.

6   Implementation of the Proposed Approach for Two Case Studies

For evaluation purpose, the proposed sensor deployment approach has been used in two case studies. The first case consists on deploying a sensor network in an urban area, which is a small part of Quebec City (Fig. 13a). In the second case study, we consider a sensor network in a natural area is a small part of Montmorency Forest located in the north of Quebec City (Fig. 13b). Initially, the study areas were covered by 10 sensors with sensing range of 50 meters for both maps. The sensors can rotate - 90° to 90° vertically and 0° to 360° horizontally. Initially, the sensors were considered to be randomly distributed in both natural and urban study area. For the urban data set, we suppose that the sensors are deployed in a network to monitor activities in a small part of a city. Assuming this, the sensors could be video cameras or optic sensors with the ability to rotate in 2D or 3D orientations, installed a few meters above the ground. This assumption is necessary to better consider the presence of different obstacles in the sensing area.

(a)   (b)

Fig. 13. The study areas: (a) a small part of Quebec City (urban area) and (b) a small part of Montmorency Forest in Quebec (natural area)

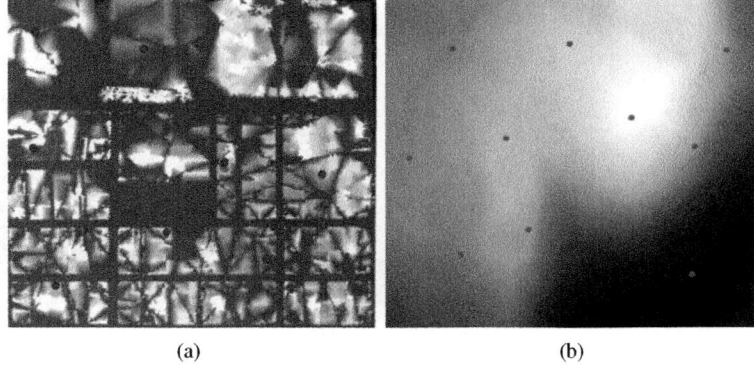

(a)   (b)

Fig. 14. Initial positions of the sensors on the DTM: (a) urban area (b) natural area

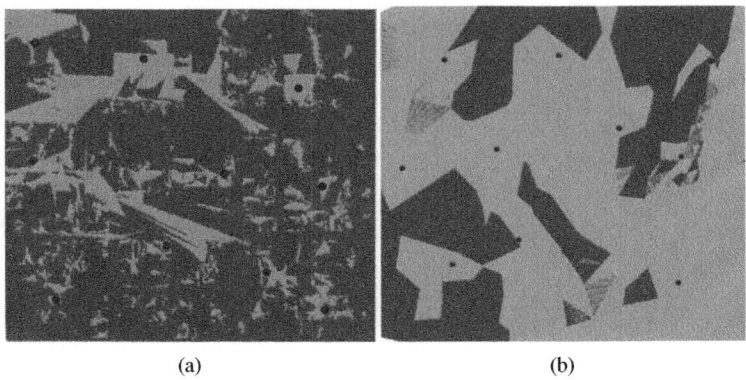

(a)   (b)

Fig. 15. Viewshed of the first sensor deployment: (a) urban area (b) natural area. Green regions are visible and red regions are invisible.

Fig. 14a and 14b show the initial position of the sensors on the DTM, of the urban and natural areas respectively, which result in viewsheds of the sensors in the environments (Fig. 15a and 15b).A pixel is assumed to be visible if it is observable by at least one sensor.

A 50 meter buffer around each sensor shows its sensing range. On the other hand, as explained in section 4, it is desired that each sensor node cover its Voronoi cell. Therefore, as shown in Fig. 16a and 16b, the current configuration is not optimal because there are areas that are covered by none of the sensors. Overlaying the buffers and the viewshed maps, the visible area in the sensing field of each sensor node is obtained (Fig 17a and 17b), which are 23% for the initial deployment of the sensors in the urban area and 66% in the natural area. We called this overlaid area, the coverage of each sensor. While, the visibility, means all of the area which have the possibility to be observed by the sensor nodes without considering the sensing range of the sensors.

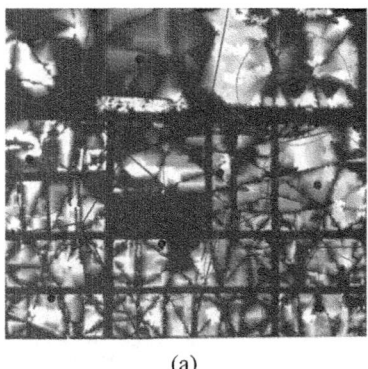
(a)

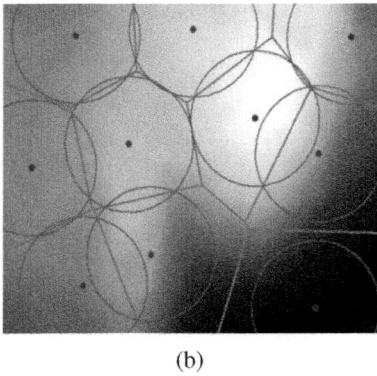
(b)

Fig. 16. Sensor's positions and their related sensing buffer and Voronoi cells in the initial deployment: (a) urban area (b) natural area

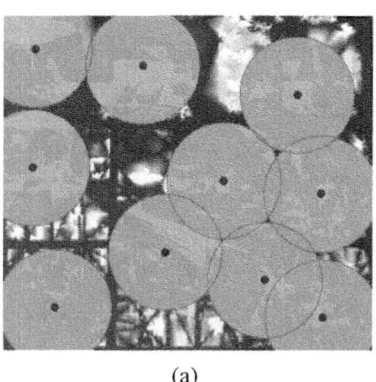
(a)

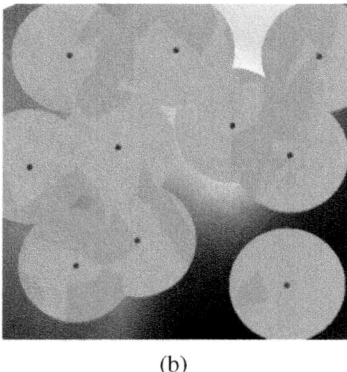
(b)

Fig. 17. The covered regions in the sensing field of each sensor node in initial deployment: (a) urban area (b) natural area. Green regions are visible and pink regions are invisible.

To increase the covered area, the VOR algorithm (section 4) is used: the sensors were moved toward the farthest Voronoi vertex, but with this restriction that the sensor stops if it reaches a position with a higher elevation than its current position.

This constraint is an extension to the VOR algorithm that allows us to better consider the topography of the terrain and the presence of various obstacles in the sensing area. This consideration will help us to significantly improve the spatial coverage of the sensor network in both case studies and also prove our initial hypothesis. Fig. 7a and 7b show the result of this movement. As Table 1 indicates, both of visibility and coverage have been relatively improved in both urban and natural areas. In urban area, the visibility has been increased 12% as well as 4% in natural area. In terms of coverage, in urban area there is 14% of coverage improvement and in natural area we can see 5% of coverage improvement.

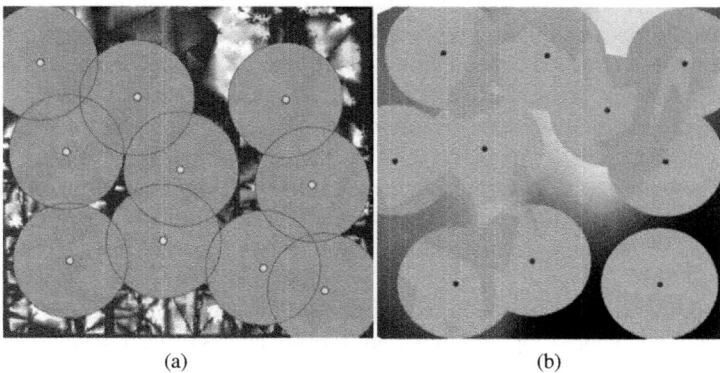

(a)   (b)

Fig. 18. The covered regions in the sensing field of each sensor node in second deployment (green regions are visible and pink regions are invisible)

Table 1. Visibility and coverage before and after optimization

Case		Visibility (no. of pixels)	Visibility (%)	Coverage (no. of pixels)	Coverage (%)
Urban area	Before optimization	23458	22	16810	23
	After optimization	37463	34	25174	37
Natural area	Before optimization	60250	67	40806	66
	After optimization	63995	71	43952	71

7   Discussion and Conclusions

This paper was focused on the coverage problem of geosensor networks. First, we have presented an overall review of the existing approaches for the optimization of the coverage of geosensor networks. Especially, algorithms that use Voronoi diagram and Delaunay triangulation were intensively investigated. As discussed in the paper, most of these methods oversimplify the coverage problem and they do not consider the characteristics of the environment where they are deployed. Spatial coverage of a sensor network is significantly related to the spatial distribution of the sensors in the environment. The coverage optimization algorithms aim at distributing sensors in the environment so that the maximum coverage is obtained.

Our extensive survey in the literature revealed that Voronoi diagram and Delaunay triangulation are well adapted for abstraction and modeling of sensor networks and their management. However, their applications are still limited when it comes to the determination and optimization of spatial coverage of more complex sensor networks (e.g., sensor networks with the presence of obstacles).

In order to overcome the limitation of these methods, a novel approach based on Voronoi diagram has been proposed in this paper. The algorithm considers spatial information in senor network deployment and coverage optimization. In order to evaluate the proposed method, two case studies were presented in the paper. The case studies provide interesting information on different challenges in the sensor network deployment both in urban and natural areas. The preliminary results obtained from these experimentations are very promising. As presented in the last section, we have observed a considerable improvement in the spatial coverage of the geosensor networks in both cases. These results are a part of an ongoing research project and more investigations will be carried out in order to improve the quality and the performance of the proposed method in the future.

Acknowledgment. This research is supported by the project entitled Integrating developmental genetic programming and terrain analysis techniques in GIS-based sensor placement systems by the Canadian GEOIDE Network of Centres of Excellence.

References

1. Aziz, N.A.A., Aziz, K.A., Ismail, W.Z.W.: Coverage strategies for wireless sensor networks. World Academy of Science, Engineering and Technology 50, 145–150 (2009)
2. Adriaens, J., Megerian, S., Potkonjak, M.: Optimal worst-case coverage of directional field-of-view sensor network. In: Proc. Sensor and Ad Hoc Communications and Networks (SECON 2006), pp. 336–345. IEEE Press (2006)
3. Ahmed, N., Kanhere, S.S., Jha, S.: The holes problem in wireless sensor networks: A survey. ACM SIGMOBILE Review 9(2) (April 2005)
4. Bahramgiri, M., Hajiaghayi, M., Mirrokni, V.: Fault-tolerant and 3-dimensional distributed topology control algorithms in wireless multi-hop networks. Wireless Networks 12(2), 179–188 (2006)
5. Berg, D.M., Kreveld, M.V., Overmars, M., Schwarzkopf, O.: Computational geometry: algorithms and applications. Springer, Heidelberg (2000)
6. Bharathidasan, A., Madria, S.: Sensor networks: an overview. IEEE Potentials 22, 20–23 (2003)
7. Chong, C.Y., Pumar, S.: Sensor networks: evolution, opportunities, and challenges. Proc. IEEE 91, 1247–1256 (2003)
8. Commuri, S., Watfa, M.: Coverage strategies in 3D wireless sensor networks. International Journal of Distributed Sensor Networks 2(4), 333–353 (2006)
9. Ghosh, A.: Estimating coverage holes and enhancing coverage in mixed sensor networks. In: Proc. 29th Annual IEEE Conference on Local Computer Networks, LCN 2004, Tampa, FL, pp. 68–76 (November 2004)

10. Ghosh, A., Wang, Y., Krishnamachari, B.: Efficient distributed topology control in 3-dimensional wireless networks. In: Proc. 4th Annual IEEE Communications Society Conference on Sensor and Ad Hoc Communications and Networks (SECON 2007), pp. 91–100 (2007)

11. Ghosh, A., Das, S.K.: Coverage and connectivity issues in wireless sensor networks: a survey. Pervasive and Mobile Computing 4, 303–333 (2008)

12. Heo, N., Varshney, P.K.: A distributed self-spreading algorithm for mobile wireless sensor networks. In: Proc. IEEE Wireless Communications and Networking Conference (WCNC 2003), New Orleans, USA, pp. 1597–1602 (March 2003)

13. Howard, A., Mataric, M.: Cover me! a self-deployment algorithm for mobile sensor networks. In: Proc. IEEE International Conference on Robotics and Automation (ICRA 2002), Washington DC, USA, pp. 80–91 (May 2002)

14. Howard, A., Mataric, M., Sukhatme, G.: An incremental self-deployment algorithm for mobile sensor networks. Autonomous Robots Special Issue on Intelligent Embedded Systems 13(2), 113–126 (2002)

15. Howard, A., Mataric, M., Sukhatme, G.: Mobile sensor network deployment using potential fields: a distributed scalable solution to the area coverage problem. In: Proc. 6th International Symposium on Distributed Autonomous Robotic Systems (DARS 2002), Fukuoka, Japan, pp. 299–308 (June 2002)

16. Hsien Wu, C., Lee, K.C., Chung, Y.C.: A Delaunay triangulation based method for wireless sensor network deployment. Computer Communications 30, 2744–2752 (2007)

17. Huang, C.F., Tseng, Y.C., Lo, L.C.: The coverage problem in three-dimensional wireless sensor networks. IEEE Globecom (2004)

18. Huang, C.F., Tseng, Y.C.: A survey of solutions to the coverage problems in wireless sensor networks. J. Internet Tech. 6(1), 1–8 (2005)

19. Huang, Q.: Solving an open sensor exposure problem using variational calculus. Technical Report WUCS-03-1, Washington University, Department of Computer Science and Engineering, St. Louis, Missouri (2003)

20. Lei, R., Wenyu, L., Peng, G.: A coverage algorithm for three-dimensional slarge-scale sensor network. In: ISPACS, pp. 420–423 (2007)

21. Linnyer, B., Vieira, L.F., Vieira, M.A.: Scheduling nodes in wireless sensor networks: a Voronoi approach. In: Proc. 28th IEEE Conference on Local Computer Networks (LCN 2003), Bonn, Germany, pp. 423–429 (October 2003)

22. Marengoni, M., Draper, B., Hanson, A., Sitaraman, R.: System to place observers on a polyhedral terrain in polynomial time. Image and Vision Computing 18, 773–780 (1996)

23. Megerian, S., Koushanfar, F., Potkonjak, M., Srivastava, M.B.: Worst and best-case coverage in sensor networks. IEEE Transaction on Mobile Computing 4(1) (Januaray 2005)

24. Megiddo, N.: Linear-time algorithms for linear programming in R3 and related problems. SIAM J. Computing 12, 759–776 (1983)

25. Meguerdichian, S., Koushanfar, F., Qu, G., Potkonjak, M.: Exposure in wireless ad-hoc sensor network. In: Proc. IEEE MOBICOM 2001, pp. 139–150 (2001)

26. Meguerdichian, S., Koushanfar, F., Potkonjak, M., Srivastava, M.B.: Coverage problems in wireless ad-hoc sensor networks. In: INFOCOM, pp. 1380–1387 (2001)

27. Meguerdichian, S., Slijepcevic, S., Karayan, V., Potkonjak, M.: Localized algorithms in wireless ad-hoc networks: location discovery and sensor exposure. In: Proc. ACM International Symp. on Mobile Ad Hoc Networking and Computing (MobiHOC), pp. 106–116 (2001)

28. Nittel, S.: A survey of geosensor networks: advances in dynamic environmental monitoring. Sensors Journal 9, 5664–5678 (2009)

29. Ruiz, L., Nogueira, J., Manna, A.L.: A management architecture for wireless sensor networks. IEEE Communications Magazine 41(2), 116–125 (2003)
30. Rung-Hung, G., Yi-Yang, P.: A dual approach for the worst-case coverage deployment problem in ad-hoc wireless sensor networks. In: Proc. 3rd IEEE International Conference on Mobile Ad-hoc and Sensor Systems, pp. 427–436 (2006)
31. Sharifzadeh, M., Shahabi, C.: Supporting spatial aggregation in sensor network databases. In: Proc. 12th Annual ACM International Workshop on Geographic Information Systems, pp. 166–175 (2004)
32. Skyum, S.: A simple algorithm for computing the smallest enclosing circle. Information Processing Letters 37, 121–125 (1991)
33. So, A., Ye, Y.: On Solving Coverage Problems in a Wireless Sensor Network Using Voronoi Diagrams. In: Deng, X., Ye, Y. (eds.) WINE 2005. LNCS, vol. 3828, pp. 584–593. Springer, Heidelberg (2005)
34. Soro, S., Heinzelman, W.: On the coverage problem in video based wireless sensor networks. In: Proc. 2nd Workshop on Broadband Advanced Sensor Networks, BaseNets 2005 (2005)
35. Szewczyk, R., Osterweil, E., Polastre, J., Hamilton, M., Mainwaring, A.: Habitat monitoring with sensor networks. Communications of the ACM 47(6), 34–40 (2004)
36. Tian, D., Georganas, N.D.: A coverage-preserving node scheduling scheme for large wireless sensor networks. In: Proc. 1st ACM International Workshop on Wireless Sensor Networks and Applications, pp. 32–41 (2002)
37. Veltri, G., Huang, Q., Qu, G., Potkonjak, M.: Minimal and maximal exposure path algorithms for wireless embedded sensor networks. In: Proc. ACM International Conference on Embedded Networked Sensor Systems (SenSys), pp. 40–50 (2003)
38. Wang, B., Lim, H.B., Ma, D.: A survey of movement strategies for improving network coverage in wireless sensor networks. Computer Communications 32, 1427–1436 (2009)
39. Wang, G., Cao, G., LaPorta, T.: A bidding protocol for deploying mobile sensors. In: Proc. 11th IEEE International Conference on Network Protocols (ICNP 2003), Atlanta, USA, pp. 80–91 (November 2003)
40. Wang, G., Cao, G., LaPorta, T.: Movement-assisted sensor deployment. In: Proc. IEEE Infocom (INFOCOM 2004), Hong Kong, pp. 80–91 (March 2004)
41. Welzl, E.: Smallest enclosing disks (balls and ellipsoids). In: New Results and New Trends in Computer Science, pp. 359-370 (1991)
42. Worboys, M., Duckham, M.: Monitoring qualitative spatiotemporal change for geosensor networks. International Journal of Geographical Information Science 20(10), 1087–1108 (2006)
43. Zhou, Z., Das, S., Gupta, H.: Connected K-coverage problem in sensor networks. In: ICCCN 2004, pp. 373–378 (2004)
44. Zou, Y., Chakrabarty, K.: Sensor deployment and target localization based on virtual forces. In: Proc. IEEE Infocom (INFOCOM 2003), San Francisco, USA, pp. 1293–1303 (April 2003)
45. Zou, Y., Chakrabarty, K.: Sensor deployment and target localization in distributed sensor networks. Transactions on IEEE Embedded Computing Systems 3(1), 61–91 (2004)
46. 21 ideas for the 21st century. Business Week, pp. 78–167 (August 1999)

Rainfall Distribution Based on a Delaunay Triangulation Method

Nicolas Velasquez, Veronica Botero, and Jaime Ignacio Velez

National University of Colombia, Street 80 No 65-223 Medellin, Colombia.
nvelasqg@unal.edu.co

Abstract. Many rainfall-run-off distributed models need rainfall data as input on a pixel by pixel basis, for each time interval. Due to the large amount of pixels that can make up a basin (proportional to the map scale), a fast and efficient method must be devised in order to obtain the rainfall field for each time interval (e.g. 20 minutes). Most models use interpolation methods such as the Inverse Distance Weighted. However, we propose the use of a Delaunay Triangulation using the incremental algorithm developed by Watson where the rainfall stations are used as the vertices of the triangles that represent a three dimensional plane of the rainfall. Once the equation of the plane is known, a rainfall value for each pixel is calculated. We compare both methods and evaluate the sensitivity to changes in time and spatial scales separately.

Keywords: Delaunay, Rainfall distribution, Time interval, Map Scale, Interpolation, IDW.

1 Introduction

In many parts of the world, rainstorm events cannot be measured directly in a distributed field due to the lack of adequate equipment, such as radars. Such information is very useful to understand diverse phenomena, for example, the spatial distribution of precipitation. In distributed hydrological models, precipitation makes part of the entry data (Velez [7], Julien [4], Chow et al [2]).

Despite not being able to rely on instruments able to conduct distributed measurements of precipitation, many places can count on pluviometric stations able to measure such events in a punctual manner and with good temporal resolution (e.g. 10 min). In this way one can capture the temporal variability of the rainstorm, and the data can be used as a starting point for the estimation of distributed precipitation fields in short periods of time.

In our case, we try to validate the construction of precipitation fields in both tropical zones with rugged topography and extra tropical zones, where rain, in short periods of time like the ones encompassed by a rainstorm, acquires a high spatial variability (Poveda [5]). This spatial-temporal variability makes the interpolation of data difficult and requires fast algorithms to produce rain data that can be used as entry data for the distributed hydrological models that in turn process great amounts of information.

M.L. Gavrilova et al. (Eds.): Trans. on Comput. Sci. XIV, LNCS 6970, pp. 173–187, 2011.
© Springer-Verlag Berlin Heidelberg 2011

Due to the agility that the interpolation method must possess, different authors (George [3]) conduct the interpolation of rain through interpolation methods such as Spline and IDW (inverse of weighted distance), methods which have some problems. One of them consists in that, to conduct the interpolation of the rain over any cell inside the river basin, all precipitation stations found within the zone are used, introducing much noise coming from remote zones where the phenomenon in the same time (Δt) has a very different behavior. Equally, due to the fact that all registers within this time interval are being weighted, a very smooth surface can be produced, which is not very common in the behaviour of a rainstorm.

Our proposal is to conduct the interpolation through the use of three dimensional planes obtained using a Delaunay triangulation and to compare their performance against the IDW method. Two sensitivity analysis of the method were conducted. The first analysis consisted in removing rainfall stations to determine how well the method behaves when few stations are available. The second analysis evaluates the behaviour of both methods using different time intervals (Δt). The triangulation method proposed by Watson (Watson [8]) has been programmed in **Fortran** 90 to obtain the Delaunay Triangulation.

2   Data Used

2.1   Spatial Data Analysis

To accomplish the exercise, a set of pluviometric stations located in the northwestern zone of the Aburra Valley were taken due to the high availability of data and the high temporal resolution they possess. The spatial distribution of the stations can be seen on Figure 1. The precipitation data from each of the stations was aggregated to obtain data every $20min$. The information for the stations was supplied by the Metropolitan Area of the Aburra Valley, who collected information during the project *OPERACION RED HIDROMETEOROLOGICA*, which took place between 2003 and 2008.

As can bee seen in Figure 1, only 14 stations were used, which were selected because they are relatively close to each other and are distributed on the slopes (stations 1, 5, 7, and 9) and the low lying areas of the Valley (stations 5, 8, and 10).

Considering the availability of data, 10 rainstorm events of 2003 were selected. That year, the shortest rainstorm lasted $240min$ and the longest reached $380min$. For the selection of the rainstorms, it was considered that the rain was registered in the majority of the stations, since in many of the evaluated events it was only registered in less than half of the stations. This gives an idea of the high spatial variability of these events in tropical zones, where macro-climatic oscillations, like the variation of the ZCIT, generates convective clouds (Poveda [5]) that, because of their shape, cover small areas and provide a great variability in rainstorm events.

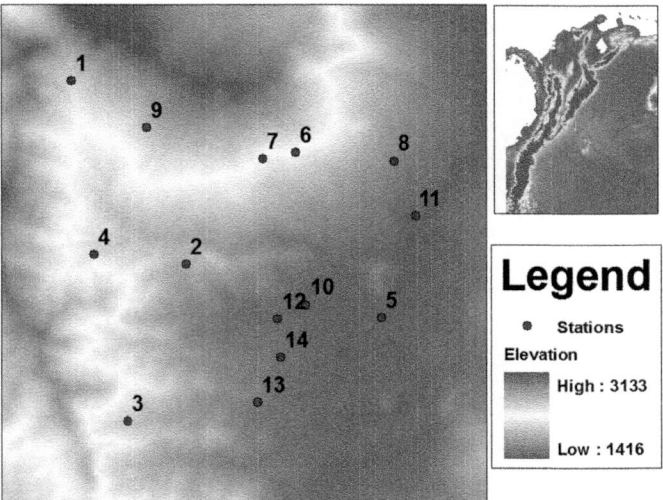

Fig. 1. Geographic location of the spatial analysis stations

2.2   Temporal Data Analysis

For the temporal sensitivity analysis we use tree storm events recorded in the Goodwin Creek, located at Batesville, Mississippi in the United States of America. Figure 2 shows the location of the second study area. This data was selected considering that it presents a detailed time scale ($\Delta t = 5min$).

The data consists of 17 stations, where all of them register the three storm events. In this case the events show less spatial variability, since all the stations register similar data.

3   Interpolation Methods

The following explains the functioning of the two methods of interpolation used, making emphasis on the method based on Tessellation or Triangulation.

3.1   Interpolation through Planes

For the definition of planes through which interpolation is done, we have used the Delaunay triangulation, which ensures that the triangulation is performed so that the closest points are joined. The algorithm used is the incremental algorithm proposed by Watson [8]. This algorithm was selected because it allows the removal and insertion of points in the triangulation without having to alter the whole calculated network, being computationally fast. In addition, its programming is relatively easy in comparison to other algorithms. The triangulation presented in Figure 3 is obtained using all the stations presented in Figure 1.

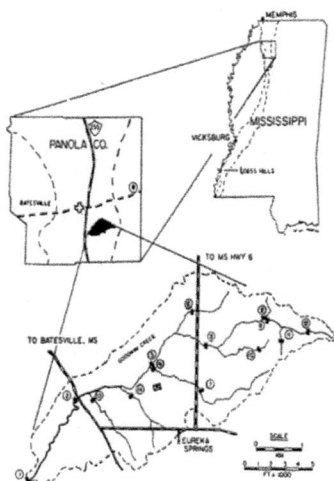

Fig. 2. Geographic location of the temporal analysis stations

As can be seen in Figure 3, each triangle covers an area. In this area, the calculation of rainfall is also done based on the triangle that covers it. For each time interval it is assumed that the rainfall is distributed according to the equation of the three-dimensional plane, where the coordinates of each station (vertex) that makes up the triangle (vertex) are the coordinates X and Y of the station, and the value of the precipitation for the time interval represents the coordinate Z. This idea is presented schematically in Figure 4.

Since each time interval is known for each plane of coordinates X, Y y Z, the value of the rainfall that is sought for a set of given coordinates can be found by solving for the determinant presented in the equation 1.

$$\begin{vmatrix} X_2 - X_1 & X_3 - X_1 & X_x - X_1 \\ Y_2 - Y_1 & Y_3 - Y_1 & Y_x - Y_1 \\ Z_2 - Z_1 & Z_3 - Z_1 & Z_x - Z_1 \end{vmatrix} = 0 \tag{1}$$

In the equation 1 it can be seen that X_x y Y_x are the coordinates of the location over which one wants to obtain the value of the rainfall, and therefore they are known, leaving as the only unknown the value Z_x, since all other values correspond to the coordinates of the rain stations and the value of the rain in the time interval (Δt) studied (see Figure 4). The value of Z_x is obtained as the projection of the coordinates X_x y Y_x in the plane, and in such a way that $Z_x = R_x$, R_x being the value of the rainfall.

The procedure described above is applied over each one of the locations where one wants to determine the rainfall in each time interval, therefore the determinant presented in the equation 1 suffers few changes in each time interval.

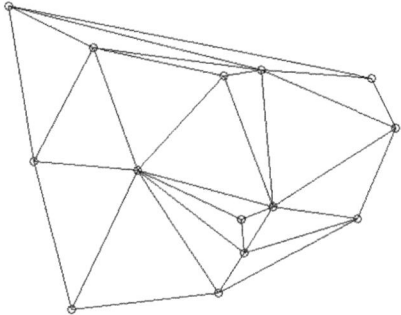

Fig. 3. Triangulation

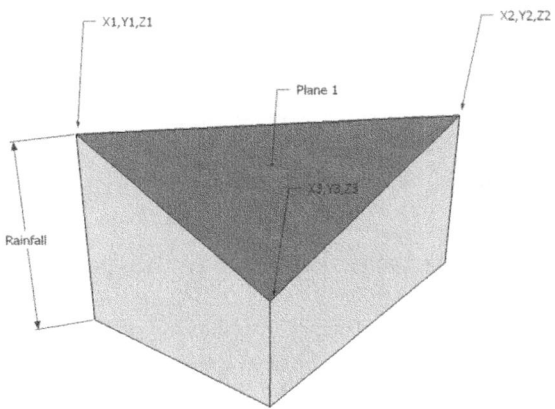

Fig. 4. Interpolation plane

3.2 IDW

The method proposed by Shepard [6] was used to do the interpolation through the weighting of the inverse of the distance. He proposed doing multivariate interpolations in two dimensions. With this method one can assign values in unknown places starting from punctual know values distributed in space. This has been a widely used method to do meteorological interpolations with variables such as rainfall, since it is an easy to implement method and has yielded satisfactory results in different rainfall interpolation works, (such as the D.W Wong [3]). In addition it has been so popular that it was proposed by the ASCE in 1996 [1].

The interpolator calculates the distance from the point in question to the known points to assign according to it, weights to the values that are found in each of the known points. The method operates under the premise that the objects that are spatially near tend to have similar behaviors. To obtain the weighting that defines the influence that each station has over the studied point, one utilizes the formula 2, in which W_i represents the calculated weighting for each station and $d(P_x, P_i)$ represents the distance between the studied point and the station i.

$$W_i = \frac{1}{d(P_x, P_i)} \tag{2}$$

Once the corresponding weights to the point in study are known, the value of the rainfall in the time interval is calculated, utilizing the equation 3, in which R_i represents the value of the rainfall for the interval at each of the stations and N is the number of stations.

$$R_x = \sum_{i=1}^{N} \frac{W_i R_i}{\sum_{i=1}^{N} W_i} \tag{3}$$

As it can be observed in the equation 3, the method proposed by Shepard, while giving much more importance to the values of the points that are nearer to the interest point, introduces noise in the interpolation when considering the value of points that are far away, so quality is lost.

4   Application of Interpolation Methods

Two exercises are proposed to study the behavior of interpolation through the method of triangular planes. The first encompasses comparing the results yielded by this method with the results yielded by the IDW method. The second consists in conducting a sensitivity analysis of the plane interpolation method to be able to discuss the advantages and limitations presented by the method.

4.1   Comparison of Methods

With the purpose of comparing both methods, from the tropical data four precipitation stations were selected as control stations and not used to realize the interpolations by the studied methods. The control stations are numbers 2, 7, 10 and 14 that are presented in Figure 1 and marked in red. These stations were selected because they are located relatively well distributed and none of them is a triangle vertex that defines the exterior limit. Figure 5 presents the triangulation obtained when these stations are removed.

The results yielded by both methods were compared to the data observed in each one of the stations for the different rainstorm events. To quantify the results of each interpolation, one calculates the percentage error produced by each method over each one of the stations for each event.

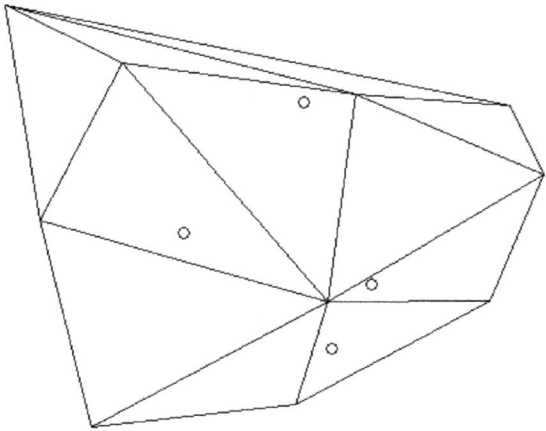

Fig. 5. Triangulation obtained without the control stations

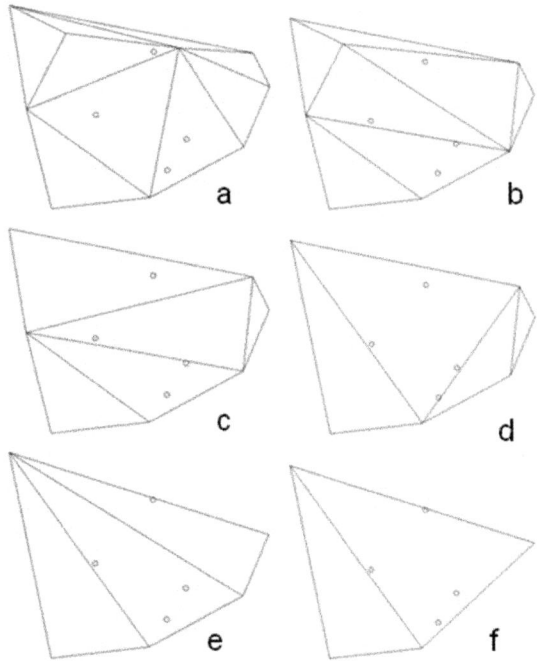

Fig. 6. Triangulation evolution removing stations. a: station 12 removed, b: stations 12 and 6 removed, c: stations 12, 6 and 9 removed, d: stations 12, 6, 9 and 4 removed, e: stations 12, 6, 9, 4 and 8 removed, f: stations 12, 6, 9, 4, 8 and 5 removed

4.2   Sensitivity Analysis

Spatial Analysis. For this analysis it was chosen to continue working with the same four control stations, since the errors obtained in this analysis can be compared to errors obtained in the method comparison.

The amount of points used in the triangulation was then analyzed, as it can be a highly important factor in the quality of the interpolation. It was considered that the more points used, the smaller the triangles that can be used for the triangulation that can better describe the spatial variation that the rainstorm possesses. The procedure conducted consisted in removing one by one the precipitation stations that made part of the triangle construction inside which the studied stations were located. The stations were removed in the following order: 12, 6, 9, 4, 8, 5. Figure 6 presents how the triangulation changes as the previously mentioned stations are removed.

One can observe in Figure 6 how, as the points are removed, one obtains a smaller amount of triangles and at the same time the remaining triangles grow bigger as each point is removed.

Temporal Analysis. The data was aggregated in 5 minute intervals, beginning with $\Delta t = 5min$ and ending with $\Delta t = 60min$. Figure 7 shows an example of the data for time intervals 5, 15, 40 and 60 minutes. The analysis consists on the evaluation of the interpolated surfaces using 9 internal stations, removing one by one. The interpolation was performed using two methods (planes and IDW) for the stations remaining after the removal of a station, and for each time scale. The errors calculated for each surface are used to quantify the overall results. Figure 8 shows the selected stations in red and the TIN generated using them.

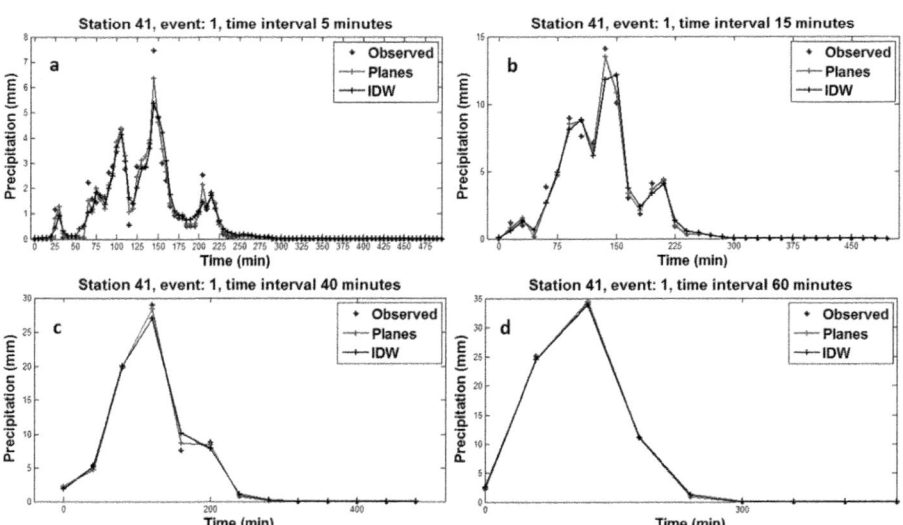

Fig. 7. Time aggregation. a: at 5 minutes, b: at 15 minutes, c: at 40 minutes, d: at 60 minutes

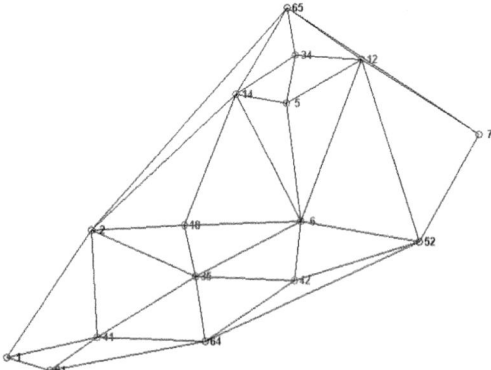

Fig. 8. TIN obtained from selected stations, and selected stations in red

5   Results

5.1   Comparison of Both Methods

In general, the plane method for the used stations and for the selected rainfall events presents a smaller error than the one obtained by the IDW method. The first method yielded an error of 31.54 percent, while the second method yielded an error of 38.98 percent, which indicates that the triangulation method allows a better interpolation.

Figure 9 presents as an example the simulation done with both methods for the first rainfall event. This figure shows how the plane method smoothes the simulation less during the different time intervals, and how in the majority of cases presents a better adjustment. In Figure 9d it is observed that both methods show high error when rain is simulated over zones that saw no rainfall during the interval. With the purpose of observing the behavior of the errors, Figures 10a and 10b are presented. Figure 10a has the average errors of the four stations for each one of the events and for both interpolation methods, while Figure 10b shows the average errors for each one of the stations.

It can be seen in Figure 10a how, for the vast majority of the events, the triangulation method presents smaller errors than the IDW methods. However, these differences are in most cases very low. On the other hand, Figure 10b shows a more marked difference between the errors produced by both methods, again with better results for the triangulation method.

5.2   Results of the Spatial Sensitivity Analysis

It was found that the interpolation error is increased as points are removed from the triangle network, since the error mentioned before of 31.54%, reaches a value of 75.43% for cases where the 6 stations are removed. In addition, it

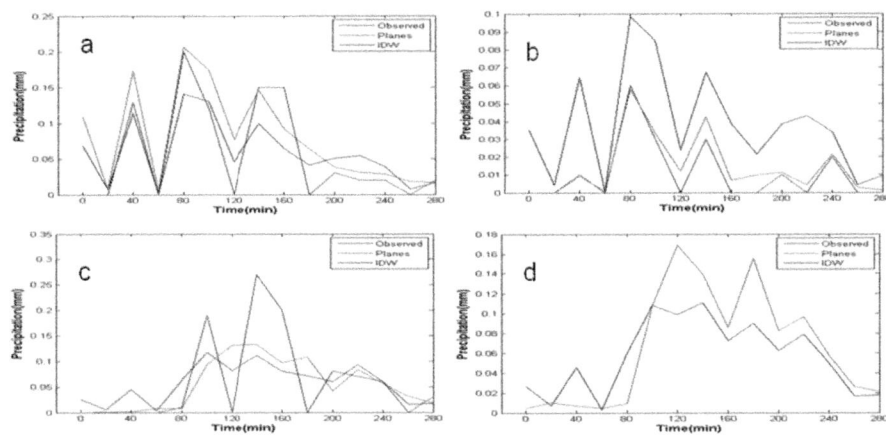

Fig. 9. Simulations for the first rain event. a: Station 2, b: Station 7, c: Station 10, d: Station 14

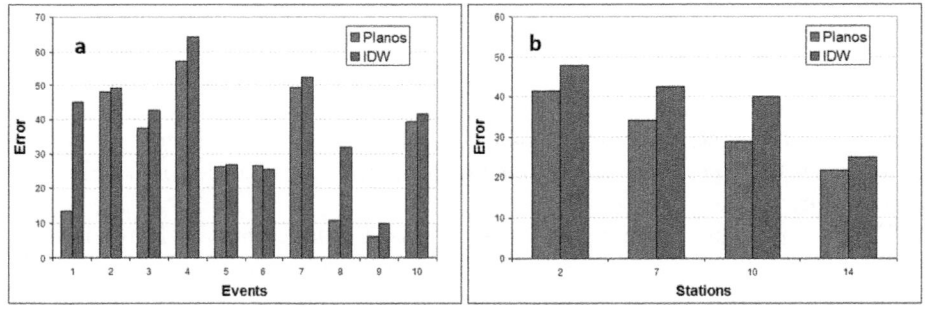

Fig. 10. Mean errors. a: different events, b: selected stations

seems the error increases with a linear tendency as the stations are removed from the triangulation. Figure 11 shows the tendency found and the value of the correlation index for the same.

To illustrate how the error varies in the interpolations as more information is used, Figure 12 shows the error as the stations are progressively removed, as presented Figure 6. The same increment error was observed for the IDW interpolation method (see Figure 13), with the purpose of making a comparison of the sensitivity of both methods to a scarcity of information. On the other hand, Figure 14 presents the increase in average error for each one of the stations versus the average distance of the stations to the closest point to each one of them, and it can be observed how this error can present a linear increase with such distance.

It can be observed in Figure 12 how, for every case there is an increase in error as the stations are removed, but this increase depends largely on how far

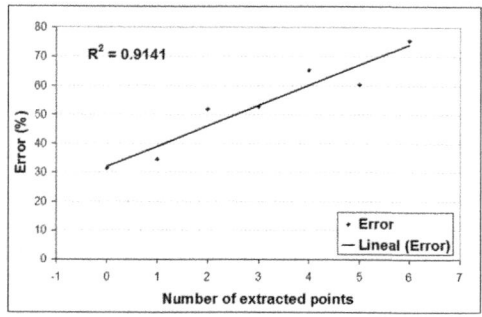

Fig. 11. Mean error change removing stations

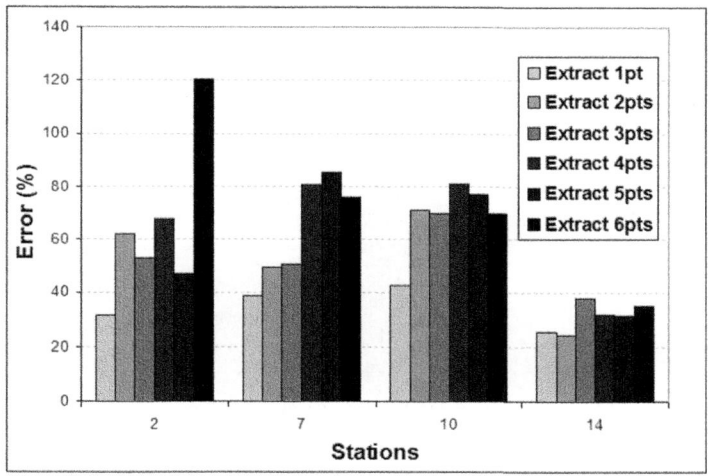

Fig. 12. Mean error change for the planes method, removing stations

away the removed station is from the one on which the interpolations are being done. This can be seen when comparing Station 2 with the other stations, where removing the 6 points produces an error much larger from the one produced at the other stations.

5.3   Results of the Temporal Sensitivity Analysis

For increased time intervals both methods (planes and IDW) increase their performance. The planes method presents a mean total error lower than the IDW method (47.2% and 52.6% respectively). Averaging the mean errors of the three storm events, the planes method presents a better performance (Figure 15d).

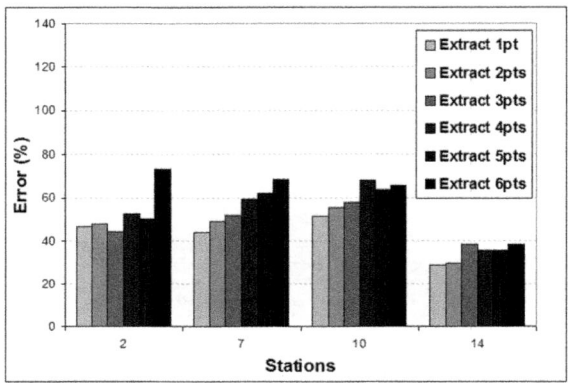

Fig. 13. Mean error change for the IDW method, removing some stations

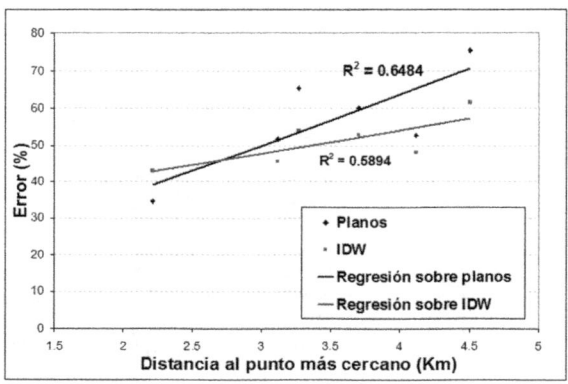

Fig. 14. Mean error change vs. the mean distance to the nearest points of the selected stations

In the first and third event, the planes method shows a significantly better performance compared to the IDW method (Figures 15a and 15c). Despite the results in the first and third event, in the second event both methods presents similar performance (Figure 15b).

6   Analysis and Conclusions

Despite having found a better performance of the proposed plane method, one must take into account that the average error is not very low, which suggests the high variability of tropical rainfall events and how this makes it difficult to obtain good precipitation fields.

Observing the results obtained by both methods (Figures 10a and 10b), it can be seen how the plane interpolation method yields better results than the IDW method. This is due to the fact that the plane method only uses the 3 closest

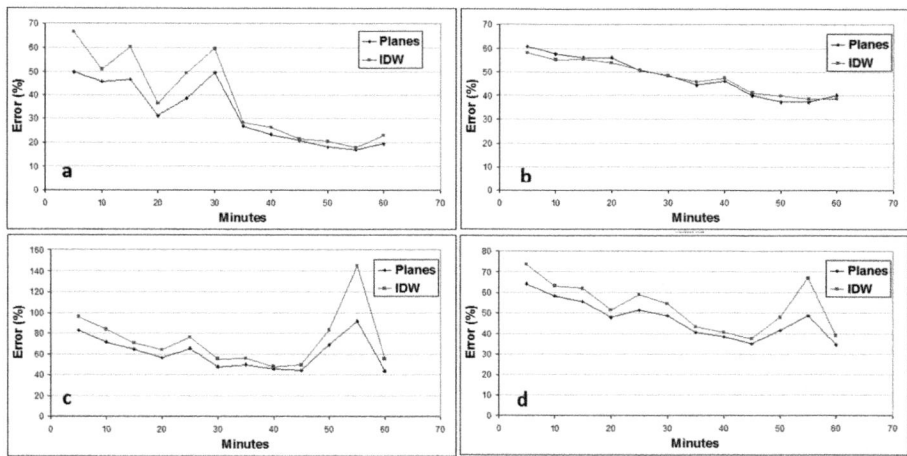

Fig. 15. Mean error for intervals. a: event 1, b: event 2, c: event 3, d: mean error of all events

stations to the point of interest, which avoids the introduction of noise from the stations that are further away, which happens in the IDW method. Likewise, the construction of the planes starting from Delaunay triangles insures that they are only relating the closest stations and that therefore there is more probability that they will behave in a similar manner during a rainfall event.

When observing the simulations done by either method for the first rainfall event selected (Figure 9), one finds that both methods display a similar behavior. Both operate mathematically differently, but work under the same premise: objects that are spatially close tend to be similar. Even though the behavior in both is similar, there exists a better adjustment in the interpolation done by the plane method. This method fits better to stations 2 and 7. However, for station 14, in this rainfall event, both methods present big differences to the observed values, seemingly because in this station there was no rainfall, and since nearby station information is used to obtain the interpolation, these introduce rainfall in this particular point and values such as the ones obtained in Figure 9a are realized. This type of problems can be very frequent in zones where rainfall is caused by convective clouds, which can be punctually located and generate effects like this one, in which inside a relatively small space there exists a high variability of rainfall. A high variability between the stations is found if the behavior of the 4 stations analyzed for the first rainfall event is compared. In one hand, stations such as 2 and 10 (Figures 9a y 9c) present much higher rainfall values than the other stations. In the other, one finds that both the beginning and the end of the events vary in the stations, and some of the peaks are out of sync.

In the sensitivity analysis, it is found that the plane method is highly sensitive to the number of stations that can be used to obtain the triangle network. This can be clearly seen in Figures 11 y 14, in which relatively high slopes are found for the calculated regressions, which begin to show the error behavior of the model even if they are calculated using few data points.

When performing comparisons of the sensitivity of both methods to the lack of entry data, it is found that the IDW method is less sensitive than the plane method. This can be clearly seen when comparing Figures 12 y 13, in which higher errors are found for all the stations in the plane method. In addition it can be observed that the change in the errors is more abrupt in this method (see Figure 12), since higher error changes can be found in each iteration where a station was removed. On the contrary, in Figure 13 softer and more continuous changes are observed. In general, it can be said that the errors of the selected stations respond in a similar manner for all cases.

Only stations that did not constitute the boundaries of the constructed network were removed for the temporal analysis, since all others were necessary for the interpolation without the use of other methods. The plane method presented a better performance for all executed temporal scales, presenting in some cases, more significant improvements over others.

For the cases of event 1 and the mean of all other events, it is clear that the higher the temporal aggregation, the mean errors are less and the performance for both methods tends to be similar. This can be attributed to the loss of variability presented by the storm as it is aggregated (see Figure 7), which facilitates the interpolation exercise.

When comparing the results obtained for the storm events presented in a tropical zone (Figure 9) and the ones presented in an extra-tropical zone (Figure 7), it can be clearly seen how the high variability in the first zone supposes more errors in the interpolation, contrary to what is observed in the Goodwin Creek stations. Here, despite that errors are produced in the interpolation, the simualtion of the behavior of the hietogram is satisfactory.

Studying the calculated regressions in Figure 14 it is found that there is a point where all the errors from the plane methods start to be larger than those obtained by the IDW method. This point corresponds to the average distance of the closest point of $2.7Km$, which can indicate up to which point it is advisable to employ one or the other method depending on the available information. Comparing the slopes of both regressions, a smaller sensitivity of the IDW method is again found, which can be explained because the method weights the available data, smoothing the results a lot and avoiding high differences in this way. It must be taken into account that regressions calculated were done to give an idea of the behavior of the errors and that they were obtained starting from a small data set.

It is then a conclusion that the planes method proposed to conduct interpolations of rainfall fields presents good results when compared a widely used method such as the IDW method, and can be used instead of it to be used inside hydrological models, since it is computationally fast and robust. In the same manner, care must be taken with the amount of available information, since depending on this and to how spatially near or far it is located, it might be preferable to use this method to the IDW method. Equally, future research must be conducted, including other interpolation models in other spatial and temporal scales, and with a much higher data amount that allows for better results.

References

1. ASCE.: Hydrology Handboo, 2nd edn. American Society of Civil Engineers (ASCE), New York (1996)
2. Chow, V.T., Madiment, D.R., Mays, L.W.: Applied Hydrology, p. 6. Mc Graw Hill (1988)
3. George, Y.L., Wong, D.W.: An adaptive inverse-distance weighting spatial interpolation technique. Computers and Geosciences 14(9), 1044–1055 (2008)
4. Julien, P.Y.: Runoff and Sediment Modeling with CASC2D, GIS and Radar Data. Parallel Session, Parallel 15 (1998)
5. Poveda, G., Mesa, O.J., Carvajal, L.F.: Introduccion al Clima de Colombia. Universidad Nacional de Colombia, Medellin (2000)
6. Shepard, D.: A two-dimensional interpolation function for irregularly-spaced data. In: Proceedings of the 1968 ACM National Conference, pp. 517–524 (1968)
7. Velez, J.I.: Desarrollo de un modelo conceptual y distribuido orientado a la simulacion de crecidas. Universidad Politecnica de Valencia Doctoral Thesis (2001)
8. Watson, D.F.: Computing the n-dimensional tessellation with application to Voronoi polytopes. The Computer Journal 24(2), 167–172 (1981)

Homotopic Object Reconstruction Using Natural Neighbor Barycentric Coordinates

Ojaswa Sharma[1] and François Anton[2]

[1] Department of Computer Science and Engineering,
Indian Institute of Technology Bombay,
Mumbai, 400076, India
ojaswa@cse.iitb.ac.in
[2] Department of Informatics and Mathematical Modelling,
Technical University of Denmark, Lyngby, 2800, Denmark
fa@imm.dtu.dk

Abstract. One of the challenging problems in computer vision is object reconstruction from cross sections. In this paper, we address the problem of 2D object reconstruction from arbitrary linear cross sections. This problem has not been much discussed in the literature, but holds great importance since it lifts the requirement of order within the cross sections in a reconstruction problem, consequently making the reconstruction problem harder. Our approach to the reconstruction is via continuous deformations of line intersections in the plane. We define Voronoi diagram based barycentric coordinates on the edges of n-sided convex polygons as the area stolen by any point inside a polygon from the Voronoi regions of each open oriented line segment bounding the polygon. These allow us to formulate homotopies on edges of the polygons from which the underlying object can be reconstructed. We provide results of the reconstruction including the necessary derivation of the gradient at polygon edges and the optimal placement of cutting lines. Accuracy of the suggested reconstruction is evaluated by means of various metrics and compared with one of the existing methods.

Keywords: Voronoi diagram, natural neighbor, Homotopy, continuous deformations, reconstruction, linear cross sections.

1 Introduction

Object reconstruction from cross sections is a well known problem. Generally a spatial ordering within the cross sections aids reconstruction. We consider the problem of reconstructing an object from arbitrary linear cross sections. Such cross sectional data can be obtained from many physical devices. An example is an acoustic probe that can obtain range information of an object by sending an acoustic pulse.

The problem of reconstruction from arbitrary cross sections has been studied by [15,9,10]. Sidlesky et al. [15] define sampling conditions on the reconstruction, while in our reconstruction algorithm, we allow the sampling to be sparse.

M.L. Gavrilova et al. (Eds.): Trans. on Comput. Sci. XIV, LNCS 6970, pp. 188–210, 2011.

Methods proposed by Liu et al. [9], and Memari and Boissonnat [10] are both based on Voronoi diagrams. Memari and Boissonnat also provide rigorous proof of their reconstruction. Our approach to reconstruction considers the "presence" or "absence" of information along any intersecting line. This is in contrast to [15], where the authors consider that a line not intersecting the object does not contribute to the reconstruction. In our algorithm, such a line is considered to contribute to the reconstruction by defining a linear section, no part of which belongs to the reconstruction.

Memari and Boissonnat [10] provide a topological reconstruction method utilizing the Delaunay triagulation of the set of segments of intersecting lines. They claim an improvement over the method by Liu et al. [9] by producing reconstructions that are not topologically effected by lines that do not intersect the object under consideration. Their reconstruction boundary, however, is a piecewise linear approximation of the boundary of the original object. In this work, we produce smooth reconstruction of the object via continuous deformations. Therefore, we anticipate better reconstruction accuracy compared to the work by Memari and Boissonnat [10].

This paper is organized as follows. Section 2 defines the reconstruction problem mathematically. We introduce the concept of homotopy continuation in section 3 followed by the main reconstruction algorithm in section 4. We discuss our Voronoi diagram based edge barycentric coordinates on convex polygons here and provide details of our homotopy based reconstruction algorithm. We provide results of the reconstruction in section 5 and analyze the accuracy of our algorithm.

2   Problem Definition

Given a set of lines $\{\mathcal{L}_i : i \in [0, n-1]\}$ in a plane, intersecting an object $\mathcal{O}$ along segments $\{\mathcal{S}_{i,j} : j \in [0, m_i - 1]\}$, the problem of object reconstruction from arbitrary linear cross sections is to reconstruct an object $\mathcal{O}$ from $\mathcal{S}_{i,j}$ such that the reconstruction $\mathcal{R}$ satisfies

$$\mathcal{L}_i \bigcap \mathcal{O} = \mathcal{L}_i \bigcap \mathcal{R}, \tag{1}$$

and that $\mathcal{R}$ is homeomorphic to $\mathcal{O}$. Further, the reconstruction should also be geometrically close to the object. We quantify the geometric closeness in our reconstruction by means of several area based ratios such as the ratio of area of reconstruction and the area of the object, and the ratio of the absolute difference of the two areas and the area of the object. Length ratio is also a good indicator of geometric closeness. Furthermore, Hausdorff distance between the two curves gives a good measure of the distance between them.

In this context we impose no restrictions on the ordering or arrangement of the intersecting lines. However, the placement of intersecting lines plays an important role in the correctness of the reconstruction. A placement that covers salient object features results in a better reconstruction. In order to quantify an optimal placement, consider a set of intersecting lines in a plane along with the object to be reconstructed. The intersecting lines partition the object into smaller regions. Considering the simply connected boundary of the object that belongs to a region (see the highlighted curve segment in Fig. 1(a)), *tortuousity* [6], which gives a simple measure of how twisted a curve is, can be computed. It is defined as the arc-chord ratio of a parametric curve $C = (x(t), y(t))$ on the interval $[t_0, t_1]$ [13]

$$\tau = \frac{L}{C} = \frac{\int_{t_0}^{t_1} \sqrt{x'(t)^2 + y'(t)^2}}{\sqrt{(x(t_1) - x(t_0))^2 + (y(t_1) - y(t_0))^2}}. \tag{2}$$

τ can be considered as a measure of straightness of a curve. According to (2), tortuosity of a circle is infinite [13] since the chord length is zero for $t_1 = t_0$. This definition can be extended for parts of the object boundary that are not intersected by any of the lines (for example, an isolated blob shown in Fig. 1(a)). In such a case, the two end points do not exist and therefore these can be set to a single point (i.e., $t_1 = t_0$) without loss of generality. Therefore, for isolated object parts that have no intersection with the cutting lines, τ is taken as infinity.

It is not difficult to see that higher the value of τ for any region, more it is susceptible to generate part of the reconstruction that is non-homeomorphic to the object. However, if the sampling is such that the intersecting lines are chosen along the medial axis of the object, then such regions can be avoided (see Fig. 1(b)). In that case, τ remains close to one for different regions. Such a sampling is illustrated in subsection 5.1 for deriving accuracy statistics for the proposed reconstruction method.

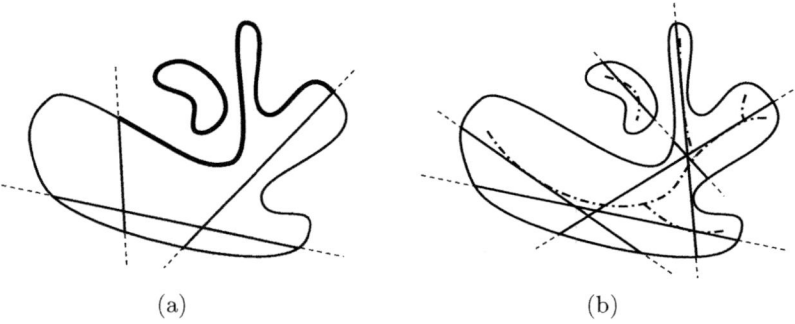

(a)   (b)

Fig. 1. Sampling condition on the intersecting lines. (a) Under-sampling. (b) Optimal sampling along medial axis.

The skeleton ($\mathcal{K}$) of the object and the skeleton ($\mathcal{K}^c$) of the complement of the object in its convex hull provide optimal placement of the cutting lines. Indeed, except if the object is a disc, the intersection of the closure of the union of $\mathcal{K}$ and $\mathcal{K}^c$ with the boundaries of the object correspond to the local extrema of curvature of the boundary object. In addition, the tangent to $\mathcal{K}$ or $\mathcal{K}^c$ at these points of intersection with the boundaries of the object correspond to the normal to the object boundary at these points in the case of regular points and the axis of the normal cone in the case of singular points (see Fig. 2). In the case of a disc, the curvature is constant, and the skeleton is reduced to a point (its center), while the complement of the disc in its convex hull is the empty space. The notion of using the skeleton of an object for optimal sampling comes from the Geometric Sampling Theorem [14]. As with 1D signals, the sampling must consider the highest frequency present in the signal, for manifolds, sampling must ensure that the regions of high curvature are measured [14]. In our case, we ensure this by sampling along the skeleton. Another desired trait

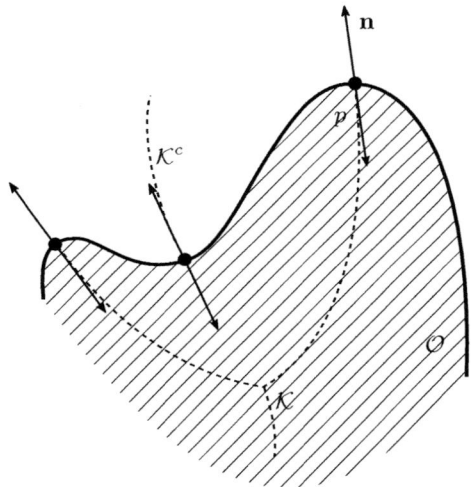

Fig. 2. Optimal placement of cutting lines. $\mathcal{O}$ is the object with $\mathcal{K}$ being its skeleton and $\mathcal{K}^c$ being the skeleton of its complement. The closure of these skeletons touches the boundary of $\mathcal{O}$ at the local extrema points.

of a reconstruction is smoothness, and we will show that the proposed method of continuous deformations results in a reconstruction that is at least C^1.

3   Homotopy Continuation

Homotopy is concerned with identification of paths between objects that can be continuously deformed into each other. The history of study of homotopy dates back in the late 1920's when the the homotopy theory was formalized.

Definition 1. *Let* $f : X \to Y$ *and* $g : X \to Y$ *be two continuous maps between topological spaces* X *and* Y. *These maps are called* homotopic, $f \simeq g$, *if there is a* homotopy *or a continuous map* $\mathcal{H} : X \times [0, 1] \to Y$ *between them, such that* $\mathcal{H}(x, 0) = f(x)$ *and* $\mathcal{H}(x, 1) = g(x)$ *for all* $x \in X$.

Therefore, we can write the homotopy $\mathcal{H}_\lambda : X \to Y$ as

$$\mathcal{H}_\lambda(x) = \mathcal{H}(x, \lambda), \tag{3}$$

and thus, $\mathcal{H}_0 = f$ and $\mathcal{H}_1 = g$. One can visualize how the deformation h continuously takes f to g (see Fig. 3) by varying the parameter λ.

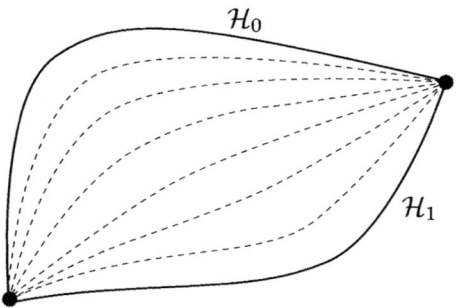

Fig. 3. Continuous deformation

One can impose additional constraints on the deformation path. For example, a specific constraint on fixed endpoints leads to homotopy of paths. For two pairs of homotopic maps $X \xrightarrow[f \simeq g]{} Y \xrightarrow[\bar{f} \simeq \bar{g}]{} Z$, the compositions $\bar{f} \circ f$ and $\bar{g} \circ g$ are also homotopic via the composition $\bar{\mathcal{H}}_\lambda \circ \mathcal{H}_\lambda$. Further, for two pairs of homotopic maps $f_i \simeq g_i : X_i \to Y_i, i = 1, 2$, the maps $f_1 \times f_2$ and $g_1 \times g_2$ from $X_1 \times X_2$ into $Y_1 \times Y_2$ are also homotopic via $\mathcal{H}_\lambda^{(1)} \times \mathcal{H}_\lambda^{(2)}$, in which case it is called a *product homotopy* [7].

Continuous deformations have been successfully used to solve non-linear system of equations that are otherwise hard to solve. A homotopy tries to solve a difficult problem with unknown solutions by starting with a simple problem with known solutions. Stable predictor-corrector and piecewise-linear methods for solving such problems exist (see Allgower and Georg [3]). The system $\mathcal{H}(x, \lambda) = 0$ implicitly defines a curve or 1-manifold of solution points as λ varies in $[0, 1]$ and x is fixed.

Given smooth $\mathcal{H}$ and existence of $u_0 \in \mathbb{R}^{N+1}$ such that $\mathcal{H}(u_0) = 0$ and $\text{rank}(\mathcal{H}'(u_0)) = N$, there exists a smooth curve $c : \alpha \in J \mapsto c(\alpha) \in \mathbb{R}^{N+1}$ for some open interval J containing zero such that for all $\alpha \in J$ (Allgower and Georg [3])

1. $c(0) = u_0$,
2. $\mathcal{H}(c(\alpha)) = 0$,
3. $\text{rank}(\mathcal{H}'(c(\alpha))) = N$,
4. $c'(\alpha) \neq 0$.

In this work, we use homotopy or continuous deformations for object reconstruction. This is discussed in the next section.

4   Reconstruction Algorithm

Starting with a set of cross sections $\{\mathcal{S}_{i,j}\}$ for lines $\{\mathcal{L}_i\}$ in a plane, we restrict reconstruction in the bounding box $\mathcal{B}_{box}$ of the cross sections. The set of lines $\{\mathcal{L}_i\}$ partition $\mathcal{B}_{box}$ into a set of convex polygons $\{\mathcal{G}_k\}, k \in [0, p-1]$. This is shown in Fig. 4 where $\mathcal{O}$ is drawn dotted, the set of lines are shown dashed with the cross sections as thick solid lines, and the boundary of the bounding box is shown dashed. Our reconstruction algorithm consists of assigning a homotopy $\mathcal{H}_k$ to every polygon $\mathcal{G}_k$. The reconstruction is then obtained as a union of reconstructions within each polygon:

$$\mathcal{R} = \bigcup_k \{(x,y) : \mathcal{H}_k = 0\}. \tag{4}$$

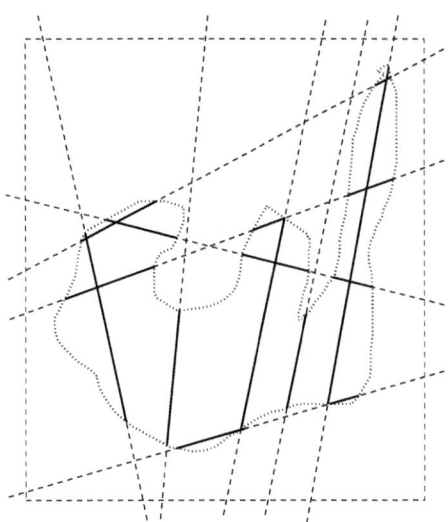

Fig. 4. A set of lines intersecting an object (dotted)

A homotopy can be seen as a smooth transition from one map to another. We can extend this definition to multiple maps by defining a homotopy in multiple variables

$$\mathcal{H}(\lambda_0, \lambda_1, \cdots, \lambda_{s-1}) = \sum_{t=0}^{s-1} f_t \lambda_t, \tag{5}$$

with

$$\sum_{t=0}^{s-1} \lambda_t = 1. \tag{6}$$

The homotopy parameters must sum up to unity in order to define the deformation $\mathcal{H}$ inside the convex hull of the domain. In our case, the domain is a polygon formed by straight line segments.

4.1   Edge Maps

Using (5), a smooth map can be defined over $\mathcal{G}_k$ for a choice of maps $\{f_t\}, t \in [0, s_k - 1]$ defined on s_k edges of $\mathcal{G}_k$. Let these maps be called *edge maps*. For continuity across all polygons, the definition of the edge maps must be consistent. Since, polygon edges are a subset of the cross section lines, it suffices to define edge maps over $\{\mathcal{L}_i\}$.

An edge map f_i should completely describe the boundary, interior and exterior of the intersection of $\mathcal{L}_i$ with $\mathcal{O}$. To define f_i, we associate a local coordinate system with each line $\mathcal{L}_i$ whose axis measures distance r along it from a chosen origin. Given abscissae $r_q, q \in [0, 2m_i - 1]$ of the intersections $\mathcal{S}_{i,j}$, we define the corresponding edge map as a piecewise quadratic polynomial

$$f_i(r) = \sum_{q=0}^{2m_i-2} \frac{\alpha_q(-r^2 + r(r_q + r_{q+1}) - r_q r_{q+1}))}{(r_{q+1} - r_q)}, \tag{7}$$

where α_q is the positive gradient $\left|\frac{\mathrm{d}f}{\mathrm{d}r}\right|_{r=r_q}$ defined as

$$\alpha_q = (-1)^{q+1}\alpha_0, \tag{8}$$

with α_0 being a chosen positive slope at r_0. Fig. 5 illustrates such an edge map.

Fig. 5. Piecewise quadratic function as an edge map

4.2   Barycentric Coordinates

It is natural to consider barycentric coordinates of a polygon as homotopy variables because of the two useful properties that they offer. Barycentric coordinates span a complete polygon and are a partition of unity. Traditional barycentric coordinates for triangles (and simplices in general) are defined by its vertices. Relevant generalizations of barycentric coordinates to n-sided polygons were provided by Wachspress [16] and later by Meyer et al. [11]. In the current context, we define barycentric coordinates in terms of the edges of a polygon rather than the vertices. Such a definition allows us to apply the concepts developed so far to associate a suitable homotopy to a polygon. We define Voronoi diagram based barycentric coordinates of edges for an n-sided polygon.

An interesting class of barycentric coordinates can be derived from Voronoi diagram of line segments and a point. Consider again a polygon $\mathcal{G}$ and a point p inside it. The Voronoi region of a point inside a polygon is a closed region of piecewise parabolic arcs as shown in Fig. 6. From the Voronoi diagram of the edges $\{e_i\}$ in a polygon, introduction of a point p steals an area from two or more existing Voronoi regions. If the stolen area for any edge e_i and p is denoted by $\mathcal{A}_i$, the barycentric coordinates for the edge can be written as

$$\lambda_i = \frac{\mathcal{A}_i}{\displaystyle\sum_{j=0}^{s-1} \mathcal{A}_j}, i \in [0, s-1], \tag{9}$$

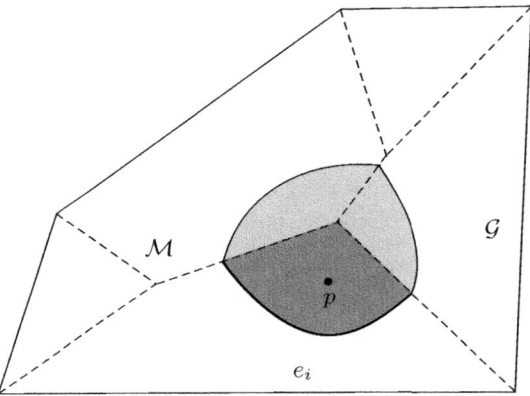

Fig. 6. Voronoi diagram of a polygon $\mathcal{G}$ and a point p inside

Area $\mathcal{A}_i$ can be computed as the area between the parabolic arc and the involved angle bisectors. Again, the barycentric coordinates defined in this way satisfy positivity, partition of unity and continuity. In the limiting case as p approaches one of the sides e_k, the stolen area $\mathcal{A}_k$ becomes very small, but simultaneously areas $\mathcal{A}_{i,i\neq k}$ become smaller (and eventually zero) by a rate higher than that of the former. Thus as p approaches e_k, $\lambda_k \to 1$, and $\lambda_{i,i\neq k} \to 0$.

4.3   Homotopy

Equipped with the above defined edge maps and barycentric coordinates for any polygon $\mathcal{G}_k$, we define a homotopy $\mathcal{H}_k$ in s_k variables as

$$\mathcal{H}_k(p) = \sum_{i=0}^{s_k-1} f_i\left(d_i(p)\right) \lambda_i(p), \tag{10}$$

where $d_i(p)$ is the distance along line $\mathcal{L}_k$ from a chosen origin O_k on it until the foot of the perpendicular from point p . We can write

$$d_i(p) = \|O_k - p_i\| + (p - p_i)^T \frac{(p_{i+1} - p_i)}{l_i}, \tag{11}$$

where p_{i+1} and p_i are two end points of an edge e_i of $\mathcal{G}_k$ lying on $\mathcal{L}_k$, and $||\mathbf{x}||$ denotes the length of a vector $\mathbf{x}$. Homotopy (10) continuously deforms edge maps f_i within the polygon and thus generates a smooth field. It can be seen as a linear combination of edge maps f_i with barycentric coordinates λ_i. We can further extend this so called *linear homotopy* to a *non-linear homotopy* as

$$\mathcal{H}_k(p, \eta) = \sum_{i=0}^{s_k-1} f_i\left(d_i(p)\right) \lambda_i(p)^\eta, \tag{12}$$

with η as a parameter.

Across polygons, the homotopy (10) is continuous and is at least C^1 smooth (see Appendix A for the proof). However, at all the intersection points $\mathcal{Q} = \{\mathcal{Q}_{i,j} : \mathcal{Q}_{i,j} \in \mathcal{L}_i \cap \partial\mathcal{O}\}$ of the lines $\{\mathcal{L}_i\}$ and the boundary $\partial\mathcal{O}$ of the object $\mathcal{O}$, the generated curve $\mathcal{H}^{-1}(0)$ is orthogonal to any line in $\{\mathcal{L}_i\}$ (see Appendix A). Therefore, the resulting reconstruction is somewhat unnatural. Given normals at the intersection points $\mathcal{Q}$ (which is the case with many range scanning physical devices), we propose a tangent alignment scheme for the resulting curve by locally warping the domain of the homotopies.

4.4   Tangent Alignment Using Local Space Rotations

Given unit normals $\widehat{\mathcal{N}}$ at intersection points $\mathcal{Q}$, the reconstruction can be constrained to be normally aligned to these normals at these points. We enforce this constraint by local space rotations around points $\mathcal{Q}$. The reconstruction

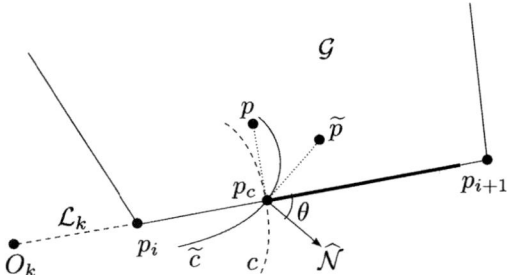

Fig. 7. Rotation of a point for tangent alignment

$c = H^{-1}(0)$ is orthogonal to the intersecting lines at $\mathcal{Q}$. Starting with a point p in the neighborhood of one of the points p_c of $\mathcal{Q}_{k,j}$ lying on $\mathcal{L}_k$, we rotate p about p_c by an angle $-\theta$ to give point $\widetilde{p}$ in the plane (see Fig. 7). The angle θ is chosen to be the signed angle between $\widehat{\mathcal{N}}$ and $\nabla\mathcal{L}_k$ at the point p_c. The resulting homotopy $\widetilde{\mathcal{H}}$ for a polygon $\mathcal{G}$ can be written as

$$\widetilde{\mathcal{H}} = \sum_{i=0}^{s_k-1} f_i\left(\widetilde{d}_i\right) \widetilde{\lambda}_i, \tag{13}$$

where,

$$\widetilde{d}_i = d_i(\widetilde{p}) = ||O_k - p_i|| + (\widetilde{p} - p_i)^T \frac{(p_{i+1} - p_i)}{l_i}, \text{ and}$$

$$\widetilde{\lambda}_i = \lambda_i(\widetilde{p}) = \frac{\widetilde{A}_i}{\sum\limits_{j=0}^{s-1} \widetilde{A}_j}, i \in [0, s-1].$$

The rotated point $\widetilde{p}$ can be written as

$$\widetilde{p} = p_c + \mathbf{R}(-\theta)(p - p_c), \tag{14}$$

where $\mathbf{R}$ is the rotation matrix

$$\mathbf{R}(\theta) = \begin{bmatrix} \cos\theta & -\sin\theta \\ \sin\theta & \cos\theta \end{bmatrix}. \tag{15}$$

The modified gradient of the homotopy field in the neighborhood of p_c can be now computed using the chain rule as:

$$\nabla\widetilde{\mathcal{H}} = \sum_{i=0}^{s-1} \left(f'_i(\widetilde{d}_i)\nabla\widetilde{d}_i\widetilde{\lambda}_i + f_i(\widetilde{d}_i)\nabla\widetilde{\lambda}_i \right). \tag{16}$$

In the limit as point $p \to p_c$ (or the orthogonal distance to $\mathcal{L}_k$, $\mu_k \to 0$), using a similar derivation as given in Appendix A, we can write

$$\lim_{\mu_k \to 0} \nabla\widetilde{\mathcal{H}} = f'_k(\widetilde{d}_i)\nabla\widetilde{d}_k \tag{17}$$

Computing the gradient of $\widetilde{d}_k$,

$$\begin{aligned} \nabla\widetilde{d}_k &= \nabla\left((\widetilde{p} - p_k)^T \frac{(p_{k+1} - p_k)}{l_k} \right) \\ &= \nabla\left((p_c + \mathbf{R}(-\theta)(p - p_c) - p_k)^T \frac{(p_{k+1} - p_k)}{l_k} \right) \\ &= \mathbf{R}(-\theta)^T \frac{(p_{k+1} - p_k)}{l_k} \\ &= \mathbf{R}(\theta)\nabla d_k \end{aligned} \tag{18}$$

Therefore,

$$\lim_{\mu_k \to 0} \nabla\widetilde{\mathcal{H}} = f'_k(\widetilde{d}_i)\mathbf{R}(\theta)\nabla d_k \tag{19}$$

We know that the gradient

$$\lim_{\mu_k \to 0} \nabla\mathcal{H} = f'_k(d_i)\nabla d_k. \tag{20}$$

Fig. 8. Weights based on higher order Voronoi diagram

From (20) and (19) it can be seen that in the limit $\mu_k \to 0$

$$\frac{\nabla \widetilde{\mathcal{H}}}{\|\nabla \widetilde{\mathcal{H}}\|} = \mathbf{R}(\theta) \left(\frac{\nabla \mathcal{H}}{\|\nabla \mathcal{H}\|} \right). \tag{21}$$

Therefore, we can achieve the desired rotation of the reconstruction curve by rotating the local coordinates around the points $\mathcal{Q}$ in the opposite direction.

4.5   Smooth Rotations of the Reconstruction Curve

In order to generate a smooth distortion $\widetilde{H}^{-1}(0)$ of the curve $H^{-1}(0)$ the neighborhoods of points $\mathcal{Q}$ must be carefully chosen. A natural neighborhood for points in $\mathcal{Q}$ is their respective Voronoi polygons. However, a constant rotation for all the points in a particular Voronoi region results in a discontinuous curve at the boundary of these polygons. Therefore, we seek a continuous weight function w_{p_i} inside a Voronoi region $\mathcal{V}_{\mathcal{Q}}(p_i)$ of any generator $p_i \in \mathcal{Q}$ such that

$$w_{p_i}(p_i) = 1, \text{ and}$$
$$w_{p_i}(\partial \mathcal{V}_{\mathcal{Q}}(p_i)) = 0. \tag{22}$$

These requirements on the weight function impose a smooth transition of rotation angles from one influence zone to another and ensure monotonically decreasing rotation angles as points get farther away from rotation centers with no rotation at the boundaries of the Voronoi regions. For any point p inside $\mathcal{V}_{\mathcal{Q}}(p_i)$, consider its nearest neighbor $p_i^{(1)}$ and the next nearest neighbor $p_i^{(2)} \in \mathcal{Q}$. Denote by $d_1(p)$ the distance between p and $p_i^{(1)}$ and by $d_2(p)$ the one between p and $p_i^{(2)}$. We can formulate the required weight function as

$$w_{p_i}(p) = \frac{d_2(p) - d_1(p)}{d_2(p) + d_1(p)}. \tag{23}$$

The first nearest neighbor $p_i^{(1)}$ is the generator point p_i of $\mathcal{V}_Q(p_i)$. The second nearest neighbor $p_i^{(2)}$ can be found by computing the second order voronoi diagram of Q. The weight function resulting from (23) is shown in Fig. 8. We outline the complete algorithm in Algorithm 1.

Input: $\{\mathcal{S}_{i,j}\}$ on $\{\mathcal{L}_i\}$, $\widehat{N}$ at Q
Output: $\mathcal{R}$
Define an edge map f_i on each $\mathcal{L}_i$ using $\{\mathcal{S}_{i,j}\}$
Partition $\mathcal{B}_{box}$ into polygonal tiles $\{\mathcal{G}_k\}$ with $\{\mathcal{L}_i\}$
Compute first order Voronoi diagram $\mathcal{V}_Q^1$ of Q
Compute second order Voronoi diagram $\mathcal{V}_Q^2$ of Q
Compute w from $\mathcal{V}_Q^1$ and $\mathcal{V}_Q^2$
for $k \in [0, p-1]$ **do**
    Compute Voronoi diagram $\mathcal{V}_k$ of $\mathcal{G}_k$
    for $\mathbf{x} \in \mathcal{G}_k$ **do**
        Let $\mathbf{x}_c \in Q$ be the generator of Voronoi polygon of $\mathbf{x}$
        $\widetilde{\mathbf{x}} \leftarrow \mathbf{x}_c + \mathbf{R}(-\theta)(\mathbf{x} - \mathbf{x}_c)$
        Compute $\{\lambda_i(\widetilde{\mathbf{x}})\}$ for all edges of $\mathcal{G}_k$ using $\mathcal{V}_k$
        Compute $d_i(\widetilde{\mathbf{x}})$ for all edges of $\mathcal{G}_k$
        Compute $\widetilde{\mathcal{H}}_k$
    end
end
$\widetilde{\mathcal{H}} \leftarrow \bigcup \widetilde{\mathcal{H}}_k$
$\mathcal{R} \leftarrow \ker \widetilde{\mathcal{H}}$

Algorithm 1. The reconstruction algorithm.

5 Results

We test our reconstruction method on a hand drawn curve intersected by a set of arbitrarily oriented lines. The lines yield a polygonal tessellation in the plane. We reconstruct the original curve from the intersections over the polygons of this tessellation.

This reconstructed curve is guaranteed to pass through the end-points of the intersections. Figs. 9(a), and 10(a) show results of our reconstruction for the linear and the non-linear homotopies respectively. The reconstructed curve is always orthogonal to the intersecting lines. The orthogonality is not apparent in Fig. 9(a) at a large scale, but is more visible for the non-linear homotopy with $\eta = 2$ in Fig. 10(a).

We also show results of the reconstructed curve after applying local rotations at points Q. The local rotations distort the original curve in the desired direction as seen in Figs. 9(b), and 10(b).

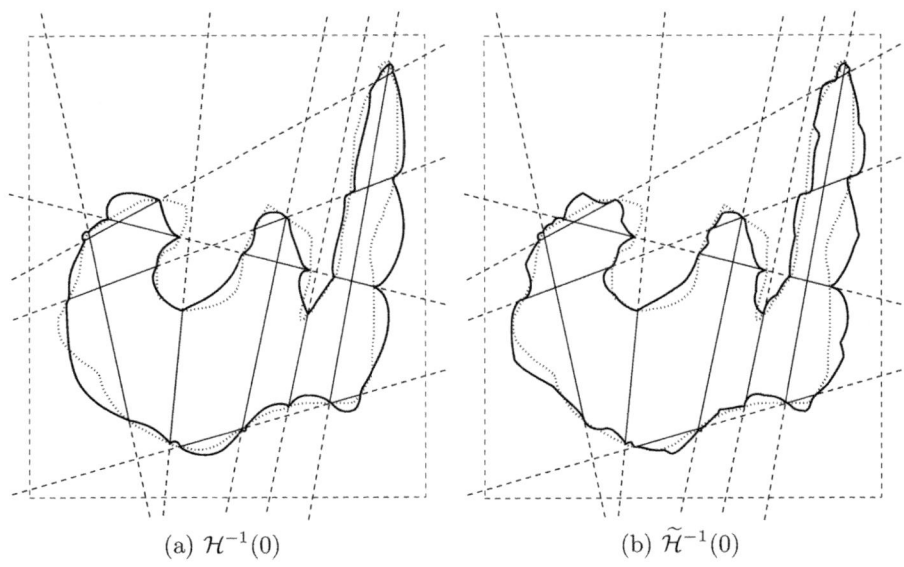

(a) $\mathcal{H}^{-1}(0)$ (b) $\widetilde{\mathcal{H}}^{-1}(0)$

Fig. 9. Reconstruction with linear homotopy

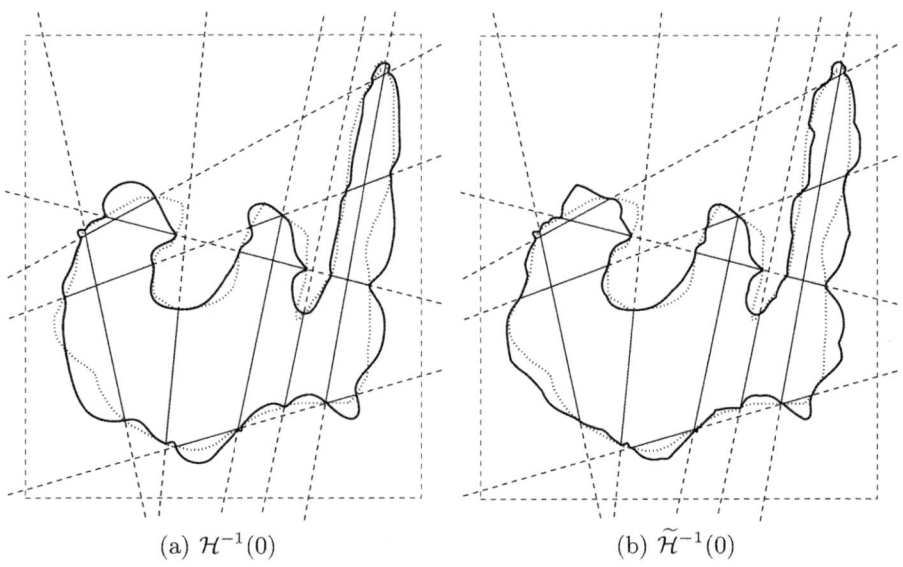

(a) $\mathcal{H}^{-1}(0)$ (b) $\widetilde{\mathcal{H}}^{-1}(0)$

Fig. 10. Reconstruction with non-linear homotopy

The next subsection discusses accuracy statistics of our method based on how much information is provided to the reconstruction algorithm in terms of the number of intersections.

5.1   Reconstruction Accuracy

A good reconstruction depends on the choice of cutting lines placed carefully to cover salient geometric features of the object to be reconstructed. In order to test the reconstruction algorithm for accuracy, we sample intersecting lines from the skeleton of the test object. We choose to sample the skeleton of the object since it captures the salient details of the object. To show the dependence of reconstruction accuracy on the number of intersecting lines, a hierarchy of skeletons is used.

A skeleton hierarchy is computed from a straight line skeleton [2]. The hierarchy provides an incremental simplification of the skeleton based on discrete curve evolution of the skeleton branches [8,5]. Curve simplification by discrete curve evolution uses local angles at every vertex of the curve to assign a relevance to each vertex. Based on the computed relevance values, the curve is evolved (simplified) by deletion of vertices. Fig. 11 shows such a skeleton hierarchy with five levels. The base skeleton is computed using the straight line skeleton module of the CGAL library [1] .

Sampled segments from the skeleton at every level are used as cutting lines for reconstruction. Thus, for every level of hierarchy of skeletons, we compute a reconstruction of the object. We next compute various error metrics for different reconstructions thus obtained. To get a comprehensive idea of the reconstruction accuracy, metrics based on area, mean distance error, and curve lengths are considered here.

Denoting the area of the model object by $\mathcal{A}_{mod}$, the area of the reconstruction at level i by $\mathcal{A}^i_{rec}$, the absolute difference of $\mathcal{A}_{mod}$ and $\mathcal{A}^i_{rec}$ is given by

$$\mathcal{A}^i_{diff} = \left(\mathcal{A}_{mod}\bigcup\mathcal{A}^i_{rec}\right) - \left(\mathcal{A}_{mod}\bigcap\mathcal{A}^i_{rec}\right). \tag{24}$$

Fig. 11 shows $\mathcal{A}^i_{diff}$ for different levels of hierarchy. Another important indicator of reconstruction accuracy is the ratio of areas $\mathcal{A}^i_{rec}/\mathcal{A}_{mod}$. Both of these measures are shown in Table 1. The tests show that with better sampling, the reconstruction accuracy increases. Also the ratio of areas indicate that the reconstructed curve has a slightly larger area than the original curve.

The *Hausdorff distance* is a good measure of the distance between two manifolds [12]. Hausdorff distance, d_H, between two curves L and L' is given by

$$d_H(L, L') = \sup_{x_0\in L}\ \inf_{x_1\in L'}\ d(x, x'), \tag{25}$$

where $d(\cdot, \cdot)$ is an appropriate metric for measuring distance between two points in a metric space. A mean value of the Hausdorff distance can be defined as [4]

$$d_m(L, L') = \frac{1}{|L|}\int_{x\in L}\ \inf_{x'\in L'}\ d(x, x')\mathrm{d}L, \tag{26}$$

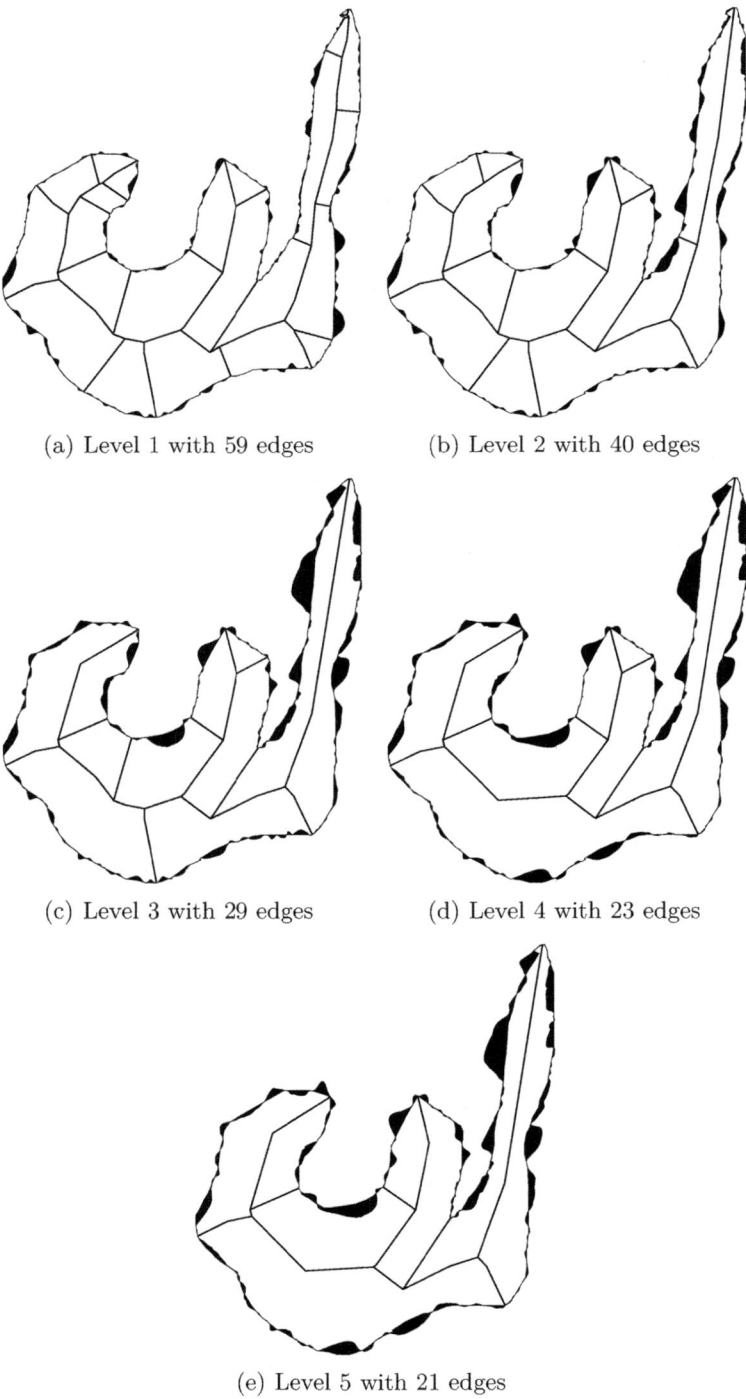

(a) Level 1 with 59 edges   (b) Level 2 with 40 edges

(c) Level 3 with 29 edges   (d) Level 4 with 23 edges

(e) Level 5 with 21 edges

Fig. 11. Skeleton hierarchy and reconstruction accuracy

Table 1. Reconstruction accuracy with respect to areas

		$\mathcal{A}_{mod} = 0.3228$			
Level	Edges	$\mathcal{A}^i_{rec}$	$\mathcal{A}^i_{diff}$	$\%\left(\dfrac{\mathcal{A}^i_{diff}}{\mathcal{A}_{mod}}\right)$	$\dfrac{\mathcal{A}^i_{rec}}{\mathcal{A}_{mod}}$
1	59	0.3266	0.0123	3.8	1.0119
2	40	0.3247	0.0168	5.2	1.0058
3	29	0.3279	0.0297	9.2	1.0160
4	23	0.3287	0.0333	10.3	1.0182
5	21	0.3283	0.0349	10.8	1.0171

where $|L|$ is the length of the curve L. A relative value of this distance is computed with respect to the bounding box of the object, as shown in Table 2. Also shown in the table is the length ratio of the reconstructed curve with respect to the original object contour. Here again, the tests show that a better sampling increases the accuracy of the reconstruction. However, we must point out that a topologically correct reconstruction is achievable with as few cutting lines as there are salient features in the object.

Table 2. Reconstruction accuracy with respect to lengths

Level	Edges	$\%\left(\dfrac{d_H}{L_{diag}}\right)$	$\dfrac{L^i_{rec}}{L_{mod}}$
1	59	0.75	1.0842
2	40	0.82	1.0919
3	29	1.02	1.1227
4	23	1.08	1.1191
5	21	1.11	1.1186

5.2 Comparison

A comparison of our reconstruction with the results of Memari and Boissonnat [10] shows some of the shortcomings of their method that can be overcome with a reconstruction using continuous deformations. Reconstruction method proposed by Memari and Boissonnat is derived from the Delaunay complex of the cross sections. The reconstruction curve (see Fig. 12) is only C^0 and misses some of the high curvature regions of the original object boundary.

We produce comparative statistics for our reconstruction with the method by Memari and Boissonnat [10]. The measures used for comparison are based on area and length of the reconstructed curve as introduced in the previous subsection. Table 3 shows reconstruction accuracy of three methods for the set of intersection lines shown in Fig. 4. Here, Homotopy$_1$ refers to the reconstruction using continuous deformations with no tangent alignment and Homotopy$_2$ refers to the reconstruction resulting from tangent alignment. Since [10] results in a

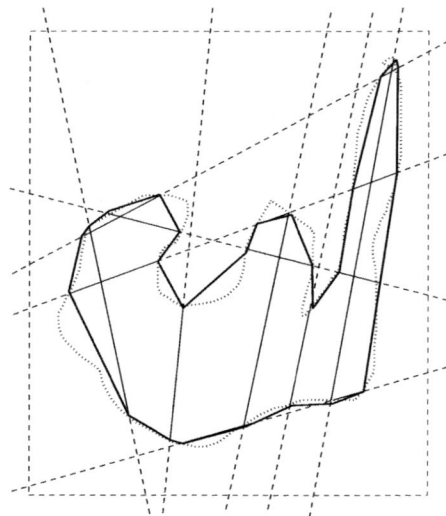

Fig. 12. Piecewise linear reconstruction using algorithm by Memari and Boissonnat [10]

piecewise linear reconstruction, the area (possibly) and length of the curve are underestimated. The ratio of absolute difference of area with the area of the model is low for Memari's reconstruction that matches up with Homotopy$_2$ reconstruction, but the relative Hausdorff measure goes bad and turns out to be more than double of that obtained with either of the Homotopy based reconstructions. Further, the relative ratio of the lengths of the reconstruction and the original object shows that homotopy based reconstructions perform better with estimating the length of the object.

Table 3. Comparison of Reconstructions

Method	$\%\left(\dfrac{A_{diff}}{A_{mod}}\right)$	$\dfrac{A_{rec}}{A_{mod}}$	$\%\left(\dfrac{d_H}{L_{diag}}\right)$	$\dfrac{L_{rec}}{L_{mod}}$
Memari [10]	14.6025	0.9652	3.9323	0.9149
Homotopy$_1$	17.5470	1.0407	1.4831	1.0256
Homotopy$_2$	14.6690	1.0425	1.3243	1.0303

6   Conclusion

In this work, we have presented a novel method of curve reconstruction from arbitrary cross sections in a planar setting. The presented algorithm uses continuous deformations to reconstruct the object smoothly. We also introduced generalized barycentric coordinates for polygons defined on its edges using the line and point Voronoi diagram. The presented method is general in nature and

can be applied to higher dimensions. We avaluate accuracy of our algorithm
based on sampling of the original object.

An interesting generalization of the problem of reconstruction from arbitrary
cross sections is 3D reconstruction from arbitrary cutting planes. These cutting
planes partition the domain of computation into polyhedra, and also embed the
intersection with the object. Similar to the approach suggested in this paper,
a function $f : \mathbb{R}^2 \mapsto \mathbb{R}$ can be embedded in a cutting plane such that $\ker(f)$
represents the boundary of this intersection. An example of one such function
is the signed distance function. Higher degree polynomial functions can be designed based on this distance function. A multi-variate homotopy can then be
constructed from the functions defined previously on the faces of a polyhedron. A
possible parameterization of such a homotopy is with respect to the orthogonal
distance of any point inside the polyhedron to the polyhedron faces (see Figure 13). Parameterization in terms of the Voronoi volume (a volume consisting
of a paraboloid face and planar faces) stolen by a point inside the polyhedron is
another choice. A union of the zero level set of the derived homotopies provides
a reconstruction surface.

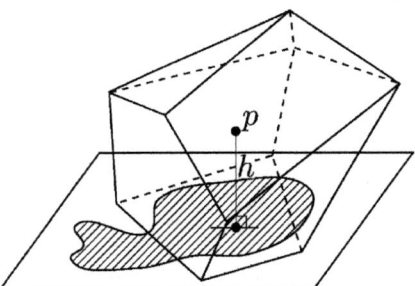

Fig. 13. Reconstruction from arbitrary cutting planes in $\mathbb{R}^3$. The polyhedron shows
one of the partitions of the domain resulting from cutting planes (one such shown here).
The shaded region depicts the cross section of the 3D object with the cutting plane.

Appendix A: Smoothness in Homotopies

Proposition 1. *For a triangle $\mathcal{T}$ in a planar triangulation , we show that the
curve $\mathcal{H}^{-1}(0)$ defined by the homotopy (10),*

$$\mathcal{H}(p) = \sum_{i=0}^{2} f_i\left(d_i(p)\right) \lambda_i(p) = 0,$$

is at least C^1.

Proof. We can prove this by showing that $\mathcal{H}^{-1}(0)$ is C^0 and C^1.

1. $\mathcal{H}^{-1}(0)$ is C^0.

From the properties of barycentric coordinates, we know that

$$\sum_{i=0}^{2} \lambda_i = 1.$$

This implies that at any edge e_k of $\mathcal{T}$, $\lambda_k = 1$ and $\lambda_{i,i \neq k} = 0$. Therefore, for any p lying on e_k, $\mathcal{H}(p) = f_k(d_k(p))$. The same holds for any other triangle in the triangulation and since the functions f_i are globally defined on lines L_i, $\mathcal{H} = 0$ is C^0.

2. $\mathcal{H}^{-1}(0)$ is C^1.

In order to show that $\mathcal{H}^{-1}(0)$ is C^1, we calculate the gradient of $\mathcal{H}$ at any e_k. The derivative of $\mathcal{H}$ is

$$\nabla \mathcal{H} = \sum_{i=0}^{2} \left(f_i'(d_i) \nabla d_i \lambda_i + f_i(d_i) \nabla \lambda_i \right) \tag{27}$$

where d_i is the distance along line $\mathcal{L}_j$ corresponding to edge e_i of the triangle. For a triangle, the proposed barycentric coordinates based on the stolen area $\mathcal{A}_i$ for any edge e_i and point p are given by (9) as

$$\lambda_i = \frac{\mathcal{A}_i}{\displaystyle\sum_{j=0}^{2} \mathcal{A}_j}, i \in [0, 2].$$

The gradient of λ_i can be calculated using the chain rule as

$$\nabla \lambda_i = \frac{\nabla \mathcal{A}_i}{\displaystyle\sum_{j=0}^{2} \mathcal{A}_j} - \frac{\mathcal{A}_i}{\left(\displaystyle\sum_{j=0}^{2} \mathcal{A}_j\right)^2} \sum_{j=0}^{2} \nabla \mathcal{A}_j$$

$$= \frac{\nabla \mathcal{A}_i}{\displaystyle\sum_{j=0}^{2} \mathcal{A}_j} - \lambda_i \sum_{j=0}^{2} \frac{\nabla \mathcal{A}_j}{\displaystyle\sum_{j=0}^{2} \mathcal{A}_j} \tag{28}$$

Using (27) and (28),

$$\nabla \mathcal{H} = \sum_{i=0}^{2} \left(f_i'(d_i) \nabla d_i \lambda_i + f_i(d_i) \left[\frac{\nabla \mathcal{A}_i}{\displaystyle\sum_{j=0}^{2} \mathcal{A}_j} - \lambda_i \sum_{j=0}^{2} \frac{\nabla \mathcal{A}_j}{\displaystyle\sum_{j=0}^{2} \mathcal{A}_j} \right] \right) \tag{29}$$

To compute the gradient of $\mathcal{H}$ at the intersection of line $\mathcal{L}_j$ and the object, we must take the derivative at a point p in the limit as p approaches line $\mathcal{L}_j$. In this limit,

$$\lambda_k \rightarrow 1$$
$$\lambda_{i,i\neq k} \rightarrow 0 \tag{30}$$

Before the gradient in the limit can be evaluated, the behavior of $\nabla\mathcal{A}_i$ and $\mathcal{A}_i$ should be analyzed. To do so, we define several quantities as following.

Consider the triangle $\mathcal{T}$ shown in Fig. 14. Let the vertices of $\mathcal{T}$ be denoted by $p_0(x_0, y_0)$, $p_1(x_1, y_1)$, and $p_2(x_2, y_2)$, and the edges by $e_0 = (p_1 - p_0)$, $e_1 = (p_2 - p_1)$, and $e_2 = (p_0 - p_2)$. Let an edge e_i of $\mathcal{T}$ be parameterized by distance α along it

$$e_i : p = p_i + \alpha\frac{(p_{i+1} - p_i)}{l_i}, \tag{31}$$

where index i is to be taken in a circular sense in the triangle. The Voronoi diagram of edges of $\mathcal{T}$ divide it internally in three regions. We consider a point p lying in the Voronoi region of an edge e_i of $\mathcal{T}$ (see Fig. 14).

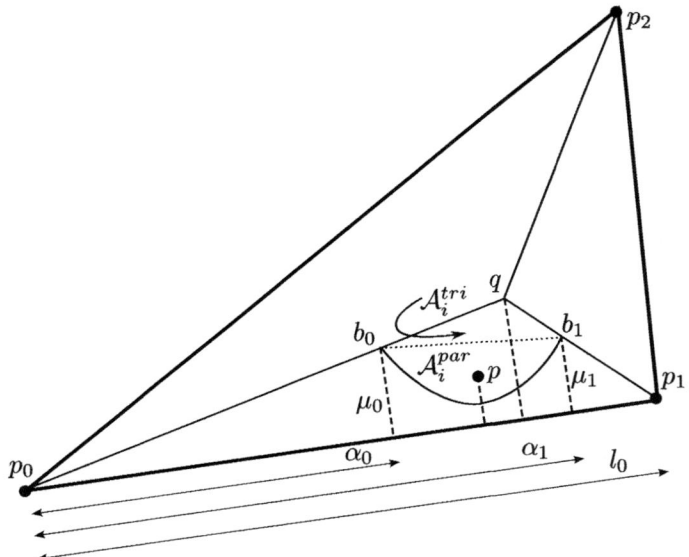

Fig. 14. Parameterization along triangle edge

A parabola with focus at point p and directrix e_i is given by

$$\mathcal{P}_i : \mu(\alpha) = \frac{(\alpha - \alpha_p)^2}{2\mu_p} + \frac{\mu_p}{2}, \tag{32}$$

where (α, μ) form an orthonormal basis, and

$$\alpha_p = (p - p_i)^T \frac{(p_{i+1} - p_i)}{l_i}, \tag{33}$$

$$\mu_p = (p - p_i)^T \mathbf{M} \frac{(p_{i+1} - p_i)^T}{l_i}, \tag{34}$$

with

$$\mathbf{M} = \begin{bmatrix} 0 & 1 \\ -1 & 0 \end{bmatrix}, \tag{35}$$

and l_i is the length of e_i.

Let the parabola $\mathcal{P}_i$ intersects the boundary of the Voronoi region at points $b_i(\alpha_0, \mu_0)$ and $b_{i+1}(\alpha_1, \mu_1)$ respectively. Denote by $q(\alpha_q, \mu_q)$ the incenter of $\mathcal{T}$. The parabola can intersect the two angle bisectors of the triangle $\mathcal{B}_i : \mu = k_0\alpha$, and $\mathcal{B}_{i+1} : \mu = k_1(l_i - \alpha)$, $k_i = \tan(\theta_i/2)$, in three possible ways

I. $\alpha_1 \leq \alpha_q$: both branches of the parabola intersect $\mathcal{B}_i$, and $\mu_m = k_0\alpha_m$, $m = \{0, 1\}$,

II. $\alpha_0 \geq \alpha_q$: both branches of the parabola intersect $\mathcal{B}_{i+1}$, and $\mu_m = k_1(l_i - \alpha_m)$, $m = \{0, 1\}$, and

III. $\alpha_0 < \alpha_q$ and $\alpha_1 > \alpha_q$: the branches of the parabola intersect $\mathcal{B}_i$ and $\mathcal{B}_{i+1}$ respectively, and $\mu_0 = k_0\alpha_0$ and $\mu_1 = k_1(l_i - \alpha_1)$.

It is sufficient to treat one of these cases here. Considering case I, α_0 and α_1 are the roots of

$$\alpha^2 - 2\alpha(\alpha_p + k_0\mu_p) + (\alpha_p^2 + \mu_p^2) = 0. \tag{36}$$

The areas to compute the barycentric coordinates can be written as

$$\mathcal{A}_i = \mathcal{A}_i^{par} + \mathcal{A}_i^{tri} \tag{37}$$

where, $\mathcal{A}_i^{par}$ is the area enclosed between the parabolic arc and line connecting b_i and b_{i+1}, and $\mathcal{A}_i^{tri}$ is the area of the triangle connecting points b_i, b_{i+1}, and q. Using (32)

$$\mathcal{A}_i^{par} = \frac{\mu_0 + \mu_1}{2}(\alpha_1 - \alpha_0) - \int_{\alpha_0}^{\alpha_1} \left(\frac{(\alpha - \alpha_p)^2}{2\mu_p} + \frac{\mu_p}{2} \right) d\alpha, \tag{38}$$

and

$$\mathcal{A}_i^{tri} = \frac{\mu_0 + \mu_q}{2}(\alpha_q - \alpha_0) + \frac{\mu_1 + \mu_q}{2}(\alpha_1 - \alpha_q) - \frac{\mu_0 + \mu_1}{2}(\alpha_1 - \alpha_0). \tag{39}$$

Therefore,

$$\mathcal{A}_i = \frac{\mu_0 + \mu_q}{2}(\alpha_q - \alpha_0) + \frac{\mu_1 + \mu_q}{2}(\alpha_1 - \alpha_q) - \int_{\alpha_0}^{\alpha_1} \left(\frac{(\alpha - \alpha_p)^2}{2\mu_p} + \frac{\mu_p}{2} \right) d\alpha \tag{40}$$

Note that (40) holds for all the three cases above. Simplifying (40), we get

$$2\mathcal{A}_i = (\mu_0 + \mu_q)(\alpha_q - \alpha_0) + (\mu_1 + \mu_q)(\alpha_1 - \alpha_q) - \left[\frac{(\alpha - \alpha_p)^3}{3\mu_p} - \mu_p\alpha\right]_{\alpha_0}^{\alpha_1}$$

$$= (\mu_0 + \mu_q)(\alpha_q - \alpha_0) + (\mu_1 + \mu_q)(\alpha_1 - \alpha_q)$$
$$- \frac{(\alpha_1 - \alpha_p)^3 - (\alpha_0 - \alpha_p)^3}{3\mu_p} - \mu_p(\alpha_1 - \alpha_0) \tag{41}$$

Distance d_i in (10) can be written as

$$d_i(p) = ||O_j - p_i|| + \alpha_p, \tag{42}$$

where we know that p_i lies on $\mathcal{L}_j$ and O_j is the chosen origin on $\mathcal{L}_j$. We note the following derivatives

$$2\nabla\mathcal{A}_i = \nabla\alpha_0(-\mu_0 - \mu_p) + \nabla\alpha_1(\mu_1 + \mu_p) + \nabla\mu_0(-\alpha_0 + \alpha_q) + \nabla\mu_1(\alpha_1 - \alpha_q)$$
$$+ \frac{(\alpha_1 - \alpha_p)^3 - (\alpha_0 - \alpha_p)^3}{3\mu_p^2}\nabla\mu_p - \frac{(\alpha_1 - \alpha_p)^2(\nabla\alpha_1 - \nabla\alpha_p)}{\mu_p}$$
$$+ \frac{(\alpha_0 - \alpha_p)^2(\nabla\alpha_0 - \nabla\alpha_p)}{\mu_p} - (\alpha_1 - \alpha_0)\nabla\mu_p - \mu_p(\nabla\alpha_1 - \nabla\alpha_0) \tag{43}$$

$$\nabla d_i = \nabla\alpha_p = \frac{(p_{i+1} - p_i)}{l_i} \tag{44}$$

$$\nabla\mu_p = \mathbf{M}\frac{(p_{i+1} - p_i)}{l_i} \tag{45}$$

$$\nabla\alpha_m = \frac{\nabla\alpha_p(\alpha_m - \alpha_p) + \nabla\mu_p(\mu_m - \mu_p)}{(\alpha_m - \alpha_p - k_0\mu_p)}, m = \{0,1\} \tag{46}$$

$$\nabla\mu_m = k_0\nabla\alpha_m, \ m = \{0,1\} \tag{47}$$

In the limit $\mu_p \to 0$ for some edge e_k of $\mathcal{T}$,

$$\alpha_m \to \alpha_p, \ m \in \{0,1\}$$
$$\mu_m \to k_0\alpha_p \ m \in \{0,1\} \tag{48}$$

Consequently, $\mathcal{A}_i \to 0$, but $\mathcal{A}_{i,i\neq k}$ becomes 0 much faster than $\mathcal{A}_k$. Therefore, using (30) and (48), the gradient (29) in the limit is

$$\lim_{\mu_p \to 0} \nabla\mathcal{H} = f_k'(d_k)\nabla d_k + f_k(d_k)\left[\frac{\nabla\mathcal{A}_k}{2} - \frac{\nabla\mathcal{A}_k}{2}\right]$$
$$\sum_{j=0}\mathcal{A}_j \quad \sum_{j=0}\mathcal{A}_j$$

$$= f_k'(d_k)\nabla d_k. \tag{49}$$

Gradient of line e_k is

$$\nabla e_k = \mathbf{M}\frac{(p_{i+1} - p_i)}{l_i}. \tag{50}$$

We note that

$$
\left\langle \lim_{\mu_p \to 0} \nabla \mathcal{H}, \nabla e_k \right\rangle = f_k'(d_k) \frac{(p_{i+1} - p_i)^T}{l_i} \mathbf{M} \frac{(p_{i+1} - p_i)}{l_i}
$$
$$
= 0.
$$

A similar result can be shown for the gradient of the homotopy on the other side of $\mathcal{L}_j$. This implies that the reconstructed curve is orthogonal to the intersecting lines from either side.

Therefore, the curve reconstruction is at least C^1.   □

References

1. CGAL, Computational Geometry Algorithms Library, http://www.cgal.org
2. Aichholzer, O., Aurenhammer, F., Alberts, D., Gärtner, B.: A novel type of skeleton for polygons. Journal of Universal Computer Science 1(12), 752–761 (1995)
3. Allgower, E.L., Georg, K.: Numerical continuation methods: an introduction. Springer-Verlag New York, Inc., New York (1990)
4. Aspert, N., Santa-Cruz, D., Ebrahimi, T.: Mesh: Measuring errors between surfaces using the hausdorff distance. In: Proceedings of the IEEE International Conference on Multimedia and Expo., vol. 1, pp. 705–708 (2002)
5. Bai, X., Latecki, L., Liu, W.: Skeleton pruning by contour partitioning with discrete curve evolution. IEEE Transactions on Pattern Analysis and Machine Intelligence 29(3), 449–462 (2007)
6. Dougherty, G., Varro, J.: A quantitative index for the measurement of the tortuosity of blood vessels. Medical Engineering & Physics 22(8), 567–574 (2000)
7. Jänich, K.: Topology, Undergraduate texts in mathematics (1984)
8. Latecki, L.J., Lakämper, R.: Polygon Evolution by Vertex Deletion. In: Nielsen, M., Johansen, P., Fogh Olsen, O., Weickert, J. (eds.) Scale-Space 1999. LNCS, vol. 1682, pp. 398–409. Springer, Heidelberg (1999)
9. Liu, L., Bajaj, C., Deasy, J.O., Low, D.A., Ju, T.: Surface reconstruction from non-parallel curve networks. Computer Graphics Forum 27, 155 (2008)
10. Memari, P., Boissonnat, J.D.: Provably Good 2D Shape Reconstruction from Unorganized Cross-Sections. Computer Graphics Forum 27, 1403–1410 (2008)
11. Meyer, M., Lee, H., Barr, A., Desbrun, M.: Generalized barycentric coordinates on irregular polygons. Journal of Graphics, GPU, and Game Tools 7(1), 13–22 (2002)
12. Munkres, J.: Topology, 2nd edn. Prentice Hall (1999)
13. Patasius, M., Marozas, V., Lukosevicius, A., Jegelevicius, D.: Model based investigation of retinal vessel tortuosity as a function of blood pressure: preliminary results. In: 29th Annual International Conference of the IEEE Engineering in Medicine and Biology Society, pp. 6459–6462 (2007)
14. Saucan, E., Appleboim, E., Zeevi, Y.Y.: Geometric approach to sampling and communication. Arxiv preprint arXiv:1002.2959 (2010)
15. Sidlesky, A., Barequet, G., Gotsman, C.: Polygon Reconstruction from Line Cross-Sections. In: Canadian Conference on Computational Geometry (2006)
16. Wachspress, E.L.: A rational finite element basis. Academic Press (1975)

Round-Trip Voronoi Diagrams and Doubling Density in Geographic Networks

Matthew T. Dickerson[1], Michael T. Goodrich[2],
Thomas D. Dickerson[3], and Ying Daisy Zhuo[1]

[1] Dept. of Computer Science, Middlebury College, Middlebury, VT, USA
{dickerso,yzhuo}@middlebury.edu
[2] Dept. of Computer Science, Univ. of California, Irvine, Irvine, CA, USA
goodrich@ics.uci.edu
[3] St. Michael's College, Colchester, VT, USA
tdickerson@smcvt.edu

Abstract. Given a geographic network G (e.g. road network, utility distribution grid) and a set of sites (e.g. post offices, fire stations), a *two-site Voronoi diagram* labels each vertex $v \in G$ with the pair of sites that minimizes some distance function. The *sum* function defines the "distance" from v to a pair of sites s, t as the sum of the distances from v to each site. The *round-trip* function defines the "distance" as the minimum length tour starting and ending at v and visiting both s and t. A *two-color* variant begins with two different types of sites and labels each vertex with the minimum pair of sites of different types. In this paper, we provide new properties and algorithms for two-site and two-color Voronoi diagrams for these distance functions in a geographic network, including experimental results on the *doubling distance* of various point-of-interest sites. We extend some of these results to multi-color variants.

Keywords: Voronoi diagrams, road networks, round-trip, point-of-interest, two-color.

1 Introduction

Given a set S of d-dimensional points, called *sites*, the *Voronoi diagram* of S is defined to be the subdivision of $\mathbf{R}^d$ into cells, one for each site in S, where the Voronoi cell for site $p \in S$ is the loci of all points closer to p than to any other point in S. The Voronoi diagram is an important and well-studied geometric structure, now more than 100 years old [3,4,14,16,24,25]. For example, when coupled with point location data structures and algorithms, the Voronoi diagram provides a solution to Knuth's well-known *post office problem* [18]: given a set of n post offices, create a structure that allows us to identify, for each house, its nearest post office.

As useful as is the geometric version of the Voronoi diagram defined above for many real world applications, for other applications it is inappropriate because it assumes that distance can be measured via a straight line in Euclidean space— or, in the case of post offices, on a Euclidean plane. That is, it applies a *geometric*

M.L. Gavrilova et al. (Eds.): Trans. on Comput. Sci. XIV, LNCS 6970, pp. 211–238, 2011.

structure to a *geographic* problem. In the real world, by contrast, postal delivery cars travel along roads, and not on straight lines (across fields, mountains, rivers, and through forests.) Post offices as well as mailboxes or addresses are points in a geographic network—the graph defined by the set of roads in a given geographic region—not points in $\mathbf{R}^2$. Thus, a Voronoi diagram defined geometrically is, in practice, a poor solution to Knuth's post office problem, especially in geographic regions with natural obstacles, like lakes, rivers, bridges, and freeways, that make the Euclidean metric a inappropriate distance function for determining a nearest post office. A more practical and realistic solution to Knuth's post office problem should use a version of the Voronoi diagram that is defined for geographic networks, or non-negatively weighted graphs.

Furthermore, traditional Voronoi diagrams define the distance only from a single vertex to a single site. However the "cost" related to a particular vertex in a geographic network may be dependent on its relationship to several sites. For example, if one likes to visit three different grocery stores each week for maximum savings, then one would want to associate each vertex on a graph with the *three* grocery store sites that together minimize the sum distance, and so we want to compute the 3-site Voronoi diagram using the sum function. If one visits two grocery stores on each trip, then the goal is to find the sites that minimize the round-trip distance function—which may be different sites than those that minimize the sum function—and we want a round-trip distance function Voronoi diagram to evaluate each vertex. If one visits a post office, grocery store, and department store each week, then optimal sites can be determined by a multi-color Voronoi diagram.

1.1   Graph-Theoretic Voronoi Diagrams

A *geographic network* is a graph $G = (V, E)$ that represents a transportation or flow network, where commodities or people are constrained to travel along the edges of that graph. Examples include road, flight, and railroad networks, utility distribution grids, and sewer lines. We assume that the edges of a geographic network are assigned weights, which represent the cost, distance, or penalty of

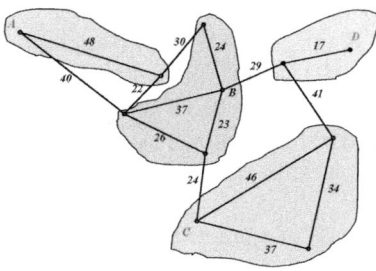

Fig. 1. An example graph-theoretic Voronoi diagram with sites A,B,C,D

moving along that edge, or some combination of these and other factors, such as scenic or ecological value. The only requirement we make with respect to these weights is that they be non-negative. In this paper, we also restrict our attention to undirected geographic networks.

Since all our edge weights are non-negative, and the edges are undirected, a shortest path exists between each pair of vertices in G. The distance, $d(v, w)$ for $v, w \in G$, is defined as the *length* of a *shortest* (i.e., minimum weight) path between v and w. This distance function, d, is well-defined, and $d(v, w) = d(w, v)$. Moreover, since by definition of "shortest", it follows immediately that for any vertices $v, w, x \in G$, $d(v, x) \leq d(v, w) + d(w, x)$; that is, the *triangle inequality* holds for this path distance d.

This observation allows us to define the Voronoi diagram of a geographic network. Formally, we define a geographic network, $G = (V, E)$, to be a set V of *vertices*, a set E of *edges* (which are unordered pairs of distinct vertices), and a *weight function* $w : E \to \mathbf{R}^+$ mapping edges E to non-negative real numbers. In a road network, this weight function could represent either distance along a road (that is, the Euclidean length of an edge) or the travel time. In the Voronoi diagram problem, we are also given a subset $K \subset V$ of special vertices called *sites*. These are the "post offices" in Knuth's post office problem, but of course they could also be any points of interest (or POIs) such as a schools, hospitals, fire stations, or grocery stores. Each site $v \in K$ is uniquely labeled with a natural number $n(v)$ from 0 to $|K| - 1$, so that we can refer to sites by number. The numbering is also used to resolve ties so that the ordering of sites by distance can be uniquely defined.

The standard first-order graph-theoretic *Voronoi diagram* [21] of G is a labeling of each vertex w in V with the number, $n(v)$, of the vertex v in K that is closest to w. All the vertices with the same label, $n(v)$, are said to be in the *Voronoi region* for v. Intuitively, if a site v in K is considered a post office, then the Voronoi region for v consists of all the homes that *ought* to be in v's zip code. (Note: if we want to consider each house on a block as separate entity with potentially a different closest post office, rather than model the entire block as a single vertex, we should use an individual vertex in V for each house, most of which would be degree-2 vertices in G.) (See Fig. 1.)

We also use these numbers, $n(v)$, to break ties in distances, which allows us to speak of unique closest sites in K for each vertex in V. That is, if we have two distinct sites $v, w \in S$ and a third vertex $x \in V$ such that $d(v, x) = d(w, x)$, then we say that x is closer to v if and only if $n(v) < n(w)$, and otherwise x is closer to y. For example, consider two distinct sites $v, w \in S$ with $n(v) = 0$ and $n(w) = 1$, and a third vertex $x \in V$, and with two edges (v, x) and (w, x) such that $d(v, x) = d(w, x)$. Then x is in the Voronoi region for site v. This becomes particularly important if there is a fourth vertex y and a third edge (x, y). Without the tie-breaking rule using $n(v)$, the entire edge (x, y) would be equidistant to v and w, and if there were no other alternate paths then any entire subgraph of G connected at x would be equidistant from v and w.

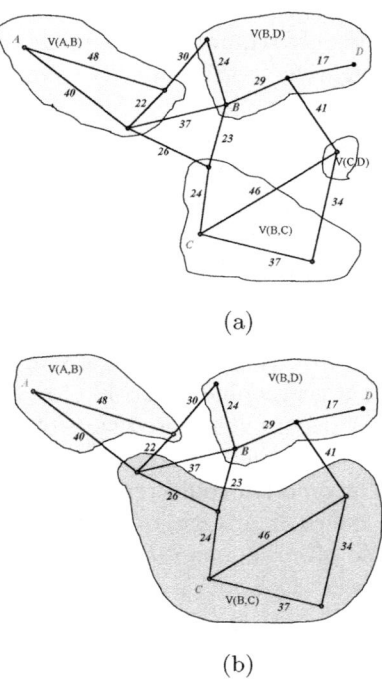

(a)

(b)

Fig. 2. (a) An example graph-theoretic two-site sum function Voronoi diagram of the same graph from Fig. 1. (b) An example graph-theoretic two-site round-trip function Voronoi diagram of the same graph as in Fig. 1.

Mehlhorn [21] shows that the graph-theoretic Voronoi diagram of a graph G, having n vertices and m edges, can be constructed in $O(n \log n + m)$ time. A similar algorithm is given by Erwig [15]. At a high level, these algorithms perform n simultaneous runs of Dijkstra's single-source shortest-path algorithm [13] (see also [10,17]).

In this paper, however, we are not interested in these types of *single-site* Voronoi diagrams.

1.2   Round-Trip Distance

In a number of applications, we may be interested in labeling the vertices of a geographic network, G, with more information than just their single nearest neighbor from the set of sites, K. We may wish, for instance, to label each vertex v in G with the names of the C closest sites in K, for some $C \le |K|$. For example, the sites in K may be fire stations, and we may wish to know the three closest fire stations for each house in our network, just in case there is a three-alarm fire at that location.

For many applications, "closest" among a set of neighbors should instead be defined by the *round-trip* or *tour* distance. (For $C = 2$, and for point sites on the

Euclidean plane, this distance was referred to in [7] as the "perimeter" distance, in reference to the perimeter of a triangle, since shortest paths are straight edges. In this paper, although we also focus on the case $c = 2$, we will refer to this function as the "round-trip" distance in reference both to road networks and to the more general case of $C \geq 2$.) (See Fig. 2.) In this notion of distance, we want to take a single trip, starting and ending at our "home" location and visiting two (or more) distinguished sites. Such distances correspond to the work that would need to be done, for example, by someone who needs to leave their house, visit multiple sites to run a number of errands, and then return home. Some hypothetical examples include the following:

- Some legal documents require the signatures of multiple witnesses and/or no-
  taries in order to be executed, so we may need to travel to multiple locations
  to get them all.
- Some grocery stores place a limit on the amount of special "loss leader" sale
  items one can purchase in a single visit, so we may need to visit multiple
  stores to get enough of such items needed for a big party.
- A celebrity just out of rehab may wish to get multiple community service
  credits in a single trip, for instance, by tutoring students at an educational
  institution and speaking at an alcoholics anonymous meeting at a religious
  institution, all on the same day.

In each case, we are likely to want to optimize our travel time to visit all the sites of interest as quickly as possible.

Alternatively, we may have a number of different kinds of sites, such as gas stations, grocery stores, and coffee houses, and we are interested in the three that are closest to each house, in terms of how one could visit all three types of sites in a single trip from home. Thus we are also interested in *multi-color* Voronoi diagrams, where each type of site (such as coffee houses and grocery stores) is represented with a different color. While an individual is not likely to use a computer implementation of a round-trip search algorithm to make such local decisions (as he or she might use Google Maps to find the fastest route to a distant city), retail and service corporations do make location decisions based on such information: maximizing the size of the population for which the location can be efficiently visited.

Figure 3 shows a gray-scale image of the 2-color round-trip Voronoi diagram of the road network in the state of Vermont featuring hospitals as one "color" site, and religious institutions as the other set of sites. For each pair of sites including one hospital and one religious institution, there is a region (in most cases empty) of the set of all vertices having that pair of sites closer than any other pair. Each non-empty region is shaded with a particular shade of grey. (This Voronoi diagram was generated using the algorithm developed in this paper.)

1.3   Related Prior Work

Unlike prior work on graph-theoretic Voronoi diagrams, there is a abundance of prior work for geometric Voronoi diagrams. It is beyond the scope of this paper

Fig. 3. A 2-Color Round-Trip Voronoi Diagram for the Road Network of the State of Vermont (Using Hospitals and Churches)

to review all this work and its applications. We refer the interested reader to any excellent survey on the subject (e.g., see [3,4,16,24]) and we focus here on previous work on multi-site geometric Voronoi diagrams and on graph-theoretic Voronoi diagrams.

Lee [20] studies k-nearest neighbor Voronoi diagrams in the plane, which are also known as "order-C Voronoi diagrams." These structures define each region, for a site p, to be labeled with the C nearest sites to p. These structures can be constructed for a set of n points in the plane in $O(n^2 + C(n - C) \log^2 n)$ time [9]. Due to their computational complexity, however, order-C Voronoi diagrams have not been accepted as practical solutions to C-nearest neighbor queries. Patroumpas *et al.* [22] study methods for performing C-nearest neighbor queries using an approximate order-C network Voronoi diagram of points in the plane, which has better performance than its exact counterpart.

Two-site distance functions and their corresponding Voronoi diagrams were introduced by Barequet, Dickerson, and Drysdale [7]. (See also [8] for a visualization of this structure.) A two-site distance function is measured from a point to a pair of points. In Euclidean space, it is a function $D_f : \mathbf{R}^2 \times (\mathbf{R}^2 \times \mathbf{R}^2) \to \mathbf{R}$ mapping a point p and a pair of points (v, w) to a non-negative real number. In a graph, it is a function mapping a vertex p and a pair of vertices (v, w)—usually sites—to a $\mathbf{R}$. Two-site distance functions D_f are symmetric on the pair of points

(v, w), though not necessarily on p, thus $D_f(p, (v, w)) = D_f(p, (w, v))$. (For some two-site distance functions, it is also true that $D_f(p, (v, w)) = D_f(v, (p, w))$ but this is incidental.) The sum function, D_S, results in the same Voronoi diagram as the 2-nearest neighbor (order-2) Voronoi diagram, but the authors considered a number of other combination rules as well including *area* and *product*. The complexity of the *round-trip* two-site distance function Voronoi diagram was left by [7] as an open problem, and remains open.

As we mentioned above, single-site graph-theoretic Voronoi diagrams were considered by Mehlhorn [21], who presented an algorithm running in $O(n \log n + m)$ time. More recently, Aichholzer *et al.* [2] study a hybrid scheme that combines geometric distance with a rectilinear transportation network (like a subway), and Abellanas *et al.* [1] study a similar approach where the subway/highway is modeled as a straight line. Bae and Chwa [5,6] study hybrid schemes where distance is defined by a graph embedded in the plane and distance is defined by edge lengths.

As far as multi-site queries are concerned, Safar [23] studies k-nearest neighbor searching in road networks, but he does so using the first-order Voronoi diagram, rather than considering a multi-site Voronoi diagram for geographic networks. Likewise, Kolahdouzan and Shahabi [19] also take the approach of constructing a first-order Voronoi diagram and searching it to perform C-nearest neighbor queries. Instead, de Almeida and Güting [11] compute C-nearest neighbors on the fly using Dijkstra's algorithm. None of these methods actually construct a multi-site or multi-color graph-theoretic Voronoi diagram, however, and, to the the best of our knowledge, there is no previous paper that explicitly studies multi-site or multi-color Voronoi diagrams on graphs. In [12], Dickerson and Goodrich study two-site Voronoi diagrams in graphs, but without employing any techniques that could improve running times beyond repeated Dijkstra-like algorithms.

1.4   Our Results

In this paper, we focus on two-site and two-color Voronoi diagrams on graphs using the round-trip function D_P for defining these concepts, although we also discuss the sum distance function, D_S, as well. In particular, for a vertex p or a point p on an edge e and a pair of sites v, w, our two-site distance functions are defined as follows:

$$D_S(p, (v, w)) = d(p, v) + d(p, w)$$
$$D_P(p, (v, w)) = d(p, v) + d(p, w) + d(v, w)$$

The sum function can easily be extended from 2 to k sites: $D_S(p, (v_1, \ldots, v_k)) = \sum_{1 \le i \le k} d(p, v_i)$. Note that with k-site distance functions, we also have a similar rule for breaking ties in distances. In the case that $D(p, (v_1, \ldots, v_k)) = D(p, (w_1, \ldots, w_k))$ for sites v_i, w_i and function D, as a means of breaking ties we consider p closer to whichever of $(v_1, \ldots, v_k)$ and $(w_1, \ldots, w_k)$ has a smaller lexicographical ordering of indices.

We prove several new properties of two-site round-trip distance function Voronoi diagrams on geographic networks, and make use of these properties to provide a new family of algorithms for computing these diagrams. We extend our proofs for the two-color variant, which is arguably more applicable than the one-color variant. (Though as noted above, there are cases when one might wish to visit several grocery stores on one trip, it is easier to imagine a case where we want the shortest tour visiting both a grocery store and a post office.)

One property we explore relates to the *doubling densities* of various types of POI sites on a geographic networks. The doubling density of a class of sites from a vertex v is the number of sites of that type within twice the distance from v as the closest site to v of that type. The run-times of our algorithms depend in part on the average doubling density of various sites from other sites. They also depend on the related density of the total number of edges within twice the distance from one site to the nearest other site of that type. (This latter property could be thought of as a different kind of doubling density.) We will prove a property that allows us to prune our search based on doubling distances, and will also provide experimental results about the doubling densities of various POIs on a set of states.

The algorithms have run times whose expected case is asymptotically faster than the algorithm of [12] under realistic assumptions of how sites are distributed in the network.

Finally, we show how to extend two-site Voronoi diagrams to multi-site and multi-color diagrams, under the sum function, while only increasing the running time by a factor of C, where C is the multiplicity we are interested in.

2   Constructing Graph-Theoretic Voronoi Diagrams

In this section, we review the approach of Mehlhorn [21] and Erwig [15] for constructing a (single-site) graph-theoretic Voronoi diagram of a graph G, having n vertices and m edges, which runs in $O(n \log n + m)$ time, and, for completeness, we also review the two-site sum function algorithm of [12], but with one minor correction.

Given a geographic network, $G = (V, E)$, together with a set of sites, $K \subseteq V$, and a non-negative distance function on the edges in E, the main idea for constructing a graph-theoretic Voronoi diagram for G is to conceptually create a new vertex, a, called the *apex*, which was originally not in V, and connect a to every site in K by a zero-weight edge. We then perform a single-source, shortest-path (SSSP) algorithm from a to every vertex in G, using an efficient implementation of Dijkstra's algorithm. Intuitively, this algorithm grows the Voronoi region for each site out from its center, with the growth for all the sites occurring in parallel. Moreover, since all the Voronoi regions grow simultaneously and each region is contiguous and connected by a subgraph of the shortest-path tree from a, we can label vertices with the name of their Voronoi region as we go.

In more detail, the algorithm begins by labeling each vertex v in K with correct distance $D[v] = 0$ and every other vertex v in V with tentative distance

$D[v] = +\infty$, and we add all these vertices to a priority queue, Q, using their D labels as their keys. In addition, for each vertex v in K, we label v with the name of its Voronoi region, $R(v)$, which in each case is clearly $R(v) = n(v)$. In each iteration, the algorithm removes a vertex v from Q with minimum D value, confirming its D label and R label as being correct. It then performs a *relaxation* for each edge (v, u), incident to v, by testing if $D[v] + w(v, u) < D[u]$. If this condition is true, then we set $D[u] = D[v] + w(v, u)$, updating this key for u in Q, and we set $R(u) = R(v)$, to indicate (tentatively) that, based on what we know so far, u and v should belong to the same Voronoi region. When the algorithm completes, each vertex will have its Voronoi region name confirmed, as well as the distance to the site for this region. Since each vertex is removed exactly once from Q and each key is decreased at most $O(m)$ times, the running time of this algorithm is $O(n \log n + m)$ if Q is implemented as a Fibonacci heap. In addition, note that this algorithm "grows" out the Voronoi regions in increasing order by distance from the apex, a, and it automatically stops the growing of each Voronoi region as soon as it touches another region, since the vertices in an already completed region are (by induction) closer to the apex than the region we are growing.

2.1   Two-Site Distance Functions on Graphs

In this section, we discuss algorithms for two-site Voronoi diagrams, which we then generalize in a subsequent section to multi-site and multi-color Voronoi diagrams. The advantage of this approach is that it highlights the additional complications needed to go from single-site to two-site Voronoi diagrams, while also showing the perhaps surprising result that we can construct two-site Voronoi diagrams with only a small, constant factor blow up in the running time. As mentioned above, the two-site sum function Voronoi diagram is equivalent to the second order two-nearest neighbor Voronoi diagram, which identifies for each vertex v in our graph, G, the two nearest sites to v. We state and prove the equivalence of these two types of Voronoi diagrams in the following simple lemma, the proof of which holds for both Voronoi diagrams in the plane and on weighted undirected graphs.

Lemma 1. *If v and w are the two closest sites to a vertex p in G, then the pair (v, w) minimizes $D_S(p, (v, w))$.*

Proof. Suppose that v and w are the two closest sites to a vertex p in G, but that the pair (v, w) did not minimize $D_S(p, (v, w))$. Without loss of generality, let $d(p, v) \leq d(p, w)$. By our assumption, there exists a vertex x such that $d(p, x) < d(p, w)$. It follows immediately that $D_S(p, (v, x)) < D_S(p, (v, w))$ which is a contradiction.

It follows that the two-site Voronoi diagram is equivalent to the two-nearest neighbor Voronoi diagram for a set of points in the plane or a graph.

So we are ready to formally define our construction problem.

Problem 1. Given a graph $G = (V, K, E)$ of n vertices V, m edges E, and a subset $K \subset V$ of s special vertices called "sites", compute the two-site Sum function Voronoi diagram of G; that is, label each vertex $v \in V$ with the closest pair of sites in K according to the two-site Sum distance function.

Intuitively, the algorithm of [12] for constructing a two-site Voronoi diagram under the sum function is to perform a Dijkstra single-source shortest-path (SSSP) algorithm from each site, in parallel, but visit each vertex twice—once for each of the two closest sites to that vertex.

More specifically, we begin by labeling each vertex v in K with correct first-neighbor distance $D_1[v] = 0$ and every other vertex v in V with tentative first-neighbor distance $D_1[v] = +\infty$, and we add all these vertices to a priority queue, Q, using their D_1 labels as their keys. We also assign each vertex $v \in V$ (including each site in K) its tentative second-neighbor distance, $D_2[v] = +\infty$, but we don't yet use these values as keys for vertices in Q. In addition, for each vertex v in K, we label v with the name of its first-order Voronoi region, $R_1(v)$, which in each case is clearly $R_1(v) = n(v)$. In each iteration, the algorithm removes a vertex v from Q with minimum key. How we then do relaxations depends on whether this key is a D_1 or D_2 value.

- Case 1: The key for v is a D_1 value. In this case we confirm the D_1 and R_1 values for v, and we add v back into Q, but this time we use $D_2[v]$ as v's key. We then perform a *relaxation* for each edge (v, u), incident to v, according to the following test:
  Relaxation(v, u):
  > **if** u has had its R_2 label confirmed **then**
  >> Return (for we are done with u).
  >
  > **else if** u has had its R_1 label confirmed **then**
  >> **if** $R_1(v) \neq R_1(u)$ and $D_1[v] + w(v, u) < D_2[u]$ **then**
  >>> Set $D_2[u] = D_1[v] + w(v, u)$
  >>> Set $R_2(u) = R_1(v)$
  >>
  >> **if** $D_2[v] + w(v, u) < D_2[u]$ **then**
  >>> Set $D_2[u] = D_2[v] + w(v, u)$
  >>> Set $R_2(u) = R_2(v)$.
  >
  > **else**
  >> **if** $D_1[v] + w(v, u) < D_1[u]$ **then**
  >>> Set $D_1[u] = D_1[v] + w(v, u)$
  >>> Set $R_1(u) = R_1(v)$.

  In addition, if the D_1 or D_2 label for u changes, then we update this key for u in Q. Moreover, since we are confirming the D_1 and R_1 labels for v, in this case, we also do a reverse relaxation for each edge incident to v by calling **Relaxation**(u, v) on each one.
- Case 2: The key for v is a D_2 value. In this case we confirm the D_2 and R_2 values for v, and we do a relaxation for each edge (v, u), incident to v, as above (but with no reverse relaxations).

When the algorithm completes, each vertex will have its two-site Voronoi region names confirmed, as well as the distance to each of its two-nearest sites for this region.

The correctness of this algorithm follows from the correctness of the SSSP algorithm and from Lemma 1. The SSSP algorithm guarantees that vertices will be visited in increasing order of distance from the origin(s) of the search. Lemma 1 states that the closest two sites to v are also the closest pair according to the sum two-site distance function.

For the analysis of this algorithm, first note that no vertex will be visited more than twice, since each vertex is added to the queue, Q, twice—once for its first-order nearest neighbor and once for its second-order nearest neighbor. Moreover, once a vertex is added to Q, its key value is only decreased until it is removed from Q. Thus, this algorithm requires $O(n \log n + m)$ time in the worst cast when Q is implemented using a Fibonacci heap, where n is the number of vertices in G and m is the number of edges. By the same reasoning, the priority queue Q won't grow larger than $O(n)$ during the algorithm, so the space required is $O(n)$.

3   Properties of Round-Trip Voronoi Diagrams on Graphs

Using the sum distance function for a two-site graph-theoretic Voronoi diagram allows us to label each vertex in G with its two nearest neighbors. Such a labeling is appropriate, for example, for fire stations or police stations, where we might want agents from both locations to travel to our home, or if we need to take separate trips to different sites. If instead we want to leave our home, travel to two nearby sites on the same trip, and return home, then we will need to use the round-trip function. Before presenting a new algorithm for this function, we first prove several properties of the round-trip distance function diagram on graphs.

Our first lemma is relatively straightforward, but provides an important property for pruning searches in our algorithms.

Lemma 2. *Let v be any vertex in a geographic network G, (s,t) a pair of sites in G minimizing the round-trip distance function D_P from v. Then for any sites p, q in G:*

$$d(v,s) \leq (d(v,p) + d(v,q) + d(p,q))/2$$
$$d(v,t) \leq (d(v,p) + d(v,q) + d(p,q))/2$$

Proof of Lemma 2: By assumption, $D_P(v,(s,t)) \leq D_P(v,(p,q))$. By the triangle inequality, $2d(v,s) \leq d(v,s) + d(v,t) + d(s,t)) = D_P(v,(s,t))$. Combining these, we get, $d(v,s) \leq \frac{1}{2}D_P(v,(p,q)) = \frac{1}{2}(d(v,p) + d(v,q) + d(p,q))$. The argument for $d(v,t)$ is symmetric. **End Proof.**

What this means is that if we know of some tour from vertex v through sites p and q—that is, we have a *candidate* pair (p,q) to minimize the round-trip distance from v— then our algorithms can safely ignore any other site s that is further from v than $\frac{1}{2}D_P(v,(p,q))$ because s cannot be a part of a pair that minimizes the round-trip distance from v.

This lemma combined with the triangle inequality $d(p,q) \leq d(v,p) + d(v,q)$ leads to the following corollary, which is a weaker condition, but one easily implementable as a pruning technique on a SSSP search.

Corollary 1. *Let p, q be the two sites closest to some vertex v under normal graph distance, and (s,t) the pair of sites minimizing the round-trip function D_P from v. Then $d(v,s) \leq d(v,p) + d(v,q)$ and $d(v,t) \leq d(v,p) + d(v,q)$.*

The following *double distance* lemma provides a similar but less obvious condition that can also be used for pruning.

Lemma 3. *(Doubling Distance Property) For any pair of sites s,t in a geographic network G, if there exists any other sites $p, q \in G$ such that $d(s,t) > 2d(s,p)$ and $d(s,t) > 2d(t,q))$, then (s,t) cannot minimize the round-trip distance function for any vertex $v \in G$.*

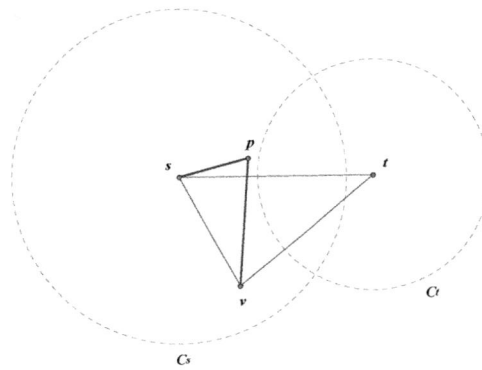

Fig. 4. Illustrating the proof of Lemma 3. The edges represent shortest paths, not single edges.

Proof of Lemma 3 (by contradiction): (See Figure 4.) Assume that there is some vertex v such that (s,t) is the closest pair of sites in the round-trip distance—that is, v is in the Voronoi region for (s,t). Assume also that there are sites $p, q \in G$ such that $d(s,t) > 2d(s,p)$ and $d(s,t) > 2d(t,q))$. Without loss of generality, let $d(v,s) \leq d(v,t)$. (Otherwise reverse the role of s and t.) We now consider the round-trip distance $D_P(v,(s,p))$. Applying the triangle inequality, we get: $D_P(v,(s,p)) = d(v,s) + d(s,p) + d(p,v) \leq d(v,s) + d(s,p) + (d(v,s) + d(s,p))$. By assumption, $2d(s,p) < d(s,t)$ and $d(v,s) \leq d(v,t)$ and thus: $D_P(v,(s,p)) < d(v,s) + d(v,t) + d(s,t) = D_P(v,(s,t))$, contradicting our assumption that (s,t) is the closest pair to v. **End Proof**

Note that Lemma 3 holds even if if s and t both meet the condition of Lemma 2—that is, even if for some vertex v, s and t are both closer to v than $\frac{1}{2}D_P(v,(p,q))$ for all sites p, q, the pair (s,t) cannot minimize the round-trip distance from v. Thus if the conditions of Lemma 3 hold, it follows immediately that the Voronoi region for (s,t) is empty in the two-site round-trip distance function Voronoi diagram.

We now state a final property of round-trip function Voronoi diagrams on graphs.

Lemma 4. *Let s be any site in a geographic network G, and v, w any vertices in G such that a shortest path from s to v goes through w. If there exist any sites $p, q \in G$ such that $d(w,s) > \frac{1}{2}(d(w,p) + d(w,q) + d(p,q))$ then s cannot be part of a nearest round-trip pair to v.*

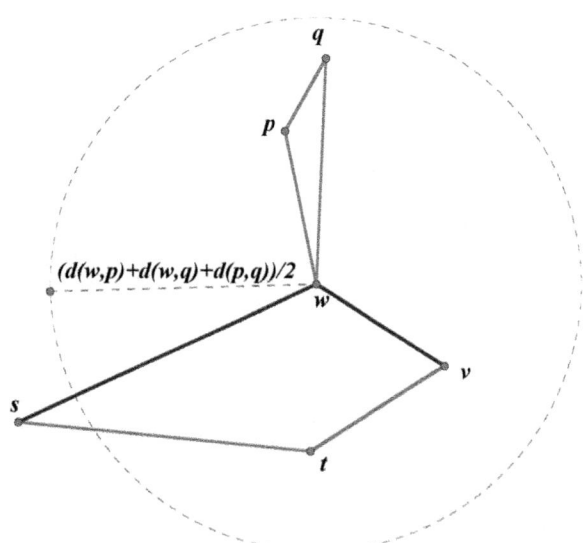

Fig. 5. Illustrating the proof of Lemma 4. The edges represent shortest paths in the graph, and not single edges. The edges (s,w) and (w,v) represent (by assumption) the *shortest* path from s to v.

Proof of Lemma 4: (See Figure 5.) For any t we know the following by the triangle inequality and the fact that w is on the shortest path from s to v:

$$d(v,t) + d(t,s) \geq d(v,w) + d(w,s) \tag{1}$$

Also by assumption

$$2d(w,s) > d(w,p) + d(w,q) + d(p,q). \tag{2}$$

Using Equations 1 and 2, we now show that $D_P(v,(p,q)) < D_P(v,(s,t))$ for s and any other site t.

$$D_P(v,(p,q)) = d(v,p) + d(v,q) + d(p,q)$$

$$D_P(v, (p, q)) \le [d(v, w) + d(w, p)] + [d(v, w) + d(w, q)] + d(p, q)$$
$$D_P(v, (p, q)) \le 2d(v, w) + d(w, p) + d(w, q) + d(p, q)$$
$$D_P(v, (p, q)) < 2d(v, w) + 2d(w, s)$$
$$D_P(v, (p, q)) < 2d(v, t) + 2d(t, s)$$
$$D_P(v, (p, q)) < d(v, t) + d(t, s) + d(v, s)$$
$$D_P(v, (p, q)) < D_P(v, (s, t))$$

End Proof.

We could rephrase this in the contrapositive, in a form similar to that of Lemma 2. Let v be any vertex in a geographic network G, (s, t) a pair of sites in G minimizing the round-trip distance function D_P from v, w a vertex on a shortest path from s to v, and p, q any sites in G, then $d(w, s) \le (d(w, p) + d(w, q) + d(p, q))/2$. What this Lemma means is that even if the pair of sites (s, t) meet the condition of Lemma 2 for some vertex v—that is, s is a candidate site to be part of the closest pair to v—if the shortest path from s to v goes through some vertex w for which s does not meet that condition, then not only is s not part of a closest pair for w, s also cannot be part of a closest round-trip pair to v. Together, these three lemmas are sufficient to prove the correctness of the algorithms in the following section.

3.1 Two-Color Variants

These lemmas can all be extended to apply to the two-color variant. The two color versions are given below. In the follow lemmas, we let $G = (V, E, S, T)$ be a geographic network, with $S \subset V$ and $T \subset V$ two disjoint sets of sites (of different colors). The two-color round-trip distance is from a vertex in V to a pair of sites (s, t) with $s \in S$ and $t \in T$. The proofs of these lemmas are directly analogous to the proofs above.

Lemma 5. *Let v be any vertex in G. Let $s \in S$ and $t \in T$ be a pair of sites such that (s, t) minimizes the two-color round-trip distance function D_P from v. Let $p \in S$ and $q \in T$ be sites in G. Then:*

$$d(v, s) \le (d(v, p) + d(v, q) + d(p, q))/2$$
$$d(v, t) \le (d(v, p) + d(v, q) + d(p, q))/2$$

Corollary 2. *Let $p \in S$ and $q \in T$ be the sites in S and T respectively that are closest to some vertex v under normal graph distance, and let (s, t) (with $s \in S$ and $t \in T$) be the pair of sites minimizing the two-color round-trip function D_P from v. Then $d(v, s) \le d(v, p) + d(v, q)$ and $d(v, t) \le d(v, p) + d(v, q)$.*

Lemma 6. *For any pair of sites $s \in S$ and $t \in T$ in a geographic network G, if there exists any other sites $p \in T$ and $q \in S$ such that $d(s, t) > 2d(s, p)$ and $d(s, t) > 2d(t, q)$, then (s, t) cannot minimize the round-trip distance function for any vertex $v \in G$.*

Lemma 7. *Let s be any site in a geographic network G, and v, w any vertices in G such that a shortest path from s to v goes through w. If there exist any sites $p \in S$ and $q \in T$ such that $d(w, s) > \frac{1}{2}(d(w, p) + d(w, q) + d(p, q))$ then s cannot be part of a nearest round-trip pair to v.*

4   Round-Trip Voronoi Diagram Algorithms

We now provide algorithms to compute the round-trip function Voronoi diagram for a geographic network and set of sites $G = (V, K, E)$. Specifically, the algorithm labels each vertex $v \in V$ with a pair of sites in K minimizing the two-site round-trip distance function from v.

4.1   A Brute Force Algorithm

An algorithm for this problem was first presented in [12]. The algorithm, in Step 1, performs a complete SSSP algorithm on G from each of the k sites in K. (Unlike in the Sum function algorithm above, these searches do not need to be interleaved—that is, performed in parallel—as the algorithm searches the entire graph from each site.) It records the distances from v to every site in K, and then creates a table of distances between all pairs of sites $(p, q) \in K$, allowing constant time access to $d(p, q)$. Then, in Step 2, for each vertex $v \in V$ and each pair of sites $(p, q) \in S$, the algorithm explicitly computes the round-trip distance

$$D_P(v, (p, q)) = d(v, p) + d(v, q) + d(p, q)$$

and labels each v with pair (p, q) minimizing this function.

This brute force approach uses the SSSP algorithm to efficiently compute all distances between pairs of vertices, and then explicitly compares all round-trip distances. The algorithm requires $O(k^2 n + km + kn \log n)$ time and $O(nk)$ space when we implement it using Fibonacci heaps as discussed above.

4.2   Improving the Brute Force Method: A Revised Algorithm

We now show how the properties of the previous section can be used to prune the search depth of the brute force algorithm of [12]. Our new algorithm has three steps, or phases.

Step 1 corresponds to Step 1 of the brute force approach above, except that we interleave the SSSP searches (as is done with the sum function) and we bound the number of sites that visit each vertex using some value B. In practice, this bounds the SSSP search outward from each site in K. The specific value of B–possibly determined as a function of n, m, k –will be described in the next section; the algorithm is correct regardless of the value of B, but its run time will depend on B. Ideally, the SSSP of Step 1 provides enough information for most (or all) of the vertices to determine the pair of sites minimizing the round-trip distiance. However the pruning may result in some vertices having incomplete information.

In Step 2, we need to complete information for each of these vertices that still have incomplete information by preforming an addition SSSP search outward from that vertex until it reaches all the sites satisfying Lemma 2. In particular, the smaller the value of B, the less work is done in Step 1, but the more potential work will need to be done in Step 2.

By Step 3, we have all the distance information needed to compute $D_P(v, (s, t))$ for all pairs of sites (s, t) satisfying Corollary 1, and Lemmas 3 and 4. We need only explicitly compute these distances from the information computed in Steps 1 and 2. Note that the pruning of Step 1 reduces not only the time required by each SSSP search, but also the number of explicit distances computed.

We start with the basic three-phase revised algorithm to compute the round-trip distance function two-site Voronoi diagram on a graph $G = (V, K, E)$.

- **Step 1:** Perform parallel Dijkstra SSSP algorithm from each site $p \in K$. For each vertex $v \in V$, record the distances from the first $B + 1$ sites whose SSSP search visits v. Any subsequent search (after the $B + 1^{st}$) visiting v is not recorded and the search is terminated. (As we will show, the result of Step 1 is that for each vertex, we have a list of the $B + 1$ closest sites in sorted order.

- **Step 2:** For each vertex $v \in V$, let p, q be two closest sites in K, and compute $d_v = d(v, p) + d(v, q)$. By Corollary 1, no site further than d_v from v is a candidate to be part of a pair minimizing the round-trip distance from v. So we consider two cases:
  case i: If the final site p on the sorted list of $B + 1$ closest sites to v is further from v than d_v, then we have found all sites closer to v than d_v and no work needs to be done on vertex v in this Step; the list of sites at v contains all possible candidate sites that could be part of a closest pair in the round-trip function.
  case ii: If the final site on the sorted list for v is not further than d_v, then we cannot guarantee that v was visited by all the candidate sites. In this case, we perform a SSSP algorithm from v and halt when we reach any vertex further from v than d_v. (Note that this is done also for those vertices that are also sites in K. Since d_v is at least as great as twice the distance from v to its nearest site, we will compute distances between all pairs of sites satisfying Lemma 3.)

- **Step 3:** For each vertex $v \in V$, compute $D_P(v, (s, t)) = d(v, s) + d(v, t) + d(s, t)$ for all sites s, t for which $d(v, s)$ and $d(v, t)$ are stored at v and $d(s, t)$ is stored at either s or t. (If $d(s, t)$ was not computed, then (s, t) is not a candidate pair and may be ignored.) Store at v the pair (s, t) minimizing $D_P(v, (s, t))$.

Correctness. Since the first $B + 1$ SSSP searches that reach any vertex will continue through the vertex, by induction each vertex is guaranteed to be reached by the SSSP from at least its closest $B + 1$ sites in Step 1. In Step 2, therefore, by looking at the first two and the last site in the list for each vertex v, we can determine if all sites meeting Corollary 1 have visited v. If not, then an

SSSP from v (in case ii) will reach those sites. So by Corollary 1 and Lemma 4, any sites s, t for which the algorithm does not explicitly computer $D_P(v, (s, t))$ cannot be a candidate to minimize the round-trip distance from v.

Worst Case Analysis. We now analyze the algorithm. In Step 1, we visit each vertex $B+1$ times. (If a search arrives at a vertex v that has already been visited $B + 1$ times, we count that work to the edge along which the SSSP came to v.) An edge can be traversed at most $B + 1$ times from the vertices on each end, for a total of $O(B)$ visits. So Step 1 requires $O(Bm + Bn \log n)$ time because we are overlapping B SSSP searches. We are storing $B + 1$ sites and distances at each vertex, as well as a list of $O(k^2)$ distances between each pair of sites, so the space required is $O(Bn + m + k^2)$.

In step 2, we need to store distances between pairs of sites s, t that are candidates to minimize round-trip distance for some vertex. If we use a table, we need worst case $O(k^2)$ space with $O(1)$ time access for any pair (s, t). We also store distances between vertex v and its candidate sites in sorted order; there are at most B sites per list in Step 1, and though in Step 2 the lists can grow to size $O(k)$ we only need to store one list at a time, and so space required is $O(Bn + m + k^2)$.

In step 2, if for a vertex v, the $B + 1$ vertices on its list includes all the sites within distance d_v, then we are in case i, and the total amount of work for that vertex in step 2 is $O(1)$ and in step 3 is $O(B^2)$ to explicitly compute all possible round-trip distances of candidate pairs (since for each pair of sites (s, t) the distance $d(s, t)$ has already been computed and can be retrieved in $O(1)$ time.) The total run time for these sites is thus $O(B^2 n)$.

For the rest of the vertices v, those in case ii, we must do a new SSSP from v. This requires $O(m + n \log n)$ time per vertex for the search and $O(k^2)$ time per vertex to look at all pairs of sites. Let A be the number of sites processed in case ii. The run time for all of them is $O(Am + An \log n + Ak^2)$.

The overall run time is thus $O((A + B)(m + n \log n) + B^2 n + Ak^2)$ and the space required is $O(nB + m + k^2)$.

In the next section, we formalize this and also provide some experimental data on values of A and B. First, however, we provide a further revision showing how for many real world networks such as road networks, we can make fuller use of Lemma 2 for an algorithm whose run time is significantly better.

4.3   Further Revisions: A Dynamic Variation

It is possible that we can further reduce the depth of our SSSP searches, and thus the number of candidate pairs examined in our algorithm. Lemma 2 gives a stronger condition than Corollary 1 that must be met by any site that is a candidate to minimize the round-trip distance from a vertex v.

Specifically, instead of using a static bound that prunes the depth of our searches in Step 1, and then simply computing the distance from vertex v to its two nearest sites, we would like to keep an updated minimal value of $D_P(v, (s, t))$ for *all* sites s, t whose SSSP searches have visited v. By Lemmas 2 and 4, we can then prune any search that reaches v from any site further away than the minimum value of $\frac{1}{2} D_P(v, (s, t))$.

Unfortunately, using this stronger condition requires that we dynamically update the minimum value of $D_P(v, (s, t))$ which in turn requires that we precompute or preprocess the values of $d(s, t)$ for all pairs of sites meeting the condition of Lemma 3. This leads to the following two-step algorithm.

- **Step 1:** Perform a SSSP algorithm from each site $p \in K$, terminating the search at the first vertex whose distance from p is greater than $2d(p, q)$ where q is the closest other site to p (discovered in the SSSP). Store the values of $d(p, q)$ for all pairs of sites reached in all of the searches.
- **Step 2:** Perform interleaved SSSP searches from each site $p \in K$, as in Step 1 of the previous algorithm. At each vertex v, store the sites s whose searches reach v along with the distance $d(v, s)$. Using this information and the table from Step 1, once a second site search has visited v, also compute and maintain the distance $D_P(v, (s, t)) = d(v, s) + d(v, t) + d(s, t)$ that minimizes this function among all pairs of sites s, t which have visited v (as well as the pair (s, t) minimizing that distance). Terminate the search from any site farther from v than $\frac{1}{2}D_P(v, (s, t))$ for the minimum value of $D_P(v, (s, t))$ seen so far.

Worst Case Analysis. In the worst case, Step 1 will require $O(m + n \log n)$ time and $O(n + m)$ space for each SSSP for a total of $O(km + kn \log n)$ time, plus an extra $O(k^2)$ space to store the table of distances between pairs of sites, for a total of $O(k^2 + m + n)$ space.

Similarly, in the worst case in Step 2, each of the k SSSP algorithm may require $O(m + n \log n)$ time, but since the searches are interleaved we may need extra $O(nk)$ space to have k searches active at once. We also need to compute $O(k^2)$ distances at each vertex in the worst case, but $k \leq n$ and so we have a total of $O(km + kn \log n)$ time and $O(nk + m)$ space.

As we will see in the following section, however, road networks and many types of POI sites have properties that result in a much more efficient algorithm.

4.4 The Two-Color Variant

The algorithms of the previous section can be extended to the two-color variant, where for each vertex v we want to find the pair of sites (or POIs) of two *different* types–say a grocery store and a post office–that minimize the distance of the shortest round-trip from v. The same basic approaches of both the revised algorithm and the dynamic variant of the revised algorithm work for the two-color version. Lemmas 5, 6, and 7 suffice as proof.

Other than the obvious change that the two-color versions of the algorithms compute and minimize the round-trip distances to pairs of sites of different types, there are only two other primary changes that are necessary. In the first stage, we still perform the interleaved SSSP algorithms from all sites (of both types). However at each vertex v we store separate lists of the sites of the two different types that visit v. This doubles the worst-case memory requirement.

Second, the application of Lemma 6 two-color variant is slightly different than that of Lemma 3 to the standard round-trip distance function. In the dynamic version we need to pre-compute only the distances between sites of different type. In particular, if our two sets of sites are S and T, we need to compute the distance from each $t \in T$ to all the sites in S no more than twice the distance of the closest site in S to t, and symmetrically from each $s \in S$ to all the sites in T no more than twice the distance of the closest site in T to s.

In terms of run-times, what this means for the two-color variant of the round-trip distance function Voronoi diagram is that we care about the *doubling density* of sites in T with respect to sites in S and vice versa–rather than the doubling density of sites in one set to other sites in the same set, as is the case with the standard round-trip two-site distance function.

5   Empirical Analysis on Doubling Density and Dynamic Pruning on Road Networks

The actual run time of our algorithms when they are run on real world data such as road networks with sites coming from standard points-of-interest (POI) files (fire departments, educational institutions, etc.) may be much better than the worst case asymptotic analyses presented in the previous section. In particular, the Lemmas of Section 3 enable each SSSP search in both of our algorithms to be terminated (pruned) well before a linear number of edges and vertices have been visited.

In the first algorithm, for example, the algorithm balances work between the first two phases by a careful choice of the value of B, where B is an expected number of sites that are "close" to most vertices, where "close" for a vertex v is defined by the value of d_v: the sum of the distances to the two nearest sites to v. If the distance to the two nearest sites is, in general, a good indication of how densely packed sites are in the proximity to a vertex v, then the first algorithm—the static version—will perform well.

When the first algorithm does not do well is when there are large rural or wilderness areas that have roads (vertices and edges on the network) but no sites. Consider, for example, a southwest desert area on the edge of an urban area, or a large northeastern or northwestern forest near a city. It might be 50 miles to the nearest gas station site, which sits in a city on the edge of the dessert, but there might be several dozen or even a few hundred gas stations within 100 miles of that rural vertex. In terms of the notation of the previous section, distance from v to the nearest two sites is very large, resulting in a large value of d_v, thus also resulting in a large number of sites—more than B—within a distance of d_v of v.

By contrast, the second algorithm—the dynamic variant—uses the stronger version of the lemma to handle even instances of these poorly distributed sites efficiently, though at the cost of more overhead and a possibly costly preprocessing phase. Consider a vertex v in a rural area that is a large distance from its nearest sites. In real world applications of read networks, however, it is likely that its two nearest sites p and q are close to one another, even though they are far from v.

We thus prune using the dynamically updated value of $\frac{1}{2}(d(v,p)+d(v,q)+d(p,q))$ which is never greater than $d(v,p) + d(v,q)$ and in the example described above is likely to be much smaller.

We ran a variety of experiments to determine empirical performance of both the static and dynamic versions of the algorithm, for both one-color and two-color variants of the problem. In particular, we ran experiments to determine the empirical values of A and B in the static variant, and to determine the doubling densities and the expected number of edge and vertex visits in the dynamic variant. Our experiments included 22 different U.S. states: AK, CT, DC, DE, HI, ID, IL, IN, LA, MA, MD, ME, ND, NH, NJ, NY, OH, RI, TN, UT, VT and WY. The state road networks ranged in size from Hawaii, with only 64892 vertices and 76809 edges to New York, with 716215 vertices and 897451 edges. They also varied greatly in terrain, urban areas, and presence of large areas of wilderness with sparse roads.

Between the one- and two-color variants, we also experimented on a variety of POIs as our sites including: *educational institutions, recreational sites, hospitals, shopping centers, fire stations,* and *religious institutions* accessed from a publicly available collection of POIs. (Multiple POIs of the same type at the same address were combined into one site. However POIs in close proximity but at different vertices were treated separately.) The number of sites in a file ranged to a maximum of 7640 (educational institutions in TN). In addition to being publicly available POIs, the variety of sites also made a good choice because some of them are intuitively distributed in a way that could lead to poor performance. Educational institutions—unlike, for example, post offices or fire stations—are unevenly distributed; a large campus for a single institution may contribute to the POI file numerous buildings in close proximity but with different addresses.

We report first on the empirical values for A and B, and then on *doubling densities* of these POIs for both the one- and two-color variants. We also report on the depths to which the SSSP searches need to go before they can be pruned by Lemmas 4 and 7.

5.1   Empirical Values for A and B on Road Networks

To study the distributions impacting the run time of our first algorithm, we ran ten trials that tested four northeastern states (VT,ME,NJ,NH), plus Hawaii (HI), testing each state on two available data sets of POI sites: where fire department data was publicly available (ME and VT), we used fire departments and religious institutions; in the other three states we used educational institutions and religious institutions. The size of the data sets had the number n of vertices ranging 64892 (HI) to 330386 (NJ), with the number m of edges ranging from to 76809 (HI) to 436036 (NJ), and the number k of POI sites ranging from 144 (educational sites in HI) to 759 (religious institutions in ME). For various values of B (the bound at which the first phase of the search is pruned) we computed the value of A (the number of vertices whose information is incomplete after the first phase).

Results were somewhat divergent and seemed to depend considerably on the geographic nature of the road network as described in the introduction to this section. With respect to the efficiency of our algorithm, Maine was the worst state, and performance for religious institutions in New Hampshire was also bad. For both fire departments and religious institutions in Maine, and religious institutions in New Hampshire, setting $B = 5 \log n$ results in values of $A = 11.6\sqrt{n}$ (NH, religious institutions), $A = 12.3\sqrt{n}$ (ME, religious institutions), and $A = 4.47\sqrt{n}$ (ME, fire departments). This is easily explained because Maine has thousands of square miles (known as the North Maine Woods) that are tracked by hundreds of miles of lumber roads, but without any real towns or villages, and thus no POIs. New Hampshire also has a large wilderness area in the White Mountains and northern forests. However on the edge of these large wilderness areas are population centers (e.g. Gorham and Conway, NH, Bangor, ME), with relatively dense distributions of POIs. If we generalize these results to other states comparable in nature to Maine and New Hanpshire, we have an algorithm with run time $O(mn^{\frac{1}{2}} + n^{\frac{3}{2}} \log n + k^2\sqrt{n})$ and the space required is $O(n \log n + m + k^2)$.

By contrast, for the seven other data sets (which use road networks on Hawaii, Vermont, New Hampshire, and New Jersey), if we set $B = \log n \log k$, then we find that $A = 0$—that is, the first step is sufficient to provide complete information for all vertices, and no additional work is required in the second step. The first algorithm therefore requires $O(m \log n \log k + n \log^2 n \log^2 k)$ time and $O(n \log n \log k + m + k^2)$ space, which is a significant asymptotic time improvement over the previous best known algorithm of [12].

5.2   Doubling Density

As noted above, the preprocessing in Step 1 of the dynamic variant of the algorithm must compute a table of distances between pairs of sites that define potential Voronoi regions. In the worst case, this may take $O(km + kn \log n)$ time to compute and additional $O(k^2)$ space to store the table, where k is the number of sites. However by Lemmas 3 and 6, we only need to store pairs of sites (s,t) if s is no more than twice as far from t as the nearest other site to t, or vice versa. This improved efficiency for Step 1 thus depends on a property we call the *doubling density*, which is defined as follows: *for a given vertex v and set of sites S, the doubling density of v is the number of sites in S no further from v than twice the distance to the nearest other site to v (not counting v if $v \in S$.)*

For the one-color two-site tour-distance problem, Lemma 3 indicates that the space and time required by our dynamic algorithm depend on the average doubling density from all sites in the current POI data set to other sites of the same type. For the two-color version where we have one set of sites K_1 and another set K_2, Lemma 6 tells us that the algorithm's space and time depend on the average doubling density of sites of type K_1 from sites of type K_2, and the average doubling density of sites of type K_2 from sites of type K_1. Empirical results of average doubling density for for both the one-color and two-color version are promising.

Let d be the total double density—that is, the number of "candidate pairs" of sites, one of each type or "color", that might have non-empty Voronoi regions. In the worst case, in the one-color case, d could be $\Omega(k^2)$ where k is the number of sites. However empirical results suggest that d is $O(k)$; or, rephrased, the average doubling density per site, d/k, is constant.

We ran nine trials for the one-color version using three POI types (religious and educational institutions and fire stations) on a mix of states. Results were very consistent. In all trials, the average doubling density for sites of a single type was found to be less than 4.1. As a result, for values of k tested up to 1783, the number of pairs of sites whose distances need to be stored for use in Step 2—that is, pairs that are candidates by Lemma 3 to minimize the round-trip distance for at least one vertex, and thus are candidates to have a non-empty Voronoi region–is empirically less than $5k$, or $O(k)$. These values appear to be independent of n. However the data does indicate a possible logarithmic relationship to k, the number of sites. In particular, for several different road networks and POI files, the average doubling density appears to be $\Theta(\log k)$, meaning the total number of pairs to be stored is $\Theta(k \log k)$ rather than $O(k)$, though for practical purposes on the POIs examined, the number of candidate pairs is $< 5k$. (Much larger POI files for the same sets of states would be needed to verify this result.) To avoid using a sparse direct table of size $\Theta(k^2)$, we can use a hash table to store these pairs in $O(k)$ or $O(k \log k)$ space, and provide $O(1)$ expected time access in Step 2.

Though the doubling density of sites (and, in particular, the size of the table computer in Step 1) impacts the size and run time of later steps, the run time of Step 1 itself is actually determined by the total number of edge visits in all of the SSSP searches used to compute the table. We kept track of the number of times each edge in the graph was visited in Step 1. In all trials for the one-color version, the average number of visits per edge was less than 5. In fact, while for fire department POIs the value was between 4 and 5, for religious and educational institutions, the number was less than 3 in all trials. These values appear to be true constants, independent of either n or k. As a result, all of the SSSPs in Step 1 of the second algorithm combined require only $O(m + n \log n)$ time—only a small constant factor more than a single SSSP search—and this holds whether the doubling density is $O(1)$ or $O(\log k)$).

For the two-color variant, we also ran trials on numerous states and a variety of different points-of-interest for the sites, to empirically determine the values of the double densities. The states (and district) tested included: AK, CT, DC, DE, HI, ID, MD, ME, ND, NH, RI, UT, VT, and WY. The types of POIs tested included education institutions, religions institutions, recreational centers, hospitals, and shopping malls.

Again, let d be the total double density—that is, the number of "candidate pairs" of sites, one of each type or "color", that might have non-empty Voronoi regions. In the worst case, as with the one-color case d could be $\Theta(k_1 k_2)$, or simply $\Theta(k^2)$ where k_1 and k_2 are the number of sites of each type, and k is the total number of sites. However for most combinations of types of sites, the

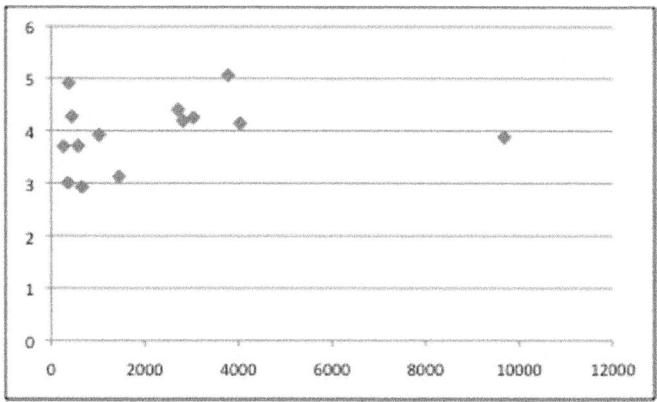

Fig. 6. The number of candidate Voronoi regions divided by k (the number of sites) as a function of k for fourteen states

average doubling density, given by the ratio d/k of the total number of candidate pairs to the total number of sites, ranged from 2.92 to 5.06. A graph of the number of candidate pairs of religious institutions and educational institutions, for the fourteen states listed above is shown in Figure 6.

For the same two types of sites, we also computed the number of times edges were visited on all of these searches, which determines the efficiency of the pre-processing step. The average number of times each edge is visited in computing this table, in the total of both types of searches, was less than 22 in all trials. See Figure 7.

Some combinations of POIs had a somewhat higher average double density. For churches and schools, the average doubling density was as high as 14. How-ever it still appears to be a constant; the doubling density does not increase as the number of vertices or sites increases, but rather the constant seems related to the types of sites and the way they are distributed related to each other. Figure 8 shows the average doubling densities for all five types of POI sites and fourteen states described above.

Thus empirical results suggest a constant average doubling density, a table of candidate pairs that is linear in the number of input sites, and a total run time of $O(m + n \log n)$ and total space of $O(n + m)$ for all the SSSP searches to compute this information.

5.3   Dynamic Pruning on Road Networks

Results for Step 2 are equally promising for the one-color variant, though not for the two-color variant. For all vertices in all trials for one-color, the total number of sites whose SSSP visited the vertex—that is, the number of sites closer to each vertex v than half of the distance of the best known round-trip distance pair yet found—is bounded by 13.

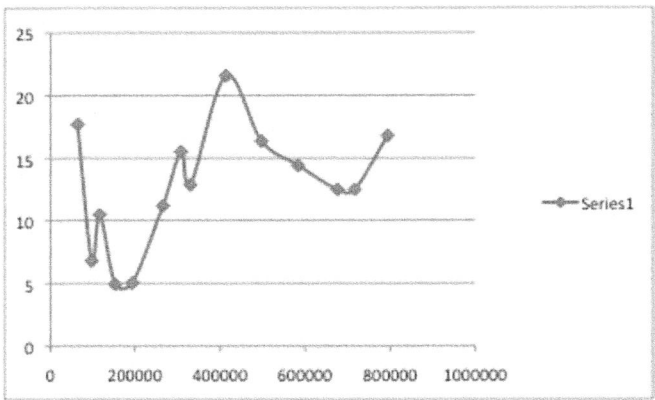

Fig. 7. Based on the doubling density, the average number of times each edge is visited (in the search for candidate pairs) visited as a function of n (for fourteen states)

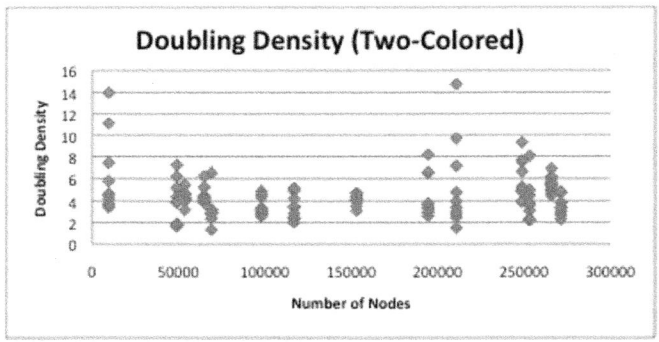

Fig. 8. The average doubling densities of five types of POIs for fourteen states

Thus the total run time of Step 2 is also $O(m + n \log n)$. Empirically, then, the overall run time of the dynamically pruned Algorithm 2 is $O(m + n \log n)$ and the space required is $O(m + n + k \log k)$ when used on road networks with standard POI files.

Results on the level of pruning are equally promising, though less immediately so; an amortized approach is required to see the efficiency. In particular, when sites of one type are much denser than sites of another—as is the case, for example, when there is a large educational campus that contributes numerous entries to a POI file in a small area—then the number of pairs of sites satisfying Lemma 2 and 5 may be large, prohibiting an early pruning of the searches and requiring distance calculations for numerous pairs of sites.

However in the two-color variant, there is no second phase of the algorithm when we must perform a SSSP search from each vertex with incomplete information. Instead, the SSSP searches from the original sites continue until all of them have been pruned. So we can bound the overall run time of these searches simply by the total number of times that a search continues through a vertex— or, equivalently, by the average number of times that each vertex is visited. In all trials except Hawaii, the average number of such vertex visits was less than 10 per vertex, which is linear on the size of the graph. (For Hawaii, which had the smallest graph, the number was 16.) That is, each vertex was visited an average of $O(c)$ SSSP searches for a total of $O(n)$ vertex visits, before further searches are dynamically pruned by Lemma 7.

Equally importantly for the efficiency of the algorithm, the number of pairs of sites of different colors that are are tested for each vertex empirically appears to converge to 10 as the number n of vertices grows. Thus, empirically the total number of distances explicitly computed to find the minimum was $O(n)$.

This immediately implies that the space complexity of all the SSSPs never exceeds $O(n)$. So empirical data suggests that Step 2 also requires $O(m+n \log n)$ time and $O(n + m)$ space.

6   Multi-site and Multi-color Sum Function Extensions

In this section, we discuss how to extend our two-site distance function algorithms to multi-site and multi-color variants, beginning with the multi-site sum function problem. The C-site distance function D_C from a vertex p to a set of C sites $v_1, v_2, \ldots, v_C$ is defined as:

$$D_S(p, (v_1, v_2, \ldots, v_C)) = d(p, v_1) + d(p, v_2) + \cdots + d(p, v_C)$$

Our next goal is to extend our sum function algorithm to solve the following problem.

Problem 2. Given a graph $G = (V, K, E)$ of n vertices V, m edges E, a subset $K \subset V$ of k special vertices called "sites", and an integer $C \leq S$, compute the C-site sum function Voronoi diagram of G; that is, label each vertex $v \in V$ with the closest C sites in K according to the C-site Sum distance function.

As we already showed to be the case when $C = 2$, it is simple to show that for general $C \geq 2$ that the C-site sum function Voronoi diagram is equivalent to the C nearest neighbors Voronoi diagram. This is sufficient argument for the correctness of the following algorithm.

6.1   The Multi-site Sum Algorithm

Our algorithm is a relatively straightforward extension of the two-site algorithm. Perform a Dijkstra single-source shortest-path (SSSP) algorithm from each site in K, in parallel. But visit each vertex only from the first C searches to reach it. Any search to reach a vertex that has already been visited C times by closer sites will not "visit" that vertex or proceed past it. Label each vertex v with the C sites from K that are the closest, namely the sites described above through which the SSSP algorithm from s first passed.

Since no vertex will be visited more than C times, no edge will be traversed more than $2C$ times (C times from each side). Essentially, the algorithm takes at most C times as many steps as the standard SSSP algorithm. The Dijkstra algorithm requires $O(m + n \log n)$ time in the worst cast when implemented using Fibonacci heaps and when visiting each vertex once. In the multi-site case, we visit each vertex C times. In order to avoid an $O(C^2)$ or $O(C \log C)$ term in our running time, we need to maintain, for each vertex, the number of nearest-neighbor sites that we have already completed. The total running time is $O(Cn \log n + Cm)$. By a similar technique we used in our two-site algorithm, we can keep a single copy of each vertex in our priority queue and just insert each vertex C times, but we keep C labels for each vertex, so the space required by our algorithm is $O(Cn)$.

6.2   The Multi-color Sum Algorithm

As discussed in our introduction, another extension of our algorithm is to a multi-color variant. Each of the K sites is colored with a color from 1 to C (for some $C \leq S$). Our goal for each vertex p is to compute the closest site of each color. The result for each vertex is a C-tuple $(v_1, \ldots, v_C)$ such that the C colors of the v_i are unique, and that minimizes the distance $D_S(p, (v_1, v_2, \ldots, v_C)) = d(p, v_1) + \cdots + d(p, v_C)$ among all such possibilities.

Again, our algorithm is a relatively straightforward extension of the standard graph Voronoi diagram algorithm. For each color c, we perform a Dijkstra single-source shortest-path (SSSP) algorithm from each site in K labeled c, visiting each vertex only from the first site to reach it. That is, we perform the Voronoi diagram algorithm once for each color. Label each vertex v with the C sites from K representing the closest site for each color. The result is that each vertex is colored with the closest site for each color; we have C overlapping Voronoi diagrams, one for each color.

Essentially, the algorithm takes at most C times as many steps as the standard Voronoi diagram graph algorithm discussed earlier. For each color, the algorithm requires $O(m + n \log n)$ time in the worst case. The total run time is therefore $O(Cn \log n + Cm)$. Since each search is conducted independently, the priority queue won't grow larger than $O(n)$ and only $O(C)$ information is stored at each vertex/edge, so the space required is $O(Cn)$.

7   Conclusion

We have given complete and efficient algorithms for multi-site and multi-color Voronoi diagrams, under both the Sum and round-trip combination functions. The only variants that we have omitted are the multi-site and multi-color Voronoi diagrams under the round-trip function. The reason we have omitted these variants here, beyond the two-site case, is that computing multi-site or multi-color distance values under the round-trip function requires that we solve miniature versions of the Traveling Salesperson Problem, which is NP-hard. For a small number p of "colors", an extension to the pruning lemmas might provide some added efficiency in minimizing the number of candidate p-tuples (p-tuples of sites with potentially non-empty p-order Voronoi regions) but the overhead of computing the least-cost tour and dynamically updating the data set would be high. So it is unlikely that we will be able to solve these variants efficiently.

References

1. Abellanas, M., Hurtado, F., Sacristán, V., Icking, C., Ma, L., Klein, R., Langetepe, E., Palop, B.: Voronoi diagram for services neighboring a highway. Inf. Process. Lett. 86(5), 283–288 (2003)
2. Aichholzer, O., Aurenhammer, F., Palop, B.: Quickest paths, straight skeletons, and the city Voronoi diagram. In: SCG 2002: Proceedings of the Eighteenth Annual Symposium on Computational Geometry, pp. 151–159. ACM, New York (2002)
3. Aurenhammer, F.: Voronoi diagrams: A survey of a fundamental geometric data structure. ACM Comput. Surv. 23(3), 345–405 (1991)
4. Aurenhammer, F., Klein, R.: Voronoi diagrams. In: Sack, J.-R., Urrutia, J. (eds.) Handbook of Computational Geometry, pp. 201–290. Elsevier Science Publishers B.V., North-Holland, Amsterdam (2000)
5. Bae, S.W., Chwa, K.-Y.: Voronoi Diagrams with a Transportation Network on the Euclidean Plane. In: Fleischer, R., Trippen, G. (eds.) ISAAC 2004. LNCS, vol. 3341, pp. 101–112. Springer, Heidelberg (2004)
6. Bae, S.W., Chwa, K.-Y.: Shortest Paths and Voronoi Diagrams with Transportation Networks Under General Distances. In: Deng, X., Du, D.-Z. (eds.) ISAAC 2005. LNCS, vol. 3827, pp. 1007–1018. Springer, Heidelberg (2005)
7. Barequet, G., Dickerson, M.T., Drysdale, R.L.S.: 2-point site Voronoi diagrams. Discrete Appl. Math. 122(1-3), 37–54 (2002)
8. Barequet, G., Scot, R.L., Dickerson, M.T., Guertin, D.S.: 2-point site Voronoi diagrams. In: SCG 2001: Proceedings of the Seventeenth Annual Symposium on Computational Geometry, pp. 323–324. ACM, New York (2001)
9. Chazelle, B., Edelsbrunner, H.: An improved algorithm for constructing kth-order Voronoi diagrams. IEEE Trans. Comput. C-36, 1349–1354 (1987)
10. Cormen, T.H., Leiserson, C.E., Rivest, R.L., Stein, C.: Introduction to Algorithms, 2nd edn. MIT Press, Cambridge (2001)
11. de Almeida, V.T., Güting, R.H.: Using Dijkstra's algorithm to incrementally find the k-nearest neighbors in spatial network databases. In: SAC 2006: Proceedings of the 2006 ACM Symposium on Applied Computing, pp. 58–62. ACM, New York (2006)
12. Dickerson, M.T., Goodrich, M.T.: Two-site voronoi diagrams in geographic networks. In: GIS 2008: Proceedings of the 16th ACM SIGSPATIAL International Conference on Advances in Geographic Information Systems, pp. 1–4. ACM, New York (2008)
13. Dijkstra, E.W.: A note on two problems in connexion with graphs. Numerische Mathematik 1, 269–271 (1959)
14. Dirichlet, G.L.: Über die Reduktion der positiven quadratischen Formen mit drei unbestimmten ganzen Zahlen. J. Reine Angew. Math. 40, 209–227 (1850)
15. Erwig, M.: The graph Voronoi diagram with applications. Networks 36(3), 156–163 (2000)

16. Fortune, S.: Voronoi diagrams and Delaunay triangulations. In: Du, D.-Z., Hwang, F.K. (eds.) Computing in Euclidean Geometry, 1st edn. Lecture Notes Series on Computing, vol. 1, pp. 193–233. World Scientific, Singapore (1992)
17. Goodrich, M.T., Tamassia, R.: Algorithm Design: Foundations, Analysis, and Internet Examples. John Wiley & Sons, New York (2002)
18. Knuth, D.E.: Sorting and Searching. The Art of Computer Programming, vol. 3. Addison-Wesley, Reading (1973)
19. Kolahdouzan, M., Shahabi, C.: Voronoi-based k nearest neighbor search for spatial network databases. In: VLDB 2004: Proceedings of the Thirtieth International Conference on Very Large Data Bases, pp. 840–851. VLDB Endowment (2004)
20. Lee, D.T.: On k-nearest neighbor Voronoi diagrams in the plane. IEEE Trans. Comput. C-31, 478–487 (1982)
21. Mehlhorn, K.: A faster approximation algorithm for the Steiner problem in graphs. Information Processing Letters 27, 125–128 (1988)
22. Patroumpas, K., Minogiannis, T., Sellis, T.: Approximate order-k Voronoi cells over positional streams. In: GIS 2007: Proceedings of the 15th Annual ACM International Symposium on Advances in Geographic Information Systems, pp. 1–8. ACM, New York (2007)
23. Safar, M.: K nearest neighbor search in navigation systems. Mob. Inf. Syst. 1(3), 207–224 (2005)
24. Sugihara, K.: Algorithms for computing Voronoi diagrams. In: Okabe, A., Boots, B., Sugihara, K. (eds.) Spatial Tesselations: Concepts and Applications of Voronoi Diagrams. John Wiley & Sons, Chichester (1992)
25. Voronoi, G.M.: Nouvelles applications des paramètres continus à la théorie des formes quadratiques. premier Mémoire: Sur quelques propriétés des formes quadratiques positives parfaites. J. Reine Angew. Math. 133, 97–178 (1907)

Author Index